Important addresses & telephone numbers

Emergency room _____
_____

Poison control center _____
_____

Ambulance _____
_____

Family doctor _____
_____

Pediatrician _____
_____

Specialist _____
_____

Dentist _____
_____

Dental emergency number _____
_____

Hospital _____
_____

Other _____
_____

TAKING CARE
OF YOUR CHILD

BOOKS BY THESE AUTHORS

The Parent's Pharmacy, by Robert H. Pantell and David Bergman

Vitality and Aging, by James F. Fries and Lawrence M. Crapo

Taking Part: A Consumer's Guide to the Hospital, by Donald M. Vickery

Living Well: Taking Care of Your Health in the Middle and Later Years, by James F. Fries

Arthritis: A Comprehensive Guide, by James F. Fries

The Arthritis Helpbook, by Kate Lorig and James F. Fries

Lifeplan: Your Own Master Plan for Maintaining Health and Preventing Illness, by Donald M. Vickery

Take Care of Yourself, by Donald M. Vickery and James F. Fries

Taking Care of Your Child

A Parent's Guide to Complete Medical Care

FOURTH EDITION

Robert H. Pantell, M.D.

James F. Fries, M.D.

Donald M. Vickery, M.D.

ADDISON-WESLEY PUBLISHING COMPANY

Reading, Massachusetts ■ Menlo Park, California ■ New York
Don Mills, Ontario ■ Wokingham, England ■ Amsterdam ■ Bonn
Sydney ■ Singapore ■ Tokyo ■ Madrid ■ San Juan ■ Paris ■ Seoul
■ Milan ■ Mexico City ■ Taipei

DEDICATION

To our parents, who first taught us about health;
To our mentors, Floyd Denny, M.D., and Helen Lamb, who first taught
    us how to care for children;
To innovators in child health, Evan Charney, Floyd Denny, Bob
    Haggerty, Barbara Korsch, Clarence McIntyre, Terry Reilly, and
    Barbara Starfield, who provide continuing inspiration; and
To our children, patients, and their parents, who continue to teach us.

Many of the designations used by manufacturers and sellers to distinguish their products are claimed as trademarks. Where those designations appear in this book and Addison-Wesley was aware of a trademark claim, the designations have been printed in initial capital letters (for example, Actifed).

Library of Congress Cataloging-in-Publication Data

Pantell, Robert H., 1945–
    Taking care of your child: a parent's guide to medical care /
  Robert H. Pantell, James F. Fries, Donald M. Vickery.—4th ed.
      p. cm.
    Includes index.
    ISBN 0-201-63293-4
    1. Children—Health and hygiene. 2. Children—Diseases.
3. Family—Health and hygiene.   I. Fries, James F. II. Vickery,
Donald M.   III. Title.
RJ61.P215 1994
618.92'0024—dc20   93-37850
                                     CIP

Cover design: Mike Stromberg
Text design: Editorial Design/Joy Dickinson
Production: Michael Bass & Associates
Electronic composition and page make-up: Publishers Design Studio, Inc.

0-201-63293-4
0-201-40782-5

First edition, 1977, twelve printings
Second edition, 1984, sixteen printings
Third edition, 1990, seven printings

2 3 4 5 6 7 8-MA-97969594
Second printing, January 1994

Instructions on the abdominal-thrust maneuver for choking (pages 253–256) have been adapted from the National Safety Council publication *Family Safety and Health*, Winter 1986–1987.

TO OUR READERS

*T*his book can be of great help to you and your family. The medical advice is as sound as we can make it. Many doctors have reviewed each section, and we have tried to capture for you the essence of the standard medical recommendations for each problem. But the advice will not always work. Like medical recommendations of all kinds, it will not always prove successful. There is also a dilemma: If we don't give you direct advice for taking care of your children, we may not be helpful; if we do, we will sometimes be wrong. So here are some qualifications:

- If your child is under the care of a doctor and if you receive advice contrary to that given in this book, follow your doctor's advice; the individual characteristics of your child's problem can then be taken into account.

- You know your child best; do not hesitate to follow your own judgment as to when to seek professional assistance.

- If your child has an allergy or a suspected allergy to a medication, check with your doctor, at least by phone. With any medication, read the label instructions carefully; instructions vary from time to time, and you should follow the latest.

- If your child's problem persists beyond a reasonable period, you should usually make an appointment with a doctor.

CONTENTS

PART II

The Growing Child 69

CHAPTER 4

Growth and Development 70

CONTENTS

PART III

The Healthy Child 155

CONTENTS

CONTENTS

CONTENTS

*A*cknowledgments

The following parents contributed to the development of this book and their help was invaluable: Anne and Dave Bergman, Val Blanchette, Nora Blay, Jani Butler, Susan Charles, Libby and Milton Clapp, Linda Collins, David Curtis, Audrie Engbretson, Barbara and Joel Falk, Lee Hall, Mary Ann and Arthur Hanlon, Lenore and Larry Horowitz, Lois Kazmer, Michaline Krey, Shirley Lewis, Irmhild and Matthew Liang, Pit Lucking, Maria Maglio, Margaret Olebe, Kay Neumann, Suzanne Pennycook, Rosa Delgadillo Reilly, Kitty and Glenn Roberts, Marni Smith, Tom Stewart, Melissa Thornhill, Heather Van Nostrand, and Anne Wilkins. Several children have also contributed material: thanks to Loren and Matthew Loney, and Ciaran and Rylan Jacka.

We would also like to thank the following individuals for their assistance and review: G. Robin Beck, M.D.; John Beck, M.D.; June Bernzweig, Ph.D.; Irene Cannon, M.D.; Mark Constantz; Doris Denney, R.Ph.; June Fisher, M.D.; Victor Fuchs, Ph.D.; Judy Geisinger, M.S.W.; Dewleen Hayes; Halsted R. Holman, M.D.; Larry Horowitz, M.D.; Charles Irwin, M.D.; Al Jacobs, M.D.; Catherine Lewis, Ph.D.; Matthew Liang, M.D., M.P.H.; Jane Morton, M.D.; Bev Kusler; Tom Mcmeekin, M.D.; Pat Moylan, R.N.; Bonnie Obrig; Margaret Olebe; Tom Plaut, M.D.; Shirley Rudd, R.N.; Alan Shapiro, M.D.; John Slane, R.Ph.; Tom Stewart, Ph.D.; Elihu Sussman, M.D.; John Wasson, M.D.; and Annie Wilkins.

We include special thanks for major conceptual guidance to David Bergman, M.D., and Marcia Pantell, M.S.W., M.P.H., and to Maureen Shannon, M.S., F.N.P., C.N.M., for her considerable contributions to the revisions of Chapters 1 and 2.

The clerical assistance of Harry Reddrick and Susan Tettelbach made this edition possible.

For continuing advice, review, and encouragement we should like to acknowledge our wives Maureen and Sarah, and our children Andrew, Elizabeth, Gregory, Gregory Michael, Matthew, Megan, and Meredith. Their daily input and effort make this as much their book as ours.

Finally, special insights and many contributions for this fourth edition were provided by the antics of Gregory and Megan, siblings who are 7½ months apart in age, and the continuing adaptation of their older brother, Matthew. Inspiration for much of the new writing came from Maureen, who as a mother nurtured, refereed, dealt with earthquakes and hurricanes, and did it all with style.

Introduction

*Y*ou *can do more for your child's health than your doctor can.*
This statement began the first three editions of *Taking Care of Your Child* and still rings true. Since the last edition, diseases such as AIDS have gained momentum. Problem behaviors such as substance abuse are taking increasing tolls. New medical technologies, including immunizations and CVS, have emerged. However, neither the diseases nor scientific breakthroughs that capture headlines are the principal determinants of whether your children remain healthy. Accidents still claim the lives or permanently injure more young children than any other cause. As parents, you have it in your power to reduce injuries and save lives by securing your child in automobile safety restraints, insisting on the use of bicycle helmets, and turning your water-heater temperature down. You also have considerable control of your child's medical destiny by being informed about the availability of proven preventive measures, such as immunizations, by being aware of the risks and benefits of procedures for which there is no universally accepted policy, such as circumcision, and by modeling good health through not smoking.
In the past few years, we have increasingly witnessed the phenomena of "hurried" and "overburdened" children. Depression, drug abuse, and suicide are major concerns for children. It is no secret that families are under increasing stress. An important part of childhood is learning to cope with stress and adapt; however, children

can become overwhelmed. Parents know their children best. Although parents cannot shield their children from today's family stresses and social pressures, parents have the responsibility to be alert to the signs of emerging difficulty and to enlist proper help.

We undertook this revision in response to the changing forces affecting children's health and to keep current with medical advances available to children. Although most of the decision charts we developed have weathered well and required little or no revision, there are constant advances in our understanding of common problems, which have enabled us to make many of the decision charts more accurate. New technologies and developments concerning some problems also prompted an update. Many sections required expansion. Finally, we have included new material to update some of the major concerns of today's family, including child care, divorce, discipline, and exposure to AIDS.

Some things, however, have not changed. Your care and judgment are still the most important ingredients for fostering a lifetime of health for your child. A sound diet, a clean environment, and sensible living habits are the best way to reach this goal. However, illnesses and injuries are a part of growing up. While some of these problems will require professional assistance, many can be handled at home. The purpose of this book is to help you manage the common problems of childhood and to enable you to make better decisions about when to see a medical professional.

Medical science encompasses a vast body of knowledge and an intricate technology. It is easy to become intimidated by the complexity of modern medicine and to feel uncomfortable making commonsense medical decisions. The most commonly encountered medical problems, however, are usually uncomplicated and inevitably get better by themselves. Each day you make sound decisions requiring good judgment in caring for your child. But there will always be a time when you are confronted with a new problem or an illness that is not typical. This book should help you in these times of uncertainty and improve your ability to judge how best to use your doctor.

There is an expression used by many medical educators: "When you hear hoofbeats, don't think of zebras." This has been our guiding principle. We have tried to focus in *Taking Care of Your Child* not on rare events that might happen, but on what probably will occur. Our own children have encountered the majority of the problems outlined in this book. Most of these problems can be successfully managed at home.

This book is divided into three sections. Parts I–III provide information helpful to understanding the several subject areas that we have found to be of greatest concern to parents. You will find discussions of pregnancy, birth, and physical development, as well as behavioral, school, and family problems, and advice for staying healthy and for stocking the medicine cabinet. We have tried to give you both sides of such controversies as home versus hospital birth, circumcision, and immunizations. Parts I–III provide background information for your role as parent, and will assist you in developing a long-range strategy aimed at ensuring a healthy life for your child. In attempting to capture the complex feelings that develop between parent and child, we have relied on parents' statements of their experiences throughout the book.

Part IV is the heart of this book; it provides specific guidance for 100 of the medical problems most likely to be encountered by infants, children, and adolescents. The charts in this section present step-by-step guidelines to help you decide on a program of home treatment or whether to make a telephone call or visit to your doctor. These charts are useful as an aid to your common sense and not as a substitute; you know your child best.

Part V provides a place to record growth, development, and medical information for each child.

As parents, we preside over the maturation of our children from complete dependence at birth to complete independence less than two decades later. In gaining independence, a young adult needs to learn how to make personal and medical decisions with confidence. Teaching children to make independent decisions about their own health is one of the most important things a parent can do. The decision guidelines in this book can help you and your child with these important skills.

The typical child sees a doctor four times a year and has many more problems managed at home. Your family's experience with these scores of inevitable illnesses and accidents provides an opportunity for your children to learn, directly or by example, skills that will last a lifetime. The information presented in this book is intended to help you increase your confidence in the commonsense decisions you must make. In this process, you may save time and money, to be sure, but the real goal is to assist you in helping your child become a healthy adult.

PART I

Birth of a Family

CHAPTER 1

A Child Is Coming

*P*regnancy is as complex as any natural event in our lives. It is exciting, exhausting, fulfilling, fatiguing, happy, sad, simple, creative, and full of many more contradictions for both mother and father. In a few pages, we cannot do justice to a subject about which poets, parents, philosophers, and doctors have written extensively. Taking care of your child, however, begins with pregnancy. In this chapter, we will cover some of the most relevant topics that may concern you at this important time.

I'm Pregnant, We're Pregnant

There is no such thing as an unnatural feeling during pregnancy. Most parents find that they experience a variety of emotions, from uncertainty and anxiety to satisfaction and exhilaration. We have tried to capture some of these feelings by collecting written accounts and recording conversations with parents. These statements, and those in the following chapters, are only a sampling. You will certainly experience feelings not found on these pages. Indeed, many people will be concerned if they initially have no feelings about such an important occurrence. This, too, is a common experience, especially before the first movements of the developing infant are felt. It is difficult to appreciate the transition in feelings that occurs by reading isolated statements.

Many of the parents who expressed tremendous anxiety about child raising during pregnancy have found that the actual task is natural and rewarding. The advice of one mother is important to remember: "In having children and raising children, nobody is likely to experience something that hasn't happened to someone else. All of the emotions, from terrible fear to joy, have been shared by millions of others many times before."

MOTHER OF TWO:
Before our first child was born, I hadn't given very much thought to what our lives would be like after the birth itself, which occupied most of my attention. If I thought at all about it, I guess I had a subconscious picture of myself in a Woman's Day *ad, one of those slightly out of focus pictures of a beautiful young mother in a pastel dressing gown nursing her baby in a pose of peaceful fulfillment. I must have imagined that she took care of her baby in her spare time. Was I surprised!*

EXPECTANT FATHER:
Libby kept asking me, "Aren't you excited about it?" She, of course, already was, and I had to admit to her disappointment that it was just hard to realize that we now had a baby on the way and I really had no solid feelings or comprehensions about the child. She did come home wearing one of those tacky T-shirts with "baby" bold across the front. My shy Libby did this. At first I wasn't too interested in being seen at the A & P with her and her new brazen shirt. But she was and still is proud of her tummy. I begrudgingly gave her my hand which she pressed firmly on her tummy. I told her not to press too hard. I could hurt the baby. I still have a feeling of slight uneasiness when pressing on her tummy.

MOTHER OF SIX:
He said, "I'm so glad you are going to have your baby at last," and I said, "But it isn't me having mine, it's us having ours," and he said, "But the baby is so much more the mother's thing." And something went cold, and I often wish I had walked away then, found someone who wanted to share it with me.

EXPECTANT FATHER:
We seem to blunder into most of the major decisions in our lives with no facts and no experience. We're going into child raising with a lot of ignorance and a lot of trepidation. My cousins asked us to baby-sit for their four kids recently and we realized for the first time we had very discordant philosophies about raising children. I always thought that I would naturally know how to raise kids—it never occurred to us we might have very different philosophies.

FATHER OF TWO:

We had been trying for several months before Lenore got pregnant. When each month came and went with no pregnancy, we were filled with both relief and anxiety, and I think it is the mixture of the two emotions that is important. When Lenore did become pregnant, my first reaction was one of great exhilaration and excitement. But there must have been subconscious underpinnings of uneasiness or uncertainty because within a very few days these feelings surfaced. The exhilaration of starting a family was tempered by the uncertainty of not knowing what kind of change in our lives was to be forthcoming. Would we lose freedom? Would our new lifestyles be as comfortable as our present ones? Would the rewards justify the work? What was parenting all about anyway? We tended to try to cram many things into the nine-month pregnancy period. We had the feeling that each trip we took, each vacation, each special activity, each late-night dinner might be the last for a very long time. This ambivalence continued throughout the pregnancy, but as Lenore changed physically, I felt more sure that this was a very positive thing. The anxieties over the unknown were still present, but somehow less important. Maybe they were replaced by the anxieties surrounding pregnancy and delivery. Would the baby be healthy? Would the delivery be easy? Would Lenore be well?

MOTHER OF TWO:

I was totally unprepared for being almost solely responsible for her care . . . Why did I waste nine months thinking about a birth that lasted only 20 hours instead of how to live with a dependent human being for another 20 years? I did not even see or hold a newborn in the time I was pregnant. I never saw breast milk or formula bottles. And none of the people who were professional helpers seemed to think it was important.

MOTHER OF ONE:

Being brought up as a woman, you see child care as your responsibility, and being brought up as a man, John doesn't even conceive of some of the things that could fall apart or could possibly go wrong.

MOTHER OF TWO:

The one big thing that was taught in the Catholic Church was that in the event of a problem at birth, you were supposed to save the child. The doctors aren't allowed to make the decision, and they come out and ask the husband and he's supposed to say, "Save the child." And I woke up upset and told David if that happened, I would really like a chance to try again.

EXPECTANT MOTHER:
We've hired a person to do full-time child care. The first month we hope to share tasks of child raising. I hope to get a feeling of what it will be like to leave the child with her before I actually do it. It really will be an experimental time. I hope it won't be a competition. Other working mothers who have had this arrangement have reassured me that the children know who mommy is.

EXPECTANT MOTHER:
I feel much more protected and cared for. Bob shares work a lot more than he did.

Taking Care of Yourself During Pregnancy

Taking care of yourself during pregnancy is just like living sensibly at other times, with one obvious exception: You are now providing for two individuals. Your habits—smoking, drinking, taking drugs or medications—now affect a second life. Most things that you have done before becoming pregnant can continue without interruption. We will spend some time now discussing exceptions to this rule and some current misconceptions.

DIET

A well-balanced diet is always advisable. Your nutritional status at the beginning of pregnancy is probably as important as what you eat during pregnancy. The exact amount that you need to eat during this time varies according to your individual requirements, but common sense tells us that extremes are harmful.

Your total pre-natal weight gain should be between 24 and 35 pounds. Women who are underweight should gain 35 pounds, whereas obese women may need to gain as little as 20 pounds. Gaining more than 35 pounds makes it harder to get back in shape after delivery. Most of the weight gained is for the infant and for the increased size of the uterus and breasts and the increase in blood volume. Less than one-fourth of the weight is for increased stores of fat and protein. During pregnancy, you will need about 15 percent more calories, and an additional 20 to 30 grams of protein, on the average, each day.

Most women do not need elaborate dietary changes and caloric bookkeeping during pregnancy; common sense and a good diet will usually suffice. The increased nutritional requirements you will have throughout your pregnancy will increase again during breast feeding by another 200 calories per day.

These are the nutrients that are most important during pregnancy:

- The majority of the extra calories you eat should come from increased **protein** (milk, meat, fish, poultry), the basic building blocks of fetal development.

- **Carbohydrates**, such as bread, potatoes, and cereals, also provide energy for the developing fetus. Restricting carbohydrates forces your body to rely on other sources of energy such as fat; too heavy a reliance on fat produces chemical by-products known as ketones, which alter your mood and are potentially harmful to both mother and fetus.

- **Fats** (butter, cheese, meat, whole milk), however, are also required for fetal development; they aid in the absorption of important vitamins.

- **Vitamins and minerals** are largely provided by fruits and vegetables. The requirements for Vitamins A and C increase considerably during pregnancy, but are usually met by an ordinary diet. In general, a proper diet will supply all nutritional requirements, but many doctors rely on vitamin supplements if they are uncertain of your diet.

    Folic acid, one of the B vitamins found in milk and green vegetables, is required for the creation of blood. The need for folic acid increases during pregnancy, hence the occasional need for pre-natal vitamins. Pre-natal vitamins differ from regular vitamin supplements by the addition of folic acid. Adequate folic acid may reduce the birth defects known as "neural tube defects"; for this purpose, taking supplemental folic acid six weeks *before* pregnancy and throughout the first trimester is important, if possible. In general, a proper diet will supply all nutritional requirements, but many doctors rely on vitamin supplements if they are uncertain of your diet.

- The body requirement for **calcium**, important for maintaining sound bones and preventing muscle spasms, also increases considerably. Calcium is best supplemented by an intake of milk, cheese, or eggs. Broccoli and oranges are also rich in calcium.

■ More **iron**, needed for blood building, is also required. Foods rich in iron include meats, cereals, and many vegetables, such as peas, spinach, lima beans, and lentils.

The human fetus is remarkable in its ability to obtain nutrition from the mother. If there is a shortage of the necessary nutrient, the fetus will receive preferential treatment in getting what is available. Improper nutrition is thus detrimental first to the mother, and then to the fetus. However, significant shortages may result in the fetus not getting enough calories.

Cravings

Many mothers report cravings for fruits and vegetables during pregnancy; these are undoubtedly the body's way of telling you what you need. Feel free to follow your cravings. Eat sensibly, and review your diet with a nutritionist or doctor should any questions arise.

Nausea

During the first three months of pregnancy, nausea and vomiting may interfere with your normal pattern of eating. Morning sickness occurs in many mothers, whereas others are most nauseated at suppertime. Having frequent small feedings is often the best way to obtain nutritional requirements in these circumstances. Many drugs used to treat nausea are potentially dangerous during pregnancy and should be avoided unless absolutely necessary and after full discussion with your doctor.

FOODS TO AVOID

Beef liver, which we warned against in our previous editions, is now safe to consume. The industry finally yielded to pressure to remove diethylstilbestrol from feed. Diethylstilbestrol was used years ago to prevent miscarriages, and mothers taking the drug produced daughters with a higher risk of developing vaginal cancer than daughters of mothers not exposed to the drug.

On December 3, 1960, the Food and Drug Administration (FDA) published an order prohibiting the use of sassafras in foods because it was found to cause cancer of the liver in animals. Sassafras may still be available in some parts of the country as a tea. It should be avoided during pregnancy. In 1977, the FDA raised concern about artificial sweeteners; they should also be avoided during pregnancy.

Pregnant women should avoid alcohol. Drinking can cause the baby to suffer fetal alcohol effects.

EXERCISE

Feel free to continue any physical exercise that you enjoyed before becoming pregnant. However, moderation is in order. This is not a good time to train for a marathon or begin an even moderately strenuous aerobics class. Skydiving and scuba diving are also not safe during pregnancy. In many parts of the country, there are now exercise classes especially designed for pregnant women.

Because of the ability of the fetus to preferentially obtain an adequate blood supply, some of your physical reserves for strenuous exercise may be lost during pregnancy. Consequently, you may tire more quickly and feel faint, particularly at high altitudes. Basically, your body will tell you when to stop. You should also use your own judgment about certain types of physical exercise that involve the possibility of abdominal injury—for instance, skiing and hockey.

TRAVELING

With one exception, there are no traveling restrictions except those dictated by common sense. As you approach the expected day of the birth, you should avoid trips that will place you out of striking distance of the site you have chosen for delivery. If you are planning on flying in your ninth month, commercial airlines request a letter in triplicate from your doctor, stating that he or she feels it is safe for you to fly.

SEX

Some women have decreased libido during pregnancy, whereas for others the opposite is true. Both are normal. Intercourse is fine if everything is progressing normally. There is some evidence that intercourse during the last several weeks of pregnancy may increase the risk of certain complications, and it is probably wise to avoid intercourse during that time. Women experiencing pre-term labor should avoid both intercourse and orgasms because of the risk of increasing uterine contractions. In addition, vaginal bleeding, possible premature labor, or premature rupture of the amniotic fluid sac at any time during pregnancy are reasons to abstain.

After delivery, an episiotomy incision that is healing may cause discomfort with intercourse for several weeks. Other than during these periods or events, intercourse is possible whenever both partners are willing and able. Other forms of sexual pleasuring should be used and explored during this time—this can help re-establish the couple's sexual intimacy.

PRE-NATAL FETAL TESTING (SURVEILLANCE)

Amniocentesis

Amniocentesis is a procedure in which amniotic fluid is removed from the uterus by means of a needle inserted through the abdominal wall. The purpose of obtaining the amniotic fluid is to evaluate the risk of a child having an inherited disease. It is usually done between the 12th and 14th week of pregnancy. Later in pregnancy, amniotic fluid may be tested to evaluate the infant's lung maturity if a premature cesarean delivery is planned. The procedure is complex, is expensive, requires a skilled doctor, and has complications. It is not for everyone.

Amniotic fluid can be evaluated for the presence of genetic diseases, such as Tay-Sachs disease, cystic fibrosis, sickle cell disease, beta-thalassemia, and muscular dystrophy or a chromosomal disease such as Down syndrome, formerly known as mongolism. In families with known metabolic diseases, the risk of a child having the disease is usually one in four if both parents carry the gene for the disease. Many metabolic diseases can be diagnosed by amniotic-fluid analysis. In the process of examining for abnormal or extra chromosomes, the sex of the child can also be determined by looking for the presence of an XX chromosome pattern (female) or an XY chromosome pattern (male). You can ask your doctor not to reveal your child's sex if you wish.

Down syndrome (mongolism) is one of the most common chromosomal disorders caused by an extra chromosome. As women become older, the chance of having a child with Down syndrome increases. Only one in 2500 women at age 25, but one in 40 women at age 45, will have an affected child. Advancing age is therefore one reason for contemplating amniocentesis. Younger women with an affected child already may be concerned about having a second child with this condition. Although there are seldom recurrences, many of these parents may be interested in considering amniocentesis.

Some chromosomal problems, such as hemophilia, occur only in males. Women carry the gene for hemophilia but cannot have the disease. Amniocentesis can identify a developing male, but only 50% of males with carrier mothers will actually have hemophilia.

Amniocentesis is relatively safe, but there are risks. Spontaneous abortions and infant blood system problems (Rh sensitization) have occurred, but in competent hands, the risk of a serious complication should be less than 1%. The risk of a spontaneous abortion after amniocentesis is approximately 1 in 200. Occasionally, no amniotic fluid is obtained, the laboratory is unsuccessful in its attempt to analyze the fluid, or there are test errors in the results. In such cases, the procedure must be repeated.

There are also many ethical considerations. Which "disorders" would warrant termination of the pregnancy, if you accept the practice of abortion? A given inherited disease can affect one child seriously and the next child minimally. What should be done with a fetus having a 50% chance of being born with a devastating disease? What do you do if you are looking for one problem and accidentally discover another minor problem? Will amniocentesis allow elimination of dread diseases or encourage abortion for even the most minor disabilities?

If you are considering amniocentesis, you should discuss it extensively at home and then with your doctor. In the end, you should make a decision consistent with your own moral framework. Doctors cannot tell you whether you should accept or reject a child with a given disorder. Complete information is often difficult to obtain about amniocentesis, but persist until you are satisfied that you have enough information to make a decision.

Chorionic Villus Sampling

This technique provides information about the genetic and chromosomal characteristics of the developing fetus. It is similar to amniocentesis. There are both advantages and disadvantages when compared to amniocentesis. On the positive side, chorionic villus sampling (CVS) provides information at an earlier stage of pregnancy (8 to 12 weeks). Consequently, if a woman chooses an early-stage abortion because of information provided by CVS, her risks from the procedure are considerably diminished. However, CVS carries a slightly higher incidence of fetal death compared to amniocentesis. Again, it is essential to learn how risky *your* procedure will be in the hands of *your* physician. Another difference between CVS and amniocentesis is that CVS is done too early in pregnancy to test for alpha-fetoprotein, a substance that is an indicator of spina bifida in developing infants.

Percutaneous Umbilical Cord Blood Sampling (PUBS)

This is a relatively new procedure in which the fetus's blood is sampled directly. The advantage is that studies can be performed far more quickly, particularly chromosomal analyses. It has risks, and is not as widely used as amniocentesis or CVS. Question your doctor about the risks as discerned from his or her experience.

Alpha-Fetoprotein

A blood test in a pregnant woman can identify an elevated level of a chemical called alpha-fetoprotein (AFP). A number of things can explain an elevated level, including twins or inaccurate dating of the pregnancy. However, about 2 to 4% of women with elevated AFP will be carrying an infant either with *spina bifida* (an opening in the back revealing the spinal cord) or lacking a brain (*anencephaly*). Further counseling and diagnostic testing, including amniocentesis and ultrasound, will be required if persistent elevation of AFP occurs.

Ultrasound
(Sonogram)

Ultrasound is used in many fields of medicine. High-frequency, pulsating sound waves are bounced off an object and form an image of this object on the screen. It is thus possible, by looking at the image, to determine the size, shape, and other important features of many parts of the body. In some obstetrical practices, ultrasounds are routinely used to confirm pregnancies or to date pregnancies. This is a practice that we do not recommend. Some parents request sonograms because they are interested in obtaining a snapshot of their baby in the uterus, or have desires to see the fetus move or to know beforehand the sex of the baby. Again, we are not in favor of these practices.

There are a number of important reasons for a sonogram to be obtained. Whenever there is the possibility of a complication in pregnancy (for example, too much or too little amniotic fluid, uterine growth that is either too little or too much), or a question of more than one baby, a sonogram should be done. Sonograms are done to confirm the date of the expected birth if the mother is unsure of her last menstrual period and there is an important medical reason to be precise about the due date. Sonograms are also appropriate if an amniocentesis is to be done. In this situation, the sonogram enables the exact location of the baby and the placenta to be known before and during the procedure.

The risks of ultrasound appear to be exceedingly low. No documented complications have been reported for abdominal ultrasound. Vaginal ultrasound is still relatively new and we encourage its use only when specifically indicated. To assume that it is a completely harmless procedure would be foolhardy, primarily because we have not gathered sufficient data over a significantly long period of time to declare ultrasound safe in its use in all pregnancies.

When a sonogram is recommended by your physician or midwife, ask him or her why it is being recommended. If the answer is that the procedure is routinely done on all patients to see how far along they are or to make sure the baby is okay, you may want to question the practitioner further about the possibility of specific complications in your pregnancy that may warrant this test. Furthermore, the cost of the procedure is considerable, adding an additional $100 to $200 per sonogram to your medical expenses.

If the procedure is being done because you are at an increased risk of having a child with particular abnormalities, it is important to make sure that you are referred to a radiologic group that has a fair degree of expertise in pinpointing such abnormalities. Also, do not rely on the interpretation of sex according to the sonogram, as errors are often committed in making this judgment.

MEDICINE

All drugs are potentially harmful, and the risk of taking drugs increases during pregnancy because of potential harm to the fetus as well as the mother. Despite this, American women consume on the average 4.5 different drugs during pregnancy, and 80% of these drugs are not prescribed by doctors. Heavy drug use is not surprising when one considers the number of drug messages we see and hear every day. Pharmaceutical preparations are the type of product most advertised on television. Resist them!

As a general rule, any drug that a mother takes will reach the fetus, and all drugs are potentially harmful to the fetus. The majority of drugs now on the market have never had their safety tested with respect to the fetus; effects on human infants are often discovered only later. Thalidomide, for example, was tested on rodents and did not produce the limb deformities it later did in human offspring. Documentation is difficult because some of the problems may occur in as few as 1 in every 10,000 to 50,000 infants. Although the risk may be small, the consequences may be severe, as with vaginal cancer following diethylstilbestrol usage. There is no need to take even a small risk unless a medication is absolutely necessary.

Aspirin is sometimes the exception to this rule. Talk to your doctor about whether to take one baby aspirin per day in the last half of pregnancy to thin your blood and prevent blood clots in the placenta. Some physicians recommend this low dose of aspirin to mothers who have experienced hypertension during previous pregnancies or who are pregnant for the first time but whose sister(s) or mother had severe pregnancy-induced hypertension (*pre-eclampsia*).

Table 1 is a list of drugs to be especially avoided, along with explanations of the problems they cause. Not all of these associations are well documented. However, when considering infant safety, we feel that drugs should be regarded as guilty until proven innocent. Many medications change in composition when stored on the shelf and become harmful only with increasing age. During pregnancy is a good time to throw out *all* medications on your shelf. Before taking any —even aspirin—while pregnant or nursing, consult your doctor. You might also try to obtain Gail Brewer's *What Every Pregnant Woman Should Know: The Truth About Diets and Drugs in Pregnancy* (New York: Viking/Penguin, 1985).

IMMUNIZATIONS AND ALLERGY SHOTS

All immunizations should be avoided during pregnancy; this is especially true for rubella (German measles) immunizations. Women receiving rubella immunizations should avoid becoming pregnant for three months afterward. Although no birth defects have been reported in the few children born to women accidentally immunized with rubella vaccine during pregnancy, the potential for problems remains. There does not seem to be any necessity for avoiding contact with children who have recently been immunized. Also, allergy desensitization should be discontinued during pregnancy.

PETS

Cats are often feared during pregnancy; some of them carry and transmit a disease called toxoplasmosis. Toxoplasmosis, like German measles, produces few symptoms in the mother (occasionally, there may be fever or lymph gland swelling), but is potentially harmful to the developing fetus. Toxoplasmosis can be transmitted in the excrement of cats. It is also found in meat, but is only transmitted in meat that has been undercooked. Toxoplasmosis is very common (30 to 40% of the population has had it). Congenital toxoplasmosis, a relatively rare form of this illness, appears in newborn children of women who become infected for the first time during early pregnancy.

TABLE 1   *Drugs to Avoid*

Medication	Potential Problem in Infants
Aspirin–phenacetin combination	Oxygen-carrying ability of blood decreases
Acetaminophen (TYLENOL, TEMPRA, VALADOL)	Kidney problems
Alcohol—chronic abuse or high doses	Seizures, low blood sugar Growth retardation Birth defects
Amphetamines	Birth defects
Anticonvulsants	
CARBAMAZEPINE	Facial defects Mental retardation
PHENYTOIN	Small size Facial defects
VALPROIC ACID	Brain and heart defects
Antidepressants (tricyclic) IMPRAMINE (TOFRANIL) AMITRYPTYLINE (ELAVIL, ETAFRON)	Birth defects
Antihistamines	
DIPHENHIDRAMINE—CHRONIC USE (BENADRYL)	Seizures
Antihypertensive drugs (CAPTOPRIL, ENALAPRIL)	Skull defects Small size
Antinausea drugs	
MECLIZINE (BONINE)	Birth defects
CYCLIZINE (MAREZINE)	Birth defects
CHLOROCYCLIZINE	
Antibiotics	
SULFA DRUGS	Yellow jaundice
TETRACYCLINE	Malformed teeth Suppressed bone growth Cataracts
KANAMYCIN	Hearing loss
STREPTOMYCIN	Hearing loss
Bronchial medications	
POTASSIUM IODIDE	Goiter
Caffeine—chronic heavy use	Growth retardation

Medication	Potential Problem in Infants
Cocaine (illegal)	Brain damage Growth retardation Limb defects Kidney problems
Hormones	
ESTROGENS (DIETHYLSTILBESTROL)	Vaginal cancer
ANDROGENS	Masculinization of daughter
PROGESTINS	Birth defects Growth retardation
LSD (illegal)	Limb defects
Marijuana—chronic use (illegal)	Unknown
Narcotics (illegal)	
(HEROIN, METHADONE)	Growth retardation, withdrawal symptoms
Nicotine (ten or more cigarettes daily)	Small size
Skin preparations	
ACCUTANE	Birth defects
Tranquilizers	
DIAZEPAM (VALIUM)	Decreased body temperature
CHLORDIAZEPOXIDE (LIBRIUM)	Seizures
BARBITURATES	Withdrawal symptoms (irritability, seizures)
CHLORPROMAZINE (THORAZINE)	Temperature regulation
PROMETHAZINE (PHENERGAN)	Bleeding
Vitamin excess	
C	Skin and intestinal problems
D	Heart valve problems
Vaginal preparations	
FLAGYL	Cancer (in rats) Mutations (in bacteria)

If you have cats and are concerned, your doctor can do a blood test that will inform you of whether you have already had toxoplasmosis; if so, you need not concern yourself about this problem during your pregnancy. The commonsense approach is for the pregnant woman to avoid the kitty litter box and not to buy a new cat. A cat that has been in the family for a while is less hazardous, so you can keep and love your old cat during pregnancy with relatively little risk.

INFECTIONS

It is impossible to avoid all infections during pregnancy. Immunizations before pregnancy can prevent some infections that are quite harmful to developing infants. *All* infection exposures, particularly illnesses with rashes, should be reported to the physician or nurse midwife supervising your pregnancy. Table 2 describes the risks of certain infections, along with helpful measures regarding them.

Breast Feeding or Bottle Feeding?

Decide how you are going to feed your baby before the birth. Breast feeding is best for some mothers, whereas others will prefer bottle feeding. Either choice is fully acceptable and will be guided by *individual* circumstances.

It is unfortunate that fashion is often responsible for making a mother feel she should choose one particular method. Bottle feeding became popular 50 years ago. The society at the time wanted to do everything "scientifically," women wanted to be free to leave the home, and the female breast was shifting from being a nursing object to being

TABLE 2   *Infections That May Occur During Pregnancy*

Disease	Risk to Fetus	Suggestion
Chicken pox	Early pregnancy—malformation Late pregnancy—severe newborn disease	Avoid exposure, immunoglobulin within three days of exposure
Fifth disease (parvovirus B-19)	Mild childhood illness that can cause newborn to swell with fluid (hydrops)	Avoid children with characteristic rash (page 412); consult physician
Herpes	Genital infection can lead to brain infection of newborn after delivery	Discuss with obstetrician
Rubella	Infection in early pregnancy causes a variety of defects	Have immunization status checked before pregnancy
Toxoplasmosis	Fetal death or malformation	Avoid cat excrement and undercooked meat

a sexual object. Fortunately, social attitudes are constantly in movement. Currently, there is widespread acceptance of mothers nursing in public. Breast feeding is now gaining in popularity and is currently considered to have many "scientific advantages."

It is important for mothers to realize that both methods supply sound nutrition and intimacy to the child. Neither choice should provoke feelings of guilt in the mother or father. Your baby is happy when you are happy. Choose the method *you* feel most comfortable with. Both breast feeding and bottle feeding require you to learn new skills. The choice is yours.

Making an informed decision requires, among other things, the availability of factual information, which we attempt to provide here. Given the proper encouragement and assistance, nearly every mother can breast feed her child. A few rare medical conditions, such as severe breast infections or infants with a cleft palate, may make breast feeding more difficult. Many pediatricians now recommend breast feeding for most women, and we do too. Here are some of our reasons.

ADVANTAGES OF BREAST FEEDING FOR THE CHILD

- Nutrition is provided in the proper proportion, if the mother is eating a balanced diet. Proteins, carbohydrates, fats, minerals, vitamins, and iron are all provided in amounts suitable for adequate growth. Although at one time iron drops were thought necessary to supplement human milk, this is no longer felt to be true. Only vitamin D, to the best of our knowledge at this time, does not seem to be present in adequate amounts in breast milk. Although exposure to sunshine will provide some vitamin D, supplemental vitamin D is necessary.

- Breast feeding is hygienic. There is less chance of contaminating the milk than with bottle feeding.

- A mother can spend time being very close to her baby.

- There may be fewer respiratory and intestinal infections in breast-fed infants. In addition, maternal immunity to a number of common viral illnesses may be transferred to the infant through breast feeding.

- There may be fewer allergic disorders of the skin, respiratory tract, and intestine in breast-fed infants.

- In poverty areas in underdeveloped countries, the death rate is lower among breast-fed infants. In the United States, there does not seem to be a difference; fortunately, infant death is unusual with either feeding method.

- Breast-fed babies tend not to become fat as frequently. The composition, and therefore the taste, of breast milk changes during the course of a feeding; thus infants will reject a breast after several minutes and move on to the next breast, even though the first breast still contains milk. This is one of nature's ways of preventing obesity in the newborn child.

- The stools of breast-fed babies are lighter in color and looser than those of babies who are fed a formula. Because the stools are soft, breast-fed babies may have fewer irritations of the rectum, and they hardly ever become constipated. This looseness should not be interpreted as diarrhea. In fact, breast-fed babies have diarrhea less often. A strange bonus—breast-fed babies' bowel movements smell better!

ADVANTAGES OF BREAST FEEDING FOR THE MOTHER

- Hormone release during breast feeding causes rapid contraction of the uterus and encourages the return to normal of the uterus after delivery.

- It is a period of relaxation for the mother. In a society in which time is money, this is nature's way of forcing you to slow down and relax.

- The cost is less. Bottles, nipples, and baby formula are more expensive than the extra nutrition required by a nursing mother.

- Breast feeding is convenient. There is no need to fuss with the paraphernalia of bottle feeding. You don't have to run to the store in the middle of the night to buy formula.

- Some reports suggest that breast feeding may reduce the frequency of blood clots in the legs, and some indicate that there may be a lower frequency of breast cancer in mothers who breast feed. These findings need more study.

MYTHS ABOUT BREAST FEEDING

- Breast feeding does *not* depend on your breast size. Large breasts contain predominantly excess fatty tissue and in no way increase the ability to breast feed.

- Inverted nipples are *not* an insurmountable problem.

- Breast feeding does *not* interfere with polio virus immunization. The trivalent vaccine used in the United States is equally effective in breast-fed and artificially fed infants.

- Breast feeding does *not* cause more frequent jaundice in newborns. In some instances a baby may be taken off breast milk for several days to speed the resolution of the normal jaundice that occurs at birth. (See **Jaundice**, page 342.)
- Breast feeding does *not* cause sagging breasts. However, for comfort it is important that you wear a brassiere that gives adequate support during your period of breast feeding.
- Although it suppresses ovulation, breast feeding is *not* a reliable method of contraception.

PROBLEMS WITH BREAST FEEDING

- The schedule of breast feeding may interfere with the mother's ability to spend time away from home. This problem can sometimes be solved by taking along a breast pump and emptying the breast when it becomes uncomfortable.
- The more frequent stools will require more frequent changing of the infant.
- The father may feel left out. If the father desires to feed the baby, breast milk can be emptied by pump or manually and then given from a bottle by the father. Or a supplemental bottle of formula may be given by the father.

CAUTIONS ABOUT BREAST FEEDING

- Most drugs taken by the mother will be passed to the infant through breast milk. Valium, sulfa drugs, and tetracycline are particularly common and should be avoided because they are hazardous to the infant. In general, never take any drug without consulting your doctor first.
- Chemical contamination of food or water supplies has resulted in contaminants being found in breast milk. PCBs (polychlorinated biphenyls), PBBs (polybrominated biphenyls), DDT, and lead can pass into breast milk. Such rare instances are not an argument for bottle feeding but an argument for tighter regulations against chemical poisons in our environment.
- Infection can be transmitted through breast milk. HIV (the virus responsible for AIDS) is transmitted through breast milk, so HIV-infected mothers should not breast feed unless there is no other acceptable source of infant nutrition. Similarly, mothers should not allow other women to breast feed their infants because of the risk of HIV or other transmissable agents.

Techniques of breast feeding are discussed on pages 60–63. Good sources on breast feeding are *Nursing Your Baby* by Karen Pryor (New York: Bantam, 1991), *Feed Me, I'm Yours* by Vicki Lansky (Deep Haven, Minn.: Meadowbrook Press, 1986), *A Practical Guide to Breast Feeding* by Janice Riordan (Boston: Jones and Bartlett, 1991), and *Breastfeeding Your Baby* by Sheila Kitzinger (New York: Alfred Knopf, 1989).

FORMULA OR BOTTLE FEEDING

There are many advantages to formula feeding. The father can help and can become involved in feeding very early. In addition, there is somewhat more freedom of movement for the mother. Also, the intimacy and relaxation in nursing a baby can be enjoyed when bottle feeding as well. Parents who choose bottle feeding may be assured that the commercial formulas available today are high quality. Great efforts have been made to produce products that are as close in composition to breast milk as is feasible. Millions of Americans who were bottle-fed are a testimony to its overall safety.

Sterilization is a thing of the past in all areas where tap water is safe to drink. If it is safe for you, it is safe for the baby. Parents wishing to prepare several days worth of feedings in advance should consider sterilizing, however, because bacteria can grow in the formula after it has been mixed.

Bottles and nipples are all the equipment you will need. Infants obtain milk naturally by a process of squeezing their lower jaws and gums. They use their tongues to push up on the nipple and keep it securely on the roof of their mouth. The usual bottle nipples permit milk flow more easily than the breast does, causing the infant to exercise the jaws less and forcing him or her to stick the tongue outward instead of upward to retard the rush of milk. Nuk nipples were the first commercial nipples that allowed the infant to closely duplicate the sucking performed on the breast. Numerous nipples with this flattened shape are now available in latex and silicone. Some nipples have been found to contain nitrosamides; be sure to boil them several times before using. Don't boil longer than a few minutes at a time. Types of formula and techniques of bottle feeding are discussed on page 63.

Babies will grow beautifully on breast or bottle milk. Mothers should choose the method with which they feel most comfortable. Many mothers are reluctant to try breast feeding because they feel anxious. This is natural, because it is something new that appears to be complicated. In fact, breast feeding is an elegant function of the human body with which many other mothers can assist you. Nursing-mothers councils and the La Leche League are always willing to help. The best place to get help is from a friend who has successfully breast fed her baby.

Childbirth Preparation Classes

Pregnancy places different physical and psychological demands on both parents. The key elements of a successful pregnancy are adequate physical and psychological preparation. Coping with some of the physical demands for mothers, such as eating to meet the increased requirements, can be straightforward. So too is exercising to strengthen back and abdominal muscles by swimming and kicking. (Standing in place for a long period of time should be avoided during pregnancy.) However, there are very few daily exercises that prepare a woman for the demands on her perineal muscles. Similarly, the breathing best suited for labor is not something women often have a chance to practice. Childbirth classes teach about the special physical demands of pregnancy and the most suitable exercises.

The purpose of childbirth classes is to help the mother and father take an active part in the process of labor and birth. These classes attempt to focus the energies during birth into productive psychological and physical activity, hence eliminating the portrait of the mother as a passive person depending on the doctor to relieve her pain and to deliver her baby.

Classes will generally discuss in detail what occurs biologically during labor and delivery. They will train you to use your body most efficiently in order to deliver your child. This will include physical and breathing exercises and demonstrations. In addition, they will teach your partner how to help. Classes generally consist of six to ten two-hour sessions in the last three months of pregnancy. Many also include a tour and explanation of local delivery facilities.

LAMAZE

The Lamaze method, probably the most discussed of the current methods, attempts to go beyond mere physical conditioning and elimination of fears. The method is based on reconditioning, or changing the responses that the mother already has. The method here is to "enlighten the woman by instructing her about the phenomenon involved in childbirth, the purpose being to convert delivery from the idea of pain to a series of understood processes in which uterine contraction is the leading phenomenon." The other purpose of Lamaze teaching is to develop a new set of conditioned reflexes in an effort to block out the old set. By coupling a new breathing method with uterine contractions, it is hoped that the uterine contractions may stimulate breathing rather than pain.

It is unfortunate that Dr. Lamaze has entitled his book *Painless Childbirth*. The purpose of prepared childbirth is to help the mother, father, and infant experience a safe, non-traumatic birth, but there is often pain. It is important that family members can feel good and have the energy to enjoy the moment and begin to develop positive feelings about their infant. Pain can be terrifying if misunderstood; it is not necessarily an evil force to be eliminated at all costs by drug or psychological anesthesia. The pain you may experience during swimming, running, or childbirth should not be hunted down and eliminated. In fact, very few natural childbirth classes have as their goal the creation of a "painless childbirth," but rather a childbirth in which the family is best prepared to deal with the event. Parents should also realize that problems such as prolonged labor will occur despite the best preparation. There are many problems that have a physiological basis and are *not* the fault of the mother.

OTHER TYPES

Many of our patients have recently been concerned that childbirth preparation classes focus almost entirely on the physical aspects of delivery. There has been increasing demand from patients for further discussion of some of the emotional aspects of pregnancy and delivery. For instance, the Bradley method has done much to bring fathers into the birth process. Several groups are now developing childbirth preparation classes that focus on the feelings that parents experience during pregnancy as well as on the reorganization of everyday life that is brought about by the birth of a new child.

We feel that childbirth classes are helpful for many; testimonials are the best evidence for their effectiveness. A recent study compared a group of women undergoing prepared childbirth with a group that had not taken the classes. There was no difference in the medical complications for mothers or for infants in either group, but prepared mothers requested considerably fewer analgesics and anesthetics. These drugs often make mothers and infants drowsy in the first few hours and diminish the quality of the interaction between them.

SOURCES

To find childbirth classes in your neighborhood, ask your friends, doctor, local hospital, or local nursing association. Many hospitals offer excellent classes about prepared childbirth, whereas other classes concentrate on how to best understand hospital routines. Be sure you know the curriculum before you enroll.

If you are having difficulty locating childbirth classes, you may write to one of the following organizations.

- American Academy of Husband-Coached Childbirth, P. O. Box 5224, Sherman Oaks, CA 91413.
- American Society for Psycho-prophylaxis and Obstetrics, Suite 410, 1523 L Street, NW, Washington, DC 20005.
- International Childbirth Education Association, P. O. Box 5852, Milwaukee, Wisconsin 53220.
- Maternity Center Association, 48 E. 92nd Street, New York, NY 10028.

Home Delivery, Hospital Delivery, or Something Else?

You have a choice of where your child is born and, in some areas, of which type of health-care provider will attend your birth—doctor, certified nurse midwife, or lay midwife. It has always been a paradox that people will spend hours doing comparative shopping for televisions and automobiles, yet automatically accept the local hospital for something as important as the birth of their child. The hospital can be an excellent place for your child to be born, or it can create problems.

HOSPITALS

How good is your local hospital? University affiliation does not necessarily ensure safety and quality. Each hospital must be judged on its own performance. How many deliveries are done in the hospital? Who attends these deliveries? Is there a pediatrician available for emergencies? How often are cesarean deliveries performed? What indications does your hospital use for a cesarean delivery? What about fetal heart monitoring? Severe fetal distress is an indication for an immediate cesarean delivery, but some people are now questioning whether the increased number of cesarean deliveries that have resulted from better monitoring is actually improving the outcome for the infants delivered. It certainly isn't for the mother. There are no perfect guides to choosing the right hospital or for answering these difficult questions.

The hospital you choose should be willing to treat the entire family with respect. They should be supportive of your wishes for *your* childbirth. Most hospitals allow fathers in the delivery room, but some do not. If you are planning on having the father present in the delivery room, be sure to inquire beforehand. Also be sure to find out what the "routine" procedures of medication and anesthesia are. Individualized care consistent with your wishes, and not "routine" care, should be provided. If you feel uncomfortable with certain hospital procedures, negotiate before delivery or look elsewhere.

AT HOME

More and more people are choosing home delivery for their children. In many countries, delivery of children at home is the rule and not the exception. The experience with home deliveries in these countries is excellent. Women are followed carefully throughout their pregnancy, and those who do not have complications are delivered at home by people with training and experience. In some countries, even first deliveries are performed at home; other countries have a policy of doing home deliveries only with the second and later pregnancies.

However, the fact that home deliveries are safe in such countries as England and Holland does not mean that they are safe in your community. If you are talking to someone who is offering to assist with your child's birth at home, ask about his or her qualifications. Is he or she a lay or nurse midwife or doctor well-trained in obstetrics? How many deliveries has the person done? How is he or she prepared to deal with an emergency in the home? What backup facilities are available? Does the person have admitting privileges at a hospital?

OTHER OPTIONS   You should also make an effort to investigate alternate facilities. Because home deliveries are becoming more popular, hospitals are feeling the stimulus to innovate that competition always brings. In most communities, hospitals have set up comfortable, "homey" rooms for childbirth. These rooms are often called "alternative birthing rooms." In the event of an emergency, the mother is wheeled down the hall to the conventionally equipped delivery room. A doctor in Idaho has rebuilt part of his house as a delivery area. There is a comfortable living room and two "delivery rooms." All the equipment of the hospital is available in the setting of a home. There are several areas of the country where more structured "out-of-hospital birthing centers" exist. Many of these centers are staffed by both certified nurse-midwives and doctors and have excellent and very close hospital access should more extensive care be required for either the mother or her baby.

Most communities have several options for delivery of your child, and you should know about them. We feel that rather than beginning with a categorical decision of "I want a home birth" or "I want a hospital birth," you should begin with the question "What is the best method of delivery for me and my child in my community?"

Additional Reading

ACOG Guide to Planning for Pregnancy, Birth, and Beyond. American College of Obstetricians and Gynecologists, Washington, D.C.: 1990.

Robert A. Bradley, *Husband-Coached Childbirth.* New York: Harper & Row, 1981.

Gail Brewer, *The Very Important Pregnancy Program: A personal approach to the art and science of having a baby.* Emmaus, Penn.: Rodale Press, 1988.

Fernand Lamaze, *Painless Childbirth: The Lamaze Method.* Rev. Ed. Chicago: Contemporary Books, 1987.

Elizabeth Noble, *Essential Exercises for the Childbearing Year.* Boston: Houghton Mifflin, 1982.

George Verrilli and Anne Mueser, *While Waiting: A Prenatal Guidebook.* New York: St. Martin's Press, 1987.

CHAPTER 2

The Big Event

J ust as in pregnancy, the big event, birth, is full of complex emotions for both mother and father. Here are some of the feelings that parents have shared with us:

MOTHER OF SIX:

Great joy sweeps over you, but not on schedule. They placed my newborn son on the bed beside me. . . . It was all so private having the baby born at home . . . even though it was a difficult birth. . . . I looked at the child and I did not recognize him. It's odd when you have felt his every move from the moment he started to move . . . when he has been a physical part of you for so long. . . . Even though you don't even know the sex of the child, you still expect to recognize him . . . and you look at this total stranger . . . you find out what his face looks like . . . a new and totally unknown face, a whole new and different person you have to get to know, . . . and with the knowing comes the loving.

The huge sweeping joy of giving birth . . . of the act itself . . . came with the second. The "for this was I born" feeling . . . a level of satisfaction unfelt before. It came again with the fourth and fifth . . . but not with the sixth. But then the sixth was another difficult delivery and I was too tired to feel much of anything except overwhelming relief that after all that had gone wrong in the pregnancy, the child was perfect . . . That was Christmas Eve . . . what a present.

MOTHER OF TWO:

Although I had excellent prenatal care and access to the most enlightened and modern practices, the emotion I felt strongest after Mary's birth was anger and resentment toward myself and members of the medical and helping professions. Why specifically—too much time was spent on reading, exercising, and talking about giving birth. I got sucked in by the movie star syndrome of the big event. How I would have natural childbirth, breathing, learning about my body, the muscles and how they worked, learning about pain, about control, learning my script, being undrugged and awake so I could experience birth, so I could see another baby coming out of my body. Hey, for nine months I read about it, talked about, drove everyone nuts (and I loved every minute of it). After laboring 19 hours at home with a sunny-side-up presentation, I finally was driven screaming and contorted with pain to the hospital, where I expected to die. After being quickly gassed, but not soon enough to feel a mid-forceps delivery, I passed out and missed Mary's birth.

I expected a dark-haired, different baby somehow bloody and face red and contorted with squalling. Mary was so unlike the monkeyish newborn I expected. Her face with tiny eyes, tiny mouth, and huge cheeks. Like a mongoloid! Yes! For a few days, I had the uneasy feeling that she was retarded. But I loved her intensely the moment I saw her, she was really special. I could spot her the minute I walked into the nursery, she was so fair and bald and her head so lumpy. I was so proud of her and myself.

MOTHER OF THREE:

It is almost impossible to have any kind of relationship with the unborn baby. When I saw Peter for the very first time, he was lying in a plastic bassinet and had stopped crying very soon with his eyes wide open, seemingly looking around. I thought he was so beautiful! The fact that he was a boy, which we had hoped for, and was so beautiful made it so easy to accept him right away. I had been looking forward to having a child very much but this seemed so special, like something you wish for but don't dare to hope for. After this initial feeling of joy, it was a matter of getting used to this little warm, cuddly, and sleepy living thing. I didn't sense any immediate attachment, it was an adjustment to something very new. The feeling of wanting to protect this new presence, nurse it, and take care of it was very strong—I guess this is what they meant by "motherly instinct."

FATHER OF THREE:
I remember very well the earlier parts of labor and then [my wife] Irmhild disappearing into the delivery room, where I was intentionally excluded. The next thing I remember was that Irmhild was on a stretcher being carted back to the ward and she held up this little brown thing and said, "Guess what?"

I said, "A girl," wishing it were a boy—and it was.

MOTHER OF FOUR:
My husband couldn't care less if it was a girl or a boy—he was so frightened something would happen to me—he only wanted to know, "Are you all right?"

MOTHER OF TWO:
I had a lot of depression because it didn't go according to the book. Being a middle-class person, I read all the books and did all the exercises and learned how to blow, but nobody prepared me for any of the things that might happen. And I felt that because I was healthy and was not 4'10" people looked at me funny when I told them I had a C-section. They seemed to be saying, "What's wrong with you? You seem to be big enough to have a baby. You need to know that it's not your fault." I felt guilty for six months because I thought, if I had only breathed right I would not have had 60 hours of labor.

MOTHER OF TWO:
After Aaron was delivered, I held him immediately. He nursed a bit, and I just held him. The nurses (one older one) grudgingly wrapped him up. My husband Terry held him and walked around with him. All in all, we had him over half an hour. When the nurse would keep trying to clean him up, he'd scream and kick. When I was alone with Aaron in my hospital room nursing him, I was thinking of my son Gabriel, and Terry. I had never spent a night away from Gabriel. I was jealous of Terry. I wanted to be there to share in telling Gabriel about Aaron. Gabriel had been so excited about the birth of his baby.

FATHER OF TWO:
The labor and delivery process ended with a feeling of great exhilaration, an indescribable feeling, unmatched by anything else I have ever known. Somehow everything was right with the world, and nothing could be wrong.

Stages of Labor

Most women begin labor in the 40th week of pregnancy, but 10 to 15% have premature labor. Many premature labors, however, do not proceed to delivery but end after a short period of time. For many days before delivery, the mother may experience contractions known as Braxton-Hicks contractions. The contractions are thought to prepare the uterus for delivery and thin and widen the mouth of the uterus (womb); this process is known as effacement and dilation.

The first stage of labor begins when contractions in the uterus become more regular and intense. It is often difficult to tell precisely when this stage of labor begins, and there are frequent false starts. Often a small amount of bloody mucus may show several days before labor. Contractions may be coming regularly, strongly, and at short intervals, and then suddenly they will cease. Labor has generally begun when contractions are ten minutes apart. These contractions may be felt in the lower abdomen or the lower back. Generally, the first stage of labor becomes shorter with each pregnancy. This stage often lasts from 6 to 18 hours in the first pregnancy, but from only 2 to 5 hours for the average second pregnancy. During the first stage, the mouth of the uterus (cervix) proceeds to open (dilate) more and more. The end of the first stage of labor is known as transition and is often the most tiring stage. In transition, the cervix dilates from 8 cm to 10 cm (3 in. to 4 in.), the size required for the infant's head to pass through the birth canal. It is for this stage that childbirth class preparation is often the most helpful because relaxation is important.

The second stage of labor is usually much shorter than the first. It usually lasts about 1 to 2 hours for the first child and 15 to 30 minutes for subsequent births. The second stage starts when the cervix is fully dilated and ends when the child is delivered. During the early phases of the second stage, some pain medications or anesthesia are usually administered.

The third stage is usually the shortest and lasts from several minutes to half an hour. During the third stage of labor, the placenta (afterbirth) is delivered. The doctor or midwife will examine the placenta carefully to ensure that no parts still remain in the uterus. Parts of the placenta remaining in the uterus must be removed because they can become a serious cause of bleeding later on.

Rupture of the Membranes

Throughout pregnancy, the developing infant rests in a pool of fluid enveloped by the amniotic sac. During the first stage of labor, this fluid cushions the child's head as the head is being pushed against the outlet of the uterus. The amniotic sac generally bursts by itself at the end of the first stage of labor. Frequently, the amniotic sac will begin to leak fluid or rupture even before labor has begun; this will usually bring about the beginning of labor. The absence of fluid does not directly interfere with labor, but an amniotic sac that has been ruptured for more than 24 hours increases the chances of infection. If labor has not begun and a large rush of fluid indicates that the amniotic sac has ruptured, contact your doctor.

The amniotic sac is often ruptured artificially at the end of the first stage of labor to facilitate the delivery. This is a painless procedure.

Medications Used During Labor

All medications that are given to the mother reach the child; no medications should be used automatically. For certain conditions, however, medication can be used with relative safety.

Sleep medications are sometimes used for women who are having false labor or who have been in and out of labor for several days and may have had little or no sleep. The fatigue of the mother may carry a greater risk to both mother and child than a mild sleeping medication would. Barbiturates such as Seconal, Nembutal, and Luminal are frequently used as sedatives.

Pain medications reduce the pain of labor but do not completely eliminate it. Strong narcotic pain relievers and tranquilizers such as Demerol and Nisentil may on occasion be used. These medications enter the infant's bloodstream and have temporary depressing effects on the infant. However, extreme pain can interfere with the progress of labor and can also affect the infant.

Sleep medications and pain relievers should be used only when necessary and after careful deliberation by the doctor. Since 1988 the American College of Obstetricians and Gynecologists (ACOG) and the American Academy of Pediatrics (AAP) have not recommended the routine use of pain medication.

Anesthesia

Anesthesia involves the complete elimination of pain by blocking the nerve impulses at either the local, spinal, or brain level. There are several kinds of anesthesia used in childbirth. Discuss them with your doctor before labor begins.

General anesthesia is used rarely in this country today, but it is sometimes used after the first stage of labor is complete. With general anesthesia, the mother has no awareness of either pain or the birth of her child. We do not believe that general anesthesia has any place in routine deliveries; it may be used in complicated deliveries where extreme relaxation of the mother's uterus and birth canal is necessary to perform a complicated obstetrical maneuver such as rotation of the infant. General anesthesia may also be used when a cesarean delivery must be performed.

Spinal anesthesia is administered at the end of the first stage of labor and results in complete absence of pain from the mother's waist down. This procedure had great popularity several years ago, but it is being used less and less today. Complications are possible in both mother and child, and post-partum headaches often trouble the mother. Again, this is a very acceptable procedure for complicated deliveries or when the mother is experiencing moderate pain.

Epidural anesthesia is similar to spinal anesthesia, but it can block the pain fibers more specifically than spinal anesthesia (for example, it can cause pain elimination specifically in the lower abdomen, perineum, and vagina). It is administered during the latter half of the first stage of labor. The possible complications associated with this type of anesthesia include slowing the woman's labor and/or lowering the mother's blood pressure.

Caudal anesthesia is administered in the last half of the first stage of labor. It is generally given in a lower part of the spine than where spinal anesthesia is given. It can be given continuously by a small tube inserted into the lower portion of the back. Caudal anesthesia has many of the same problems as spinal anesthesia, including the possibility of lowering the mother's blood pressure and causing post-partum headaches.

Paracervical anesthesia is administered by injecting a local anesthetic (such as Xylocaine) into the area surrounding the cervix by means of a long needle. The Xylocaine deadens the nerves going directly to the cervix. Xylocaine can produce a temporary decrease in infant heart rate. Because of this, paracervical anesthesia is rarely used today.

Pudendal block anesthesia eliminates pain in the woman's external genitalia. This provides partial anesthesia during the second stage of labor and during repair of spontaneous or surgical cuts of the birth canal (episiotomy). Some episiotomy repairs require an additional local injection of Xylocaine.

Pitocin and "Induced Births"

Oxytocin is a natural hormone that produces contraction of the uterus. It is produced by the body during labor to intensify the contractions of the uterus; it is also produced after labor to cause contraction of the uterus and to decrease uterine bleeding. Infant sucking on the mother's breast stimulates the production of oxytocin and hence uterine contractions.

Pitocin is synthetic oxytocin. It should not be used routinely to augment natural contractions or to produce a child at a more convenient time. It may be of great assistance to induce labor in women who are more than two weeks past their due date or where there are indications that, due to the distress of the infant, it is appropriate to induce labor. Pitocin may also help during prolonged labor or in aiding uterine contractions after labor. It can be administered either intravenously or by intramuscular injection. With certain types of anesthesia, it must be used because the anesthesia diminishes the strength of the uterine contractions. Too much pitocin can cause such painful or excessively strong uterine contractions that depressant drugs must be used to counteract the force.

Fetal Heart Monitoring

Fetal heart monitoring has long been accomplished by listening to an infant's heartbeat through a stethoscope. A strong, regular heartbeat is a sign that an infant is doing well during labor. A significant, prolonged drop in rate or intensity of the heartbeat is a sign of distress of a fetus that may, on occasion, necessitate a cesarean delivery in order to prevent damage to the infant.

Electronic monitors have replaced the stethoscope for keeping track of the infant's heartbeat. In some cases, external monitors are used to keep track of both uterine contractions and fetal heartbeat by electrodes strapped to the mother's abdomen. In other cases, internal fetal monitors are used. Electrodes are inserted during the first stage of labor and are placed directly on the infant's head. This requires rupture of the amniotic sac, if it has not already ruptured. Some of these monitors are capable of measuring the acidity (pH) of the infant's blood as well as the infant's heartbeat. The mother's uterine contractions are also measured.

The theoretical advantage of fetal monitors is the detection of early infant distress. Many women object to them on the grounds that they interfere with mobility during labor. There is, however, a more significant disadvantage to fetal monitors. With the use of fetal monitors at certain hospitals, the rate of cesarean deliveries has doubled. Yet the rate of newborn complications in hospitals not using fetal monitors appears no different from the hospitals using the monitors and having a higher cesarean-delivery rate.

The full risks and benefits of fetal monitors have yet to be clarified. It does not seem to us that they should be used routinely. The ACOG and the AAP agree, and for low-risk pregnancy recommend listening for the infant heart rate every 15 minutes in early labor and at least every 5 minutes in the second stage of labor. We do feel, however, that they can be important in high-risk or complicated pregnancies, or when the mother or baby develops a problem during labor.

Preparation Procedures

Prepping involves shaving the mother's pubic hair before delivery; whether this reduces the risk of infection to the infant is debatable. The re-growth of pubic hair can be uncomfortable and itchy and may add to the discomfort that sometimes accompanies the healing in the post-partum period. Discussion with your doctor or midwife about what will be done can avoid confusion and misunderstanding at the last moment. Preparation should certainly include a thorough washing of all parts of the body surrounding the vagina.

Enemas have also been the subject of recent debates. Some have argued that infants come in contact with a great number of bacteria in passing through the birth canal and that an enema will decrease this exposure. The true risk of an infant becoming contaminated with a mother's stool is not well understood. Many women have diarrhea for several days preceding delivery, and most women will not have a bowel movement at the time of delivery. Again, this is a subject for discussion with your doctor or midwife that should not be left until the last moment.

Delivery Room Procedures

Forceps are curved metal instruments designed to fit around a baby's head. They have been used for hundreds of years to make birth easier. We believe that, as with all potentially dangerous interventions into the birth process, their use should not be routine. They often must be used if anesthesia has interrupted the progress of labor, and they should be used if a prolonged or abnormal labor is causing fetal distress or if hemorrhaging is jeopardizing either mother or child. Sometimes they are used to rotate an infant's head so as to make the birth easier.

An **episiotomy** is an incision made in the skin between the lower end of the vagina and the anus to enlarge the vaginal opening and ease the birth of the child's head. Many obstetricians feel that an episiotomy minimizes damage to infant and mother. They argue that these straight incisions heal better and hurt less than the irregular tearing that might otherwise occur. Others argue that this skin is extremely elastic and usually tears only when improper techniques are used. They maintain that properly controlled deliveries will not lead to tearing. As with so many other areas of medical care, the answer is not to be found by siding with either group. There are indications supporting episiotomies, but not all women will need them. The need for an episiotomy diminishes with subsequent births. Discuss episiotomy with your doctor or midwife beforehand so you will know what to expect.

Breech deliveries are not really "feet first"; they are actually "hips first." Although about 3% of births are breech, most of those occur in premature or multiple (twins or triplets) births. Although it is difficult to assess precisely, breech births are more complicated and carry greater risks to the fetus. Cesarean deliveries are often performed to deliver babies from the breech position.

CESAREANS

A **cesarean delivery** (C-section) is a surgical procedure performed under general anesthesia to remove the infant from the uterus through the abdominal wall. C-sections are performed in from 5 to 15% of births. Because they are performed so commonly, parents should have an understanding of the procedure in the event that it must be done. Discuss this possibility fully with your obstetrician.

There are many reasons for doing C-sections. Sometimes the infant's head is larger than the mother's pelvis can accommodate. Or labor may suddenly stop; this often occurs spontaneously, and sometimes because of drugs. Prolonged labor usually is a physiological problem and *not* the fault of the mother doing something improperly. Less common reasons include a placenta implanted in front of the cervical opening and complications of diabetes or other illnesses. Signs of significant fetal distress or other serious problems of the developing infant are also indications for a C-section.

C-sections are relatively safe, although they carry a higher complication rate than vaginal deliveries. Because of more sensitive methods for detecting signs of fetal distress, minimum distress is being discovered more often and leading to a higher number of C-sections. The benefit of this new approach has not yet been proven.

Premature Labor and Births

Between 5 and 10% of all infants are born more than two weeks before the due date. A number of factors account for these premature births. Some are due to error in calculation of dates, and some are due to premature inducement of labor or cesarean section. Infections, longstanding illness, poor nutrition, and complications of pregnancy also can lead to premature labor. However, the majority of premature births are unexplained. Often, a mother who has had a good diet, exercised regularly, avoided drugs, and received proper medical help will spontaneously begin labor early. The reasons simply are not known.

Many women will experience premature labor contractions because of the nature of their uterus. There has been considerable progress in the management of premature labor. Because it is an extraordinarily fatiguing process if it goes on for more than a few days, bedrest is usually recommended. It can be particularly frustrating because friends react by sometimes envying you for a life of being catered to. In reality, this experience can produce sore and weakening muscles. In addition, there are the ever-present concerns of "Did I do anything wrong?" and "Is my baby going to be all right?" Because premature labor is increasing in frequency, you may find your best support coming from women who have been through the same experience.

It is quite common for women experiencing premature labor to be prescribed a medicine called **ritodrine**. Ritodrine acts to relax the muscular contractions of the uterus. It has been a very effective medicine in preventing the premature delivery of infants due to premature uterine contractions. It has not been shown as yet to have any adverse effects on the developing baby. While bedrest and ritodrine is often a successful treatment, many women still deliver prematurely.

Fortunately, the chances of a premature infant surviving and developing normally are excellent. Many premature infants are faced with temporary respiratory problems because of immature lungs (hyaline membrane disease or respiratory distress syndrome). Other premature infants have problems with infection. Competent medical care can do much to help the infant with these problems. Recent work suggests that corticosteroid hormones given to the mother early enough in labor can prevent some of the respiratory problems. There are methods for halting premature labor; contact your doctor or midwife immediately if labor begins early.

Premature infants often surprise parents by their appearance. They are certainly not as big and cuddly as expected. They appear quite frail and are often connected to a variety of complicated devices in the nursery. It may be difficult to locate the child among the machines. Nevertheless, you are just as important to your premature child as to your full-term child. Holding and cuddling the child is important and is encouraged in many nurseries. Premature infants who have greater human contact become stronger faster.

Several excellent publications are now available on this subject. We recommend *The Premature Baby Book* by Helen Harrison (New York: St. Martin's Press, 1983).

Newborn Procedures

We feel that delivery room procedures should encourage the unity of the family while ensuring the safety of the infant. More and more infants are given to their mothers immediately after birth, and often breast feeding starts in the delivery room. There are many reasons why we approve of this process. Stimulation of the nipples encourages uterine contraction and thus decreases bleeding. Holding the child and looking at it gives parents an early opportunity to experience the child and to explore feelings about this new baby. For many mothers, these are significant moments. We also encourage the father to hold the infant. Although we agree with recent advocates that the bathing of infants in warm baths immediately after birth is soothing, we feel this is optional and that the opportunity for parents to hold their new child is more important and gratifying for all.

A tremendous amount of emphasis has recently been placed on the importance of the first few moments of life. These first moments are important, but so are the first days, weeks, months, and years. If the delivery doesn't go as planned and you miss the first few minutes, don't worry needlessly. There are many more to come.

Hospitals vary in their newborn routines, and you may want to spend some time becoming familiar with the procedures of your local hospital before delivery. Here is a list of the most common routines; all of them are designed to improve the safety of your infant's stay in the hospital.

- Deliveries should be attended by an individual whose responsibility it is to care for the infant as soon as the infant is delivered. This need not be a pediatrician. Most delivery room nurses are skillful in managing the common problems of the first few moments of life. However, complicated pregnancies or complicated deliveries, such as breech births, should be attended by a pediatrician or other doctor who will care for the baby immediately after birth.

- We have never seen an infant held up by the feet at birth and smacked in order to begin crying. That is for the movies. Most infants breathe and cry spontaneously at birth. The fact that the infant's chest is forced through a narrow birth canal removes most of the fluid from the infant's lungs. Most infants will have their mouths and upper respiratory passages suctioned in order to remove the considerable amount of fluid that has recently been in the infant's lungs. This facilitates breathing.

- The delivery room should be equipped with infant warmers, or the infant should be wrapped snugly in a blanket. It takes a while for a baby's temperature regulation to begin performing smoothly. One of the worst things that can happen to infants is a rapid drop in their body temperature, and this can happen if a wet infant is left exposed at normal room temperature.

- Babies should be examined promptly after birth for any signs of distress. The initial assessment includes evaluation of color, tone, activity, respiratory rate, and heart rate. Often this initial evaluation is expressed as a number from 1 to 10 known as an **Apgar score**. This score is *not* a predictor of your child's intelligence or health. A more thorough examination is usually performed within the first few hours of life.

- All infants should have prophylactic eye medication administered; in most states, this is a legal requirement. The purpose of the eye medication is to stop infections from the gonococcus bacteria (gonorrhea). Gonococcal conjunctivitis is a major threat to the infant and is a common cause of blindness. Although most mothers are certain that they do not and have never had gonorrhea, this disease can remain hidden for many years. The most common types of eye medication are erythromycin and silver nitrate; these medications may cause several days of tearing in the newborn. Erythromycin is also effective against another more common infection by an organism known as *Chlamydia*.

- New babies should have a vitamin K injection. Before the administration of vitamin K became routine, hemorrhaging was common in the newborn. The newborn's immature liver is often unable to produce the important vitamin K, which is necessary for the production of one of the components of the blood that prevents hemorrhaging. (Although recent studies in England raised questions about vitamin K injections, careful studies have shown the preparation used in the U.S. to be safe.)

- Infants should be examined daily while in the hospital by a doctor, and daily conferences should be held with the parents in order to discuss any questions or concerns.

BLOOD TESTS

- Upon discharge from the hospital, all children should receive a blood test for phenylketonuria (PKU), a rare disease that causes mental retardation. Again, this is a legal requirement in most states. If detected at birth and treated with the appropriate diet, these children can have normal intelligence. Because many mothers are now discharged within the first 24 to 36 hours after delivery, you may need to bring your child to a doctor later in the first week to have this test performed; the test is not quite as accurate in the first 24 to 36 hours of life as after that period.

- A similar blood test should be done for hypothyroidism, a far more common problem which can also be treated immediately and, consequently, prevent the mental and growth retardation that accompanies the unchecked illness. Many other inherited metabolic disorders and blood disorders (e.g., sickle cell disease) are evaluated in newborns.

- HIV testing is currently recommended for all pregnant women living in high-risk areas or who have a high risk of contracting HIV. You should very seriously consider testing if you: (1) had a blood transfusion prior to April 1985; (2) used injection drugs; or (3) had any previous partner who used injection drugs. All women should be offered HIV testing and allowed to decide for themselves. If testing is not done during pregnancy, it can be considered for newborns, whose test results will be identical to their mothers. Important preventive, therapeutic, and experimental approaches are available for mothers and infants. Only one in three infants of HIV-infected mothers will develop HIV infection.

- "Rooming in" is a procedure used in many nurseries. Very simply, it means that the child is kept in your room. The advantages of rooming in are that your child is in your room at all times and you can provide most of the care for your child. In addition, rooming in can prevent the spread of infection from child to child during infectious epidemics in newborn nurseries. The disadvantage of rooming in is that you are providing the majority of care for your infant—an advantage for some, a disadvantage for others. Most hospitals offer daytime rooming in. The child is with the mother all day but returns to the nursery at night. If you are interested in rooming in, you should discuss this possibility with your doctor. It is not available in all hospitals.

Circumcision

The decision to have your child circumcised should be considered carefully by both parents before the birth. Circumcising a male infant involves the surgical removal of the foreskin of the penis. The child is sometimes sent home with a small plastic ring still in place around the penis; we mention this only because doctors occasionally forget to tell the parents and the plastic ring can cause considerable anxiety.

Circumcision has its historical roots in both ritual and health. Circumcision does prevent *balanitis,* an infection beneath the foreskin, or *phimosis,* the inability to retract the foreskin. These problems can cause swelling of the foreskin and even obstruction of the urinary stream. Phimosis happens only if the foreskin is not retracted and cleansed regularly; it does not occur where proper hygiene is used. Infants who are uncircumcised are also more likely to develop urinary tract infections than circumcised boys. However, the overwhelming majority of uncircumcised boys never develop medical problems.

Less than 5% of newborns have retractable foreskins; by three years, 90% have retractable foreskins. It is difficult and unnecessary to retract the foreskin in a newborn infant; retraction usually becomes easy after several months. Phimosis is frequently seen in children who are beginning to take responsibility for their own bathing but who do not retract the foreskin and wash the glans of the penis carefully.

Complications from circumcision can occur, although they are rare; these complications can include damage to the penis itself. In some children, the end of the penis has not completely developed, and the child is unable to produce a forceful stream of urine. This problem, termed *hypospadias,* is not a dangerous medical problem, but can create difficulties if boys find that they are unable to use the urinal like everyone else. It can also be a cause of infertility. One of the surgical procedures to correct this problem requires the foreskin. Unfortunately, some foreskins are removed by circumcision without first checking for this condition. No foreskin should be removed in a circumcision without first checking to make sure the child does not have hypospadias.

Other arguments for and against circumcision have been made with varying degrees of scientific support. Cancer of the penis has been said to be more common in uncircumcised males and cancer of the cervix more common in the sexual partners of uncircumcised males. If engaging in unprotected sexual intercourse, uncircumcised males are more likely to acquire infection. Sexual pleasure for the male has been said by some to be enhanced by the presence of a foreskin and by others to be diminished; neither argument has been substantiated.

Finally, there are those who worry, "Will he be like everyone else?" or "Will he be like his father?" Circumcision is an optional procedure, and there will be plenty of circumcised and uncircumcised boys around. We feel medical procedures should be for medical reasons and not to make boys look like somebody else.

Taking Your Child Home

The joy of taking your child home can be limited by many petty details. Here are some of the things you should prepare for in coming home from the hospital.

**HOSPITAL
ROUTINE**

Hospital **paperwork** must be completed. Generally, the father can attend to this, although many hospitals now have someone from the financial office drop by the hospital room so the mother can make the arrangements herself.

Although most hospitals permit unlimited **visiting** by fathers, young children usually are not allowed on maternity wards. A young child can be anxiously looking forward to mother's return only to be restrained from embracing the mother by hospital rules. The mother usually will be provided with a wheelchair while leaving the maternity ward (like it or not) and generally will not be allowed contact with her older children while leaving the maternity ward. While there are many compelling humane arguments against these procedures, the rules are there and should be anticipated.

The days of "lying in" hospital, where mothers could spend several weeks adjusting to and learning about their newborns, are long gone. In an era of cost containment, efforts are directed towards **early discharge**. On the positive side, there are better places to be than hospitals. However, the surveillance and support that were traditionally provided by maternity and nursery nurses have been, and continue to be, valuable. With the advent of early discharges we are seeing some disturbing signs of increasing number of emergency room visits and heightened parental anxiety that goes hand and hand with diminished experience. If you leave the hospital early, inquire about home nursing visits. Many hospitals and health plans offer such services. They save money for you and your health plan while bringing valuable services to you.

PRODUCT SAMPLES

Many hospitals provide a package of goodies for the parents to take home for their new child. These goodies are supplied by the companies making the products and not by the hospital or your doctor. They are advertising samples. Some of these products may be useful, but many are unnecessary junk. Although they are free, they can cost you money in the long run, and can get some bad habits started. We have spent many hours personally discarding many of the items from these packages.

A typical package will include the following:

- **Infant formula.** This is fine if you are going to bottle feed your child. Very often the formula included is the prepared liquid or a liquid concentrate. These are handy to have and have a long shelf life. If you do choose to use infant formulas, however, remember that powdered formulas are more economical. Check in your local supermarket to see which formula has the lowest local price. (It may *not* be the one you get a free sample of.) There is no clinically significant difference between any of the major commercial products now available on the market.

- **Lotions, creams, and oils.** Keep these to use on yourself if you like lotions, creams, and oils. Most infants, however, do not need such preparations. We do not recommend routine use of any skin preparation for children. While some preparations can protect skin in the diaper area from urine and consequently diaper rash, these same preparations usually should not be used after a diaper rash has developed, because they keep air away from the skin and prolong the rash.

- **Powders.** Baby powders have been used for centuries to keep babies dry. But the best way to dry the baby's bottom is to dry the baby's bottom. As soon as the child urinates, the effect of the powder is gone. Powders do help dry the perspiration that occurs underneath a warm diaper. If you choose to powder your child, place the powder in your hand and *then* place it on the child's bottom. Shaking powder from a distance of a foot or so will spread a cloud of powder around your child; this is dangerous because the infant can inhale the powder particles. Powders with talc in them have been incriminated as a cause of serious lung disease. Use these powders cautiously or avoid them.

- **Cotton-tipped swabs.** Cotton-tipped swabs can be used to help in cleaning a newborn's umbilical cord stump with alcohol. However, cotton-tipped swabs should *never* be used to clean the inside of children's ears. Wax is produced within the ear canal for a purpose and usually becomes a problem only when it is impacted in the ear canal by the use of cotton-tipped swabs. We tell parents never to put anything smaller than their elbow in a child's ear.

- **Vitamins.** All infant formulas are supplemented with all vitamins necessary for your child's proper growth and development and several more. For mothers choosing to breast feed their infants, only a vitamin D supplementation is necessary.

THE DRIVE HOME

The number-one killer of children in this country is accidents. The leading accidental killer is the automobile. The risk to your child from the automobile is greater than the risk from childhood infectious diseases and childhood cancer together. You may want to hold your baby in your arms on your way home from the hospital, but this is not in anyone's best interest. The driver will probably be excited about the baby coming home for the first time and will be less attentive to road conditions than usual. We know of several tragic accidents in this setting. Have an **infant car seat** ready to use on the way home. Discussion on infants' seats can be found on page 183.

ARRIVING HOME

Upon arriving home you need time and attention for yourself, your new infant, and your other children, who probably are anxious to see that you still love them. These are the most important items on your agenda. Contact with friends and relatives is secondary. If a friend or a relative will be staying with you, we advise that they help with the household chores and free you for time with your children. Well-meaning relatives too often come between older children and their mother. Older children need reassurance from their mother, as well as their grandmother, that mother still loves them.

Mothers who have no relative or no husband to help at home can seek help from neighbors, friends, social agencies, and employment agencies.

CHAPTER 3

*Y*our First Concerns

*I*s there life after birth? Becoming a new family is a major physical and emotional adjustment for everyone. Here are some thoughts on this new experience from parents we know.

The First Child

MOTHER OF ONE:
No one told me what I was going to feel like afterward.

FATHER OF TWO:
The first several weeks made it very clear that our lives had changed fundamentally and unalterably, partially for the better, partially not. We were tied to the breast-feeding schedule. Our freedom of movement was over. Our nights were continually interrupted. There were times I resented the change, resented the baby for bringing them about and felt guilty because it seemed unconscionable to hold such feelings toward such a small, adorable, helpless infant. On the whole, however, the changes were positive. The process of interaction with the baby was fantastic. It exceeded my expectations. Walking her to sleep became a cherished event. Feeding her her first cereal. Listening to her sounds develop. There were many things I couldn't do, but many new things I was doing for the first time. There is also a sense of fulfillment of purpose and of family in a sense that I had never understood before. We were both able to work and enjoy the baby and get out and felt strongly that our lives, although radically different, were on the whole better.

MOTHER OF TWO:

I found myself being very dependent during birth for support, also after we took Suzie home. I had no one else to talk to and was worried about whether I was overreacting to things like too much or too little crying, how much clothing, small things but so constant they made my day seem like a mass of indecisions and insecurity. Artie was really my only friend. I also was physically dependent. I was unable to cook or clean for at least a week and depended on Artie for meals, laundry, and going out. This topic was the subject of almost all of our fights, which increased sharply in number after Suzie's birth. I was spending literally 24 hours a day with the baby, while Artie was in school for most of the time. He didn't understand why I had to get away from her sometimes.

MOTHER OF THREE:

As Peter grew, I tried to remind myself over and over again that husbands and new fathers are supposed to be jealous of the baby, and I tried to take this into consideration by letting Matt be involved and giving me advice in the care of our son. The latter was the hardest and still is today. I was quite unsure about the care of our first child and relied heavily on my mother's advice, but I also asked a lot of other people. I was confident that I would someday learn.

FATHER OF THREE:

During the early months I don't remember feeling as close physically or emotionally to the baby as I do now. Was it because Irmhild had everything so under control and was breast feeding? Or was it because I was still working? I don't know. The first time I held the baby, even though I had held others before, was a fairly nervous time. He seemed very small and fragile and I was afraid of letting his neck fall down for fear of causing some whiplash injury.

MOTHER OF ONE:

After four years of peaceful dinners enjoyed after hectic days, my husband and I resented the interruption. I felt an obligation toward the baby but also sympathy for my husband. We felt guilty that we were being selfish. After discussions with friends, we decided that our feelings were shared by many couples adjusting to a new member in their family.

MOTHER OF SIX:

With the first, there is "the buck stops here" . . . the responsibility lasts for 24 hours each day, and the decisions are all yours . . . there is no escape . . . you can't put the problem aside or pass it on. . . . I believe this is the essence of the trapped feeling so common in new parents, much more than the fact that they can't go to the party because there is no sitter available tonight.

MOTHER OF TWO:

This is the first time in my life that any individual had the right—not the privilege—to call on me 24 hours a day any place—and I had the obligation to go. Nobody had ever done that to me before. I was a schoolteacher and my husband used to say, "Ha! This one doesn't go home at 4:30," and it was true—he was sympathizing. Another big shock was that mothers of little children don't get sick—you can have a 104°F temperature and vomiting, and you still have to be there. And my husband feels the same way. Whenever the children are sick at night, he's also up with them.

MOTHER OF FOUR:

The first baby is the hardest. Now that I have four I find I have time to get things done. With my first baby we were living abroad and even had two maids, but I still had no time to make even Jell-O for dessert. I asked myself what was I doing? It seemed that I had no time for a bath. Now that I have four, everything is organized, and there's plenty of time.

MOTHER OF TWO:

Coming from a large family—[I'm] the oldest—I knew the mechanics of caring for babies, but somehow I didn't remember babies as so small and weak. I had a feeling of nervous anxiety, not depression. Sue's breathing was so irregular I found myself checking her constantly, really expecting to find her dead from crib death or suffocation from not being able to lift her head up high enough. After two weeks, she snapped out of her depression and developed colic. Jim and I would sit silently through supper and listen to her scream until nine, when suddenly and what seemed miraculously she stopped. Being angry at her because we really tried all the remedies—burping stomach down, warm bath, extra burping, tipping up and down, longer feedings. Picking her up made her cry harder. I felt my husband thought I was neglecting her or being mean to her or taking the whole thing too lightly at first. Do something, don't just sit there and listen. We didn't realize it was colic, until it ended two weeks later.

MOTHER OF FOUR:

You worry about unnecessary things with the first baby; you don't with the second. You take a diaper off the top of the pile, and your husband says, "That one is dusty." With the first baby you waste a lot of unnecessary time on things like that.

MOTHER OF TWO:

Nursing, too, followed the same pattern. A lot of propaganda from the La Leche league. "People who don't nurse are downright degenerate, and those who nurse less than a year or two or three or four are shirkers and crummy mothers." I nursed Tally for six months and once again felt anger at those women who had full breasts at the right time instead of the three months of engorgement, leaking, pain. PAIN?

Maybe I was the only one, maybe I wasn't relaxed enough, not in tune with my body, something must be wrong with me. Only after talking with other mothers, some successful, some not, did I discover that the "letdown reflex" is often painful. OK. So I was willing to accept it. But why not say that nursing is really hard sometimes?

MOTHER OF ONE:
I used to drive in the car and plan for hours what would happen if we drove into a river—how I would get everyone out. I don't think I ever realized how fragile life was until I had Danny.

The Second Time Around

MOTHER OF SIX:
I guess the second time around it is all more real. . . . You know about the new person arriving . . . and the joy is easier. And then you discover they never both sleep at the same time.

MOTHER OF TWO:
This is what the days were like:

6:00 A.M.	*Feed/change Jeremy*
8:00 A.M.	*Dress Mirah/breakfast/clean up*
10:00 A.M.	*Feed/bathe Jeremy; change Mirah*
12:00 P.M.	*Change Mirah; lunch/clean up*
2:00 P.M.	*Feed/change Jeremy; change Mirah for nap*
3:00 P.M.	*End of Mirah's nap*
4:00 P.M.	*Change Mirah*
6:00 P.M.	*Feed/change Jeremy; dinner for Mirah*
6:30 P.M.	*Larry burps Jeremy/I fix dinner*
7:00 P.M.	*Dinner/clean up*
8:00 P.M.	*Bath for Mirah/dress for bed*
8:30 P.M.	*Ice cream for Mirah*
9:00 P.M.	*Bed for Mirah*
10:30 P.M.	*Feed/change Jeremy/pray he sleeps through the night*
11:30 P.M.	*Fall exhausted into bed*

If I thought of childbearing primarily in terms of its rewards before my children were born, I seem now to describe it largely in terms of its work. Both are lopsided views. At the same time that my days seem to turn into an endless series of jobs—like a string of paper clips—they also become filled with sudden pleasures—Mirah's smiles and kisses, her delight in running, climbing high on the slide, her excitement in reading books and going to the Bookmobile for book time.

MOTHER OF TWO:

The first night home I put Rick down and went to bed. I woke up a short while later because I realized I had not checked on Alan and tucked him in again—something I always did. I forgot Alan was in the house. That made me feel strange and also sad.

About three nights after Rick was born, I was playing with Alan and feeling very close to him. We were enjoying ourselves when Rick started crying. I had to leave Alan to feed Rick, and I felt myself resenting Rick because he had broken in on a special time.

When we first brought Rick home, Alan started taking out a lot of frustration on my husband—hitting, yelling, refusing to do anything. I hadn't thought he would react to Rick like that. But it only lasted two days. I guess it was better that he acted it out rather than keep it all in. I felt myself feeling less tense and pleased at how easily Rick became a family member. But one night my husband said, "You're enjoying him more than you thought you would." I guess I felt relieved at that.

I feel badly that Rick, Alan, housework, etc., have taken so much time—time that I would like to share with my husband. He never complains and just pitches in to take care of the children and to help out.

MOTHER OF FOUR:

After the first baby, I never cried once, and I thought there was no such thing as the post-partum blues. When Emily was born—I brought her home and I woke up one morning and burst into tears. . . . David was still in bed, and he put his arms around me and said, "What's wrong?" I said, "I'm very unhappy about that baby," and he said, "I know," and we both decided, wrongly, that since we had had another little girl we were very disappointed. I felt terribly guilty. It was a big thing in the Navy—everybody was very macho—and wanted little boys. I told myself, as long as it was healthy I'd be happy, and I really am happy and she really is healthy, so why in the name of heaven am I crying? I felt so mad at myself for being so unhappy over the fact that I had two daughters. It wasn't until a year later when I thought maybe it was post-partum blues. Since I hadn't experienced post-partum blues the first time, I didn't expect it the second time. Even though David said, "Her plumbing's on the inside and not on the outside," I couldn't understand why I was so sad when I really was so happy.

MOTHER OF TWO:
I felt and was completely trapped, unable to change their diapers fast enough. I felt guilty because Sarah was so jealous of her sister and I was hard pressed to devote an "hour," as my pediatrician had recommended, to her exclusively. I also felt badly that I was not so involved in the new baby's personality as I was in Sarah's. The little things like rolling over that Sarah did that thrilled Will and me seemed less exciting the second time around.

FATHER OF THREE:
Caring for kids in the first year is not too great, and we've tried to share it between the two of us, but obviously my wife gets the brunt. Doing every other diaper when I'm home, we thank God for a well-formed stool. In traveling, I am much more sympathetic to parents with children, but am still uptight when our kids cry in public.

MOTHER OF TWO:
I look back on the past three months as both swift and interminable—punctuated for so long by four-hour wakings, changings, and feedings. I have almost unconsciously learned to use my left hand while the baby is cuddled and nursing on the right. It's amazing how I can now manage to eat my dinner, not quickly it is true, during two-year-old tantrums and three-month-old cranky crying, talking to one and holding the other. I sometimes think that hardening of the eardrums must be estrogen-related.

FATHER OF TWO:
Other couples talked about it, but we never had the experience of feeling restricted by our new addition. In fact, it was pretty nice to have an excuse to stay home.

FATHER OF TWO:
The second baby was a very different experience. The peaks of exhilaration were lower although present; the valleys of anxiety were not nearly as deep although they too were present. A different kind of uncertainty confronted us. To what degree would our lives be more restricted? What additional increment of work would be involved? Was a second child a quantum leap in work and change of lifestyle from the first child? Or, would only a minor modification have to be made? Several times, I wondered if it was a mistake at all to enlarge our family. Delivery and birth were much easier with Jeremy, and he proved to be an easier baby to take care of

even than Mirah, but my God what an added increment of work. Two sets of diapers. Two car seats. Two feeding schedules. Two sleeping schedules. But not that much more restriction. We were able to accommodate to it. We resumed our other lives more easily.

FATHER OF TWO:
Sam really wasn't as much work as Mandy. We were less attentive to his every move. We didn't listen as closely to his grunts in bed. We didn't carefully measure every step we took. We were more relaxed. We were more natural. And I think the reason Sam has been easier is that we were easier on him than we were on Mandy. At no point did the addition of Sam provoke the kind of reaction that the first several weeks with Mandy did. Never did I feel, "My God, what have we done?" Yet, there are times when I still resent being tied down, get angered because there is very little quiet time and together time for Lori and me, but I wouldn't change it—not for anything.

MOTHER OF TWO:
We found that we were spending most of our time with people who had children—they understood the constant interruption and irrational and seemingly destructive behavior of our children.

MOTHER OF SIX:
We take on a 24-hour-a-day job with a pathetic lack of skills with which to cope. . . . And for the mother and father who have been accustomed to being effective and successful at their own work, the feeling of gross incompetence leads to a great resentment of the role inflicted upon them.

MOTHER OF FIVE:
It is more important for the child to feel loved than for the house to look good. The three-year-old will remember you sitting down and drawing silly faces for him with a warm glow. He won't feel the same about you mopping the floor and his not being allowed in until it is dry . . . eventually one has to, but not too often. It's now he needs you, mop the floors later if you must. . . . Anyway, drawing silly faces is much more fun than mopping floors. I remember the Montreal taxi driver who said it always worried him if he came home to a tidy house. . . . It made him wonder if his kids had had any fun at all that day.

In the first few days at home, new parents often spend a great deal of time checking their new baby. Perfectly normal infants often appear strange in some way, causing concern to new parents. The purpose of this section is to describe the sometimes surprising features found in normal newborns.

There are undoubtedly things that you will notice about your newborn that have not been covered in this brief discussion. Most doctors are more than willing to sit down and explain the concerns that parents have over their babies. Very often, however, parents forget to ask about a small item because of the excitement of having their child and because of the chaos of the hospital ward. We suggest that you write down any questions about your baby and make sure that you get a satisfactory answer to each one.

Activity

Infants spend much of their time sleeping. The newborn may spend 18 or even 20 hours in sleep, the six-month-old will spend from 16 to 18 hours sleeping, and the one-year-old child from 14 to 15 hours. You may observe several types of sleep states. In one state, the child will be very, very motionless and have regular breathing. In another sleep state, the breathing will be irregular, the eyes will be seen to be fluttering, and there will be motion of the hands.

There are also different stages of being awake, the least satisfying of which is crying. This is the infant's most obvious way of communicating needs, frustrations, and discomforts, but there are other subtle ways of communicating that parents can learn to recognize.

Infants spend much of their quiet time looking and watching. In this stage, the infant is exploring the world with his or her eyes and will stare intently into the parents' faces and follow them across the room. Babies have certain likes and dislikes in shapes, patterns, and even colors, and these preferences have been demonstrated even in the first few hours after birth. In the "active alert" state, the infant is looking around actively, as well as moving arms and legs.

Newborn babies are often disappointing to parents who expect laughing, gurgling infants. The image most of us have is actually when they are four to eight months old. Newborns are exciting but you must remember that it will be many months before they laugh and gurgle.

BREATHING

Infant breathing patterns are often irregular. There may be periods of five to ten seconds when infants will stop breathing; this is known as an **apneic spell**. Normal infants outgrow this pattern within the first few months of life. This breathing pattern is characteristic of a certain state of sleep; it is a normal phenomenon in the newborn, and should not be of concern to parents unless it persists for a number of months.

Sudden Infant Death Syndrome (SIDS)

Many parents fear their infants will stop breathing and become victims of SIDS, or sudden infant death syndrome. Although it is the leading cause of death beyond the newborn period, it is important to bear in mind that the risk is about 1 in 600 before age one. The risk declines rapidly after six months, but is higher for children with siblings who had the problems and in premature infants. There are numerous causes, hence the use of the term "syndrome."

Do not panic if your child's breathing is erratic, with episodes when the child does not take a breath for 10 to 15 seconds. Pauses longer than this should be discussed with your doctor, because there are a number of diagnostic and management strategies (including monitors and medications) currently available. Parents who have experienced a previous loss of a child should discuss the benefits and risks of an apnea monitor with their physician.

Parents can do things that may reduce the risk of SIDS. It is essential for a mother not to smoke during her pregnancy and after birth. Infants dying with SIDS are two to three times more likely to have smoking mothers.

There have been reports of SIDS in infants sleeping on polystyrene pillows, but it is unclear that pillows really pose an increased risk for this syndrome. (Infant cushions have been banned because of the risk of suffocation; see page 181.) As a general rule, babies do not need to sleep on pillows.

There is some evidence suggesting that infants sleeping on their stomachs are more vulnerable to SIDS than infants who sleep on their sides or backs. Australia and New Zealand reported decreases in SIDS of 40 to 50% after starting programs to encourage back sleeping.

However, Scotland had a 46% decrease at the same time without changing any policy. The American Academy of Pediatrics currently recommends that healthy infants sleep on either their sides or their backs. Premature infants and those with respiratory or intestinal problems may do better sleeping on their stomachs, and in such circumstances you should rely on your doctor's judgment.

Parts of the Body

The Skin

At birth, most infants are covered by a protective, white, thick material known as the **vernix caseosa**. This covering is generally washed off within the first day. Often some of the material is missed, especially behind the ears and in the ear folds. You should not be disturbed to find white, sticky material behind your child's ears. In addition, the baby's skin will frequently peel in the first week or two of life; this is another normal phenomenon of the newborn period. There are a number of other rashes that are extremely common in the newborn and not serious. (See **Baby Rashes**, page 414.)

Birthmarks

About 50% of babies will be born with a birthmark. The most common type of birthmark is called a **salmon patch**. This patch, when located in the center of the forehead, is known as an "angel's kiss"; if located on the back of the neck it is a "stork bite." Virtually all salmon patches disappear in the first few months, although occasionally a stork bite will remain.

Many babies have a dark discoloration around their buttocks; this will be found in about 95% of black babies, 80% of Oriental babies, 70% of Mexican-American babies, and 10% of white babies. It also disappears.

Strawberry marks are often barely visible at birth. If they can be seen, they are red and white. They gradually enlarge, achieving their biggest size at about six months of age. Occasionally, they are quite large. They always get smaller and should never be treated with surgery or X-rays. An *extremely* rare occurrence is when these strawberry marks trap some of the blood cells (platelets) that assist in preventing bleeding problems. In such circumstances, medicine (steroids) will be administered by doctors. Virtually all strawberry marks will disappear if left alone; large persistent ones can be surgically removed after the child is much older (over six years old).

Portwine stains are large purple marks that occur in about 3 of every 1000 babies. They are cosmetically disturbing if they are on the face. There are currently no good methods for removing the color, which will persist. X-rays are harmful and must be avoided. Adults find that cosmetics often cover these spots quite well. If a portwine stain covers an entire eyelid, there may occasionally be an associated problem of seizures, which will require medical help.

Finally, there are brown and black hairy and non-hairy **moles**. More than 2% of white babies and 20% of black babies have these moles, which do not disappear. Because there is a risk of large moles developing into cancer, many doctors recommend their removal soon after birth. You should discuss a problem of large moles with your doctor.

THE HEAD

The head is the largest part of the infant's body at birth. As such, the head will frequently show signs of damage from its passage through the birth canal. Infants often have a circular swelling on the back of their heads where a beanie or a yarmulke might rest; this swelling is known as a **caput succedaneum**. It is a result of pressure while the child's head is pushing against the mother's pelvis during labor. It generally disappears within two to three days. Another type of swelling that may appear in the same area is known as a **cephalohematoma**; this is due to a small amount of superficial bleeding from the scalp. It can be distinguished from the caput succedaneum in that it generally does not cross the midline of the scalp so that it is found only on one side or the other. Cephalohematomas generally take longer to disappear than do caputs but are of no concern unless they are of such large size that there may have been significant blood loss.

In addition, the head may have marks from the pressure of forceps, if forceps were used in the delivery. Again, these marks are very, very common and should resolve within a few days.

The size of the **fontanels**, the two soft spots on the infant's head, may vary greatly. Normal fontanels may range in size from one to three inches. When the infant is asleep they will ordinarily be flat. When the infant is crying, the fontanel will bulge upward. It also rises and falls regularly with the infant's heart rate. The front fontanel, on top of the head, closes at about 12 months of age, the smaller back one before 6 months.

THE EYES

Infants can see quite well from birth, although they do have some difficulty focusing for the first few months of life. One of the most frequently asked questions is what color the child's eyes will be. Most newborns have the same eye color—bluish gray. Eye color generally cannot be determined accurately until some time after the third month.

Yellowness of the "whites" of the eyes, also known as **jaundice**, is usual in newborns. The newborn liver is not as capable as the adult liver of handling the normal waste products of human red blood cells. This causes a certain amount of jaundice for the first few days of life. An extremely high level of jaundice can be a problem. If you are concerned about jaundice after you go home, call your doctor promptly. Sunlight has been discovered to be beneficial in reducing the amount of jaundice in newborns, but use caution because infant skin is extremely sensitive to the sun. A special type of lighting is used in most hospitals to reduce the jaundice level.

Occasionally, the eyelids may be swollen because of pressure placed on them in the birth canal. This swelling usually resolves by three days of age.

Excess **tearing** in newborns is usual. Tears are absorbed by tear ducts located in the inner aspect of the eyes; excess tearing on one side may signal a blocked tear duct on that side. Frequently, excess tearing on both sides in the first few days of life is the result of the silver nitrate drops placed in the newborn's eyes to protect against infection.

Doctors usually examine the newborn's eyes carefully at birth, including a test for vision. Frequently, however, doctors arrive on the scene when the infant is sleeping or crying. The test for vision is not adequate under these conditions, and it is mothers who usually confirm that their infants have good vision. Infants will follow your face and your eyes as you move from side to side. Doctors examine the lens of the eye and make sure that there are no cataracts present. In addition, they use an ophthalmoscope to examine the back of the eye, known as the retina, for signs of infection. They also examine the retina for any blackening, which will alert the doctor to the presence of a retinoblastoma, an extremely rare but treatable tumor of the eye in newborns.

Crossed eyes are common in newborn babies and are discussed further on page 355.

THE NOSE

Most textbooks tell us that infants must breathe through their noses, and much attention has been paid to clearing the nostrils of excess mucus and debris in order to make breathing easier. Seldom does mucus completely block the nostrils, and even then we have found that infants adapt quickly to mouth breathing, so parents need not worry excessively about clearing the nose. Babies also have protective reflexes that enable them to turn away from potentially suffocating situations. Newborns also spend a great deal of their time sneezing; the sneeze is not an indication of a respiratory infection or an allergy. Sneezing is a normal reflex of newborns.

THE MOUTH

Frequently, parents will notice little white spots on the roof of the infant's mouth directly in the midline, known as **Ebstein's pearls**. These pearls are normal findings in almost all newborns.

Many parents, especially years ago, used to worry about their infant being tongue-tied. This means that the frenulum, the piece of tissue on the bottom of the tongue that attaches to the floor of the mouth, is short. A short frenulum will *not* interfere with the child's speech. However, it will interfere with a child's ability to stick his or her tongue out as far as friends can, will interfere with the ability to catch M&Ms thrown in the air, and occasionally may interfere with the ability to lick an ice cream cone down to nothing as fast as friends can. Inability to do these acts can be troublesome, and a frenulum can be cut in order to enable the child to participate in these childhood activities. This is a simple procedure, but should not be carried out until the child is much older and actually experiences some disability.

THE FACE

The face is subjected to a considerable amount of pressure coming through the birth canal. This pressure may cause a dotted, purple rash over the face known as **petechiae**. Petechiae are caused by ruptures in very tiny blood vessels that occur because of the high pressure. These purple dots will generally resolve within two weeks.

In addition, infants often develop white dots with red bases all over their faces. This is known frequently as **newborn acne**. If there is a widespread rash, your doctor may tell you that your child has *erythema toxicum neonatorum*. Again, this is a harmless rash of the newborn period; see **Baby Rashes** on page 414.

THE HANDS

Infants ordinarily keep their fists clenched. You may notice that your baby has long fingernails. An infant's fingernails are generally soft, but nonetheless can scratch the infant's skin. If the fingernails appear long or scratch marks seem to be appearing on the infant (or on you when the infant is held close to you), the nails may be cut. Almost any nail-clipping instrument or even teeth will do so long as you are careful.

THE CHEST

Many parents are surprised to find that their infants have swollen breasts, boys as well as girls. This is a response to the mother's hormones, which have crossed the placenta and are found in the fetal blood. The swelling of the breasts usually disappears within the first month of life but may last for two to three months. Occasionally, a breast discharge may occur.

THE ABDOMEN

The **umbilical cord** is of frequent concern to parents; this cord consists of a white gelatinous material through which run two arteries and one vein. The cord will dry up in a period of about one week, but the decaying process usually smells. The smell comes from ordinary bacteria on the cord. Put a little alcohol on the cord; this will destroy the bacteria and also hasten the drying process. The cord will eventually become smaller, develop a brownish color, and fall off in one to three weeks. Often there is a slight amount of bleeding after the cord falls off. If an area of redness develops on the skin surrounding the cord, your doctor should be consulted.

Swelling or hernias of the navel are very common in black babies and far less common in white babies. Almost all of these disappear by themselves. The hernia is really a separation in the muscles of the abdominal wall. See the discussion of hernias on pages 175–177.

THE GENITALIA   The genitalia of both boys and girls are swollen in the newborn period because they respond to maternal hormones. Boys will have a swollen, enlarged scrotum, whereas girls will have enlarged genital lips and an enlarged clitoris. The swelling should subside within several weeks. Vaginal bleeding and discharges occur frequently; see page 529 and page 534. The foreskin in most newborn boys cannot be retracted.

THE FEET   An infant's feet assume many unusual positions. It is easy to understand why if you think of the cramped quarters in which the infant has recently been living. Most babies have feet that are turned in with the soles facing each other; the feet will assume a more normal position within a few months. Parents can check the normality of the feet by wiggling them about. It should be possible to wiggle an infant's foot into all of the positions that your own foot will assume.

Feeding

The feeding of infants and children is more than just satisfying the hunger urge. There are, in fact, three goals:

- To meet the nutritional requirements of children so that they can grow adequately.
- To help children to develop muscle and coordination skills. As infants become older, they learn to feed themselves. This requires complicated skills, such as reaching the tongue around a spoon, grasping the spoon, and eventually coordinating movement of the spoon with its portion of food into the mouth.
- To assist infants and children in developing social skills. Eating is a social activity that involves interaction with other family members.

BREAST FEEDING

With proper guidance and encouragement, most mothers can master the technique of breast feeding. Some mothers, through no fault of their own, will be unable to breast feed. An individual mother stands the best chance of success with supportive help. Many doctors are supportive, but few have had the experience to be truly helpful. Often the maternity nurses are helpful in early instructions. The best source of help in breast feeding is a friend or neighbor who has successfully breast fed her baby. The LaLeche League and the local Nursing Mothers Association are also good sources for advice on techniques.

Do not let anyone undermine your desire to breast feed. Also, do not let anyone push you into nursing longer than you want. Rely on your own judgment after you have listened to advice. You know what is best for you and your baby. Judge for yourself how long you want to breast feed. Two weeks, two months, and two years are all acceptable choices.

Preparation

Recommendations vary widely about the proper preparation of the breast for breast feeding. By and large, we do not feel that very much preparation before birth is necessary. We suggest a commonsense approach. During the last few months, the breast should be washed with a damp washcloth without the use of any soap preparations. Most creams and lotions should be avoided, although the use of lanolin on the areola (the dark area around the nipple) is acceptable. Exposure to air or massage of the nipples several months before birth sometimes helps. The purchase of several supportive nursing bras is a good investment for most mothers.

Milk Content

Infants can be put to breast within moments of birth. For the first several feedings, the breasts produce a material known as **colostrum**. The colostrum is rich in antibodies that serve to protect the child against infections in the coming months. There is no need to worry about the number of calories in your milk in the first several days because infants are born with an excess of weight that will tide them over until the high-calorie milk comes in. Between the third and fifth day milk begins to come in abundantly.

Although you may have heard that the iron content in breast milk is low, the amount of iron in breast milk is absorbed so well by the baby's intestines that supplemental iron drops are seldom necessary.

Remember that you are no longer eating just for yourself, but also for your infant! Be sure to drink two quarts of liquids a day: a quart of milk and a quart of any other favorite non-alcoholic drink.

Techniques

Nursing can be accomplished lying on your side or sitting up. Leaning forward a little in the sitting position helps. Placing the nipple near the baby's mouth will stimulate a reflex, the rooting reflex, which will help the baby find the nipple. Don't nudge him or her with your finger or the baby will root toward the finger. You may need to use your finger to hold the top part of your breast away from the infant's nose to allow breathing. Babies will let you know if they are having trouble breathing by pushing off the breast or opening their mouths and crying.

Feeding a small, premature baby can present special problems, but they can often be overcome. Indeed, many nurseries feel that breast milk is more important for the premature infant than for the full-term baby. Special bottles with "premature" nipples are available. Until the premature child is strong enough to suck and control the flow of milk, you can pump your breasts and bottle feed with these small nipples.

Mothers also have reflexes that assist them in breast feeding. The sight of their child, hearing their child cry, or even another child's cry will begin the flow of milk. Besides the rooting reflexes that help them find the nipple, infants have sucking reflexes that form the basis for their feeding. After an infant has been sucking for a while, a vacuum is created, and a tight seal is formed by the mouth around the nipple. Irritation to the nipples can be avoided by breaking the vacuum by inserting your finger between the infant's gums before removing the infant from your breast.

Timing

Newborns will require feeding every two to three hours. (Every four hours is unusual!) Frequent, prolonged feedings may help prevent engorgement. You should begin by allowing the child to suck several minutes on each breast and gradually build up to about 15 minutes on each breast by the third day of life. Infants will reject a partially full breast because the taste of the milk changes somewhat during feeding.

They should be allowed to reject the breast and go on to the other one. If your breast is so full that the infant is having difficulty grasping the nipple, express some milk by using your thumb and index finger to compress the breast at the areola.

As the child becomes older, the interval between feedings can become greater. Infants often begin to sleep through the night at a weight of about 12 pounds. Many mothers who breast feed find that the father can begin giving a supplemental bottle within the first two months. The father should usually give the supplemental bottle at the same time each day, although this is not a rigid rule. To prevent breast engorgement, mothers may pump their breasts when the father will be giving the supplemental bottle. Some parents prefer to have the mother pump her breasts and have the father bottle feed the breast milk. Other women find that their breasts adjust within a short period of time to the uneven schedule that is created by the father's feeding.

Babies have their own style of eating. Some are ravenous eaters, others are picky, while still others seem to fall asleep shortly after they begin nursing. They will all feed differently, but all thrive beautifully. Your baby will let you know his or her style early.

For further reading on breast-feeding techniques, see: *Nursing Your Baby* by Karen Pryor (New York: Bantam 1991); *Breastfeeding Your Baby* by Sheila Kitzinger (New York: Knopf, 1989); and *A Practical Guide to Breast Feeding* by Janice Riordan (Boston: Jones and Barlett, 1991).

The following helpful hints are from mothers who have breast fed and enjoyed their experiences.

MOTHER OF ONE:
The more relaxed you are about it, the easier it is. In the early weeks, there is bound to be some tension, but it can be minimized and with it the process becomes natural and easy. Try to find a comfortable place in the house where you will have the things you need close by and won't have to get up to interrupt the feeding. Either having the telephone within reach or taking it off the hook is helpful.

MOTHER OF TWO:
Set the feeding time aside for you and the baby to relax together. Keep interruptions and distractions to a minimum (this may be hard if you have other children). It is a good time to listen to music, watch TV, or read. It may be your only chance to sit and do these things. It is important to realize that it is okay to say to friends you are feeding the baby and excuse yourself from the demands of the phone, doorbell, and visitors until you are finished. [This is also good advice for mothers who choose to bottle feed.]

MOTHER OF TWO:
Wear loose clothing, especially things you can pull up from the bottom which will give you more privacy in public than things that button and zip up in front. You can also take along a small blanket or shawl to shield you and the baby.

MOTHER OF ONE:
Lying down is a very comfortable position for the mother, especially in the early weeks and during late night and early morning feedings.

MOTHER OF TWO:
By hand-expressing a little milk or putting the baby to breast you can relieve some of the pressure of engorgement. If it really becomes bad, hot compresses work. But take heart; generally it will occur only the first month or two until you and the baby are synchronized. At some point in time, you will be able to stop the middle-of-the-night feeding. This is important for the mother's health (and sleep). He may cry a lot the first night, but gradually the crying will decrease over the next few nights.

BOTTLE FEEDING

Whole or skim cow's milk has too much calcium and phosphorus for infants and should not be given before the fourth month. **Evaporated milk** (not to be confused with condensed milk, which is sweetened) is the cheapest commercially available formula. One can (or 13 ounces) of evaporated milk is added to 17 ounces of water. One tablespoon of sugar or corn syrup is then added to provide additional calories and to prevent the constipation that frequently results without its use. Should you choose not to add sugar or syrup, add only 15 to 16 ounces of water rather than 17 ounces to the evaporated milk.

Most parents use prepared formulas, including Similac, Enfamil, and SMA. These differ only slightly, and your decision should be determined by prices in your local area. The most economical way to prepare the formula is to use the powdered form (generally, two scoops of formula to 4 ounces of water). While mixing the concentrated liquid requires less shaking, it also requires more money. Care should be taken to mix the formula according to instructions; using too little water can cause serious problems for the infant. The temperature of the formula need not be higher than body temperature; room temperature is fine.

Children who are bottle fed need an iron supplement. Most commercial formulas (SMA, Enfamil, and Similac) are available with or without supplemental iron. Children should not be on an exclusive diet of whole cow's milk, or they will become moderately iron-deficient by the age of one year. Iron is present in a variety of other food sources, especially cereals, meats, and many vegetables.

The newborn needs about 50 calories for each pound of weight (or 110 calories per kilogram) daily. Requirements drop to about 45 calories per pound by a year. Formulas and breast milk provide 20 calories per ounce; thus, a seven-pound newborn needs about 350 calories (7×50) each day, or 17½ ounces ($350 \div 20$) of milk. Such calculations are seldom necessary, because infants will let you know when they are hungry or full.

HOW MUCH FOOD, AND HOW OFTEN?

Most newborns will not take more than three ounces at one time and hence require about six feedings a day. Every-four-hour feedings are seldom achievable. Your child will let you know when he or she is hungry. Don't be surprised if feedings are as frequent as every two hours or if the time interval varies daily. As infants become older, they can eat more and be fed less often. The middle-of-the-night feeding usually falls by the wayside when the infant begins sleeping through the night by about the fourth month. By one year of age, most infants will be satisfied with three meals and an occasional snack.

SOLID FOODS

Fruit juices can be served within the first two to three weeks of age. However, doctors and nutritionists disagree about the best age for beginning solid feedings. Almost all infants are capable of digesting whole foods at birth but lack of appropriate tongue coordination may interfere with feeding. Most nutritionists and doctors feel that there is no need to begin solids until six months of age. Many parents, however, claim that children sleep through the night better if given a solid feeding late in the evening. Children with a strong family tendency toward allergic disorders may do better if the addition of solid foods is deferred until six months of age. If the baby doesn't seem satisfied with formula alone, it is possible to add rice cereal to the formula in order to thicken the consistency; rice cereal is highly unlikely to produce any allergies.

Infants generally do well on most strained foods or baby foods between 3 and 6 months of age. These foods tend to be expensive, and generally have a fair amount of added water, so there has been recent interest in the home preparation of baby foods. (To make baby foods that will be acceptable to your child, consult one of the many books available. There are also a number of machines available that help in preparing baby foods.) Between 6 and 9 months of age, many children are given "junior" foods. Finally, between 9 and 12 months of age, most children are capable of eating finger foods or table foods.

The best guideline for feeding your child after the first year is to feed the child small portions of everything you eat. Foods from all of the basic food groups should be included. These include the milk and cheese group; the meat-fish-chicken group or their protein equivalent in either eggs, cheese, or beans; the fruit group; the yellow-and-green vegetable groups; and the bread, potatoes, and cereal group.

FEEDING ABILITY　Most children have a vigorous sucking action at birth, although many do not perfect this skill until they are four weeks of age. At four months of age, tongue control begins to develop, but children cannot yet remove food from a spoon. At this age, the infant will swallow a fair amount of air and burping will be necessary periodically. At four months, the infant begins to show signs of waiting for food. The child's arm will often move at the sight of food. Tongue protrusion begins developing to the point where the tongue will project out after the spoon. At this stage, food should be placed well back on the tongue; as more control develops, the child will become able to remove the food with the tongue. Between five and six months, lip control develops and the lips can be brought to a cup. Hand development is improving by this age, and the child can grab a bottle with both hands. Soon the hands can be brought to the mouth, and the ability to grasp a spoon with the hand begins to develop.

Between 6 and 9 months of age, the child learns to remove food from the spoon and might finally be able to drink from a cup. The child learns to eat a cracker without any assistance. Between 9 and 12 months of age, the infant achieves better finger control and is able to use fingers to obtain small pieces of food. He or she will try to use a spoon alone, but more often than not the spoon is still a better toy than a feeding instrument. A cup can be held, but the contents will be easily spilled. During this age, the child first begins to become choosy about food. Many myths have developed about the tastes of children. Although infant taste buds are quite developed, infants may be more sensitive to food texture than to taste and are likely to reject a food because of texture.

The period between 12 and 15 months of age is one of the more trying times for the parents. Children begin throwing food and utensils and become more assertive in the eating and rejecting of certain foods. Between 15 and 18 months of age, the muscle skills become more sophisticated, and a cup full of liquid can be handled without spilling. Between 18 months and two years of age, the child handles a cup well and can drink through a straw. The social aspects of feeding become more evident at this age as the child begins to say "eat," begins to name certain foods, and also knows when food is "all gone." At least half of a child's first six words are usually related to food. At this age, children begin feeding their favorite stuffed animals or dolls. By three years of age, a child should not be spilling too much and is able to coordinate talking and eating.

SOCIAL EATING

Feeding is an integral part of a child's social development. During mealtime, children learn how to interact with family, and they learn how to choose their foods. Although it may seem unbelievable to many parents, nine-month-old children have been shown to select a nutritionally balanced meal when presented with a wide variety of foods, and these children did not become either overweight or underweight. It was learned that a child's taste differs from day to day. A food rejected one day may make an entire meal the next. We are not fond of the practice of using foods, such as desserts, to reward children, and feel that a child should not be forced to eat a disliked food as a punishment.

Feeding time can become a battleground. Some of these battles may be avoided if parents realize that nutritional requirements change dramatically at one year of age. Infants gain about two pounds a month in the first six months of life and about one pound a month in the next six months, but then only four to five pounds a year up to the age of six. It is easier to tolerate a child's pickiness knowing that nutritional needs are being met. Children won't leave the table hungry; their appetite is a good guide.

As children become older, the dinner hour becomes the one time when all family members are assembled in one spot. This is an important social time. Watching television or reading during the dinner hour is wasting time that could be better spent with the family. Good communication during the dinner hour is an important measure for promoting mental as well as nutritional health.

COMMON FOOD CONCERNS

Junk food is an all-encompassing term that generally refers to candies, snacks, sodas, and the products of fast-food houses. The nutritional value of items called "junk food" varies tremendously. Some are quite nutritious, and some are nearly devoid of nutritional value. All are expensive. Use your common sense, read the labels, and do not use junk foods as rewards. Food can mold a child's behavior and parents should not encourage the use of expensive food with minimal nutritional benefits. Parents are often concerned that their children seem to eat only junk food. However, children do not do the shopping, and it is usually within a parent's power to control consumption of these items.

Fast-food restaurants can be fun for children, and the nutritional offerings vary. Use these opportunities to help encourage children to select balanced meals rather than give in to the common temptation of ordering the special, high-cholesterol combo.

Dry cereals range from products that are of exceptional value to products that are nearly worthless. The sugar content of dry cereals varies from 7 to 70%. The composition of dry cereals is printed on the package box, and you should read before you buy.

There are over 2700 known **additives** to the foods we consume. Two-thirds of these are of no known benefit; they include food colorings and artificial flavorings. The harm from useless additives is unknown but is potentially unlimited. While some food additives, such as preservatives, are necessary, the majority are not. Because any foreign substance can cause an allergic response in a sensitive individual, there are undoubtedly many allergic reactions that are caused by these additives. Research is currently under way to determine whether food additives are responsible for behavioral problems as well as allergic problems. Evidence is suggestive, but not conclusive, that food additives may affect behavior in some children.

Additional Reading

Marshall Klaus and Phyllis Klaus, *The Amazing Newborn*. Reading, Mass.: Addison-Wesley, 1985.

PART II

The Growing Child

CHAPTER 4

Growth and Development

O ne of the most pleasurable rewards of being a parent is watching your children grow. The excitement of witnessing your child's first steps or hearing your child's first words is difficult to exaggerate. As the years roll by, scrapbooks become as important to parents as the writings of their favorite author. Home movies are in a league above and beyond this year's Academy Award winners. Your favorite van Gogh is no more priceless than your favorite crayon creation.

Parents with no interest in history suddenly become meticulous historians of their child's life. The precise timing of the first step, the first word, the first birthday party, and the first date is remembered after the date of the Magna Carta has long been forgotten. As other historians do, they remember battles—the battle of the bottle, the battle of the toilet. Momentous pacts and treaties also have their day. Parents also grow. They learn to deal with the trauma of beheaded toy bears, bogey men in the dark, departing friends, and illnesses in brothers, sisters, and pets. In just a few years, parents watch their children grow from complete dependence to total independence.

Watching children grow is fun, but sometimes it can be worrisome. Parents are easily and naturally concerned that their children are not developing properly, and sometimes parents worry unnecessarily. As a society, we often seem pre-occupied with predicting the future success of our offspring. A child throwing a ball at an early age should prompt excitement, but not necessarily anticipation of a career in sports. And children who read at an early age may be destined for mechanical or artistic interest rather than headed toward a scholarly life.

Much of the work on child development of decades ago is still being used to predict a child's future performance. There are no crystal balls. We discuss patterns of growth and development in this chapter, as well as some of the tools used to measure and predict this process. Part of the purpose of this chapter is to point out the limitations of these tools.

Growth

YOUR CHILD'S WEIGHT

The physical growth of infants is remarkable. They usually double their weight in the first four to five months of life and triple their weight by the time they are one year of age. These are average weight gains. There are wide variations that are normal. Their height increases by a full 50% by the end of the first year. At birth, a child's head size is already nearly 60% of the adult size; by the age of three years, the head size will be nearly 90% of its adult size. Except in cases of rare diseases, physical growth is determined by two factors: proper nutrition and heredity.

Proper nutrition begins during pregnancy and continues throughout infancy, childhood, and adult life. The measure of growth that is most sensitive to nutrition is, of course, weight. Children largely make up their own minds about when and what to eat. So there are periods when children will refuse to eat a great deal of food. Remember that if your child will not eat carrots or spinach today, it will have no effect on his or her weight, and certainly will have no effect on height or brain growth. Prolonged periods of not eating will eventually affect weight, and then height, and ultimately brain growth, but malnutrition of this severity is extremely uncommon in this country.

The eating habits of children concern most parents. In discussing feeding problems with parents, we have found that all children either eat too much or too little and not enough of the right kind of food. It is hard for children to eat too little if presented with food. Hunger cravings are biologically determined, and the survival instinct is strong. Normal children always eat enough.

More often, the problem is that children eat too much. A child in the first year who eats too much gets chubby, but also undergoes several invisible changes. An excess number of fat cells may develop within his or her body. According to one theory, once these fat cells have multiplied, they send out messages that help to determine appetite cravings. Although losing weight in the future will decrease the size of these fat cells, the *number* of fat cells will remain the same. Excessive eating also can stretch the size of the stomach. Hunger pains are caused when the walls of the stomach contract against one another. Stomachs that are large and stretched require more food to decrease the empty feeling of hunger. Other factors appear to determine eating patterns. For example, thin people seem to be able to tell precisely when they have had enough and will not eat any additional food. Fat people, on the other hand, tend to continue eating even beyond the point when hunger has been satisfied.

Children learn eating patterns from their parents, and proper weight is partly culturally determined. In some countries, being heavy is considered a sign of prosperity, and the same holds for some sub-cultures in the United States. A wide range of weights must be considered as normal. Being ten pounds overweight or ten pounds underweight for your height, when compared with the average weight, makes little health difference. But as you get beyond a few pounds overweight, you must pay a price. That price is poorer health.

YOUR CHILD'S HEIGHT

The genetic factors important to physical growth are reflected in the height of the parents. Tall parents tend to have tall children, and short parents tend to have short children; it is as simple as that. However, an interesting phenomenon called "regression toward the mean" has been observed. Tall parents have children that are taller than average, but shorter than the parents. Shorter-than-average parents will have children who are shorter than average, but who tend to be taller than the parents.

The upward growth trend is not a steady trend. For long periods of time, children may not grow at all; at other periods they may exhibit rapid growth spurts.

Your doctor will record the growth of your child on a growth chart similar to the charts shown in Part V of this book. After recording a series of heights and weights at different ages, you can visualize the growth of the child. If growth is extremely fast or stops for a long period, it will show up on the graph and be noted much more easily than with the traditional marks on the closet wall. Instructions for using those charts are given in Part V, together with charts for your children. Don't worry too much about how your child compares to the "normal" lines. At some time during growth and development, most children will be high or low in something because of growth spurts or lags. Usually, the growth rate is roughly parallel to the lines on the growth chart and does not deviate more than two or three lines from the original. A prolonged period of delayed growth requires investigation. The charts will help you with the problems of **overweight** (page 320) and **underweight** (page 323).

Hormone Treatments

We are pleased that tall women are no longer considered "unusual." Height has always seemed a trivial characteristic by which to judge any person. Unfortunately, many tall women can remember being teased because of their height when they were younger. This attitude is no longer common (particularly with the trend requiring fashion models to be nearly as tall as professional basketball players); but many parents, understandably, wish to protect their daughters from such experiences. Other parents have been concerned that their daughter's height might interfere with a ballet career. Hormones that will prematurely stop bone growth are available, but they are dangerous and have numerous side effects. To be effective, they are best given before puberty when career decisions about ballet are seldom seriously made. Once growth is stopped, it cannot be restarted. Except for *very* unusual circumstances, we do not recommend the use of these hormones.

Many parents are concerned that their child will be handicapped because he or she is too short in a culture that places a value on being tall. Some have heard that **growth hormone** is a drug for increasing height. Growth hormone is produced by the pituitary gland and enhances growth. This hormone is effective in children who have a deficiency of growth hormone and has recently been found to be effective in other children with extreme growth problems. Although genetic engineering is about to make this hormone widely available, its use should be restricted to severe growth retardation because of potential side effects.

GROWTH AT PUBERTY

Initially, growth is controlled by two hormones: growth hormone and thyroid hormone. During puberty, additional hormones are responsible for growth. These hormones are also responsible for the development of sexual characteristics in boys and girls. Although hormones are primarily responsible for sexual behavior in animals, they play only a minor role in human beings; sexual behavior in adolescents and adults is far more a psychological than a hormonal response. Even in most animals, mature sexual behavior requires far more than the presence of hormones.

As an adolescent reaches puberty, the pituitary gland increases the secretion of a hormone called follicle-stimulating hormone (FSH). FSH in women stimulates the ovaries to produce estrogen (female hormone) and in men stimulates the development of sperm. In males, another pituitary hormone stimulates the increase in testosterone (male hormone) from the testicles. Testosterone can also be produced by the adrenal glands in men and women.

Boys

In boys, there is no single dramatic event that makes it clear that puberty is on its way. There is a typical sequence of changes, however. These changes may not be immediately obvious because they tend to occur over a fairly long period of time. These changes usually begin between the ages of 9½ and 14, but some of the changes are so subtle that they may not be noticed until a later age.

In most boys, enlargement of the testicles begins at about age 11 and continues until about age 18. Penis enlargement usually begins about a year later and continues until about age 16. Pubic hair generally does not begin to appear until about age 13½, although some 11-year-old boys show signs of early pubic hair. If there are signs of pubic hair development before age 10, a call to the doctor is in order. Testosterone—which is responsible for the enlargement of the penis, testicles, and growth of pubic hair—is also responsible for the development of axillary and facial hair, voice changes, adult body odor, acne, and increasing muscle mass at about age 15.

Half of all boys will experience some enlargement of one or both breasts. Commonly, there may be a lump or tenderness under one or both nipples. This can be easily explained on the basis of hormonal changes, but it is often very embarrassing. Your son will need your help and support without having undue attention drawn to this condition. Obesity may accentuate this problem by giving the appearance of very large breasts while at the same time making the penis and testes appear small. Occasionally, breast enlargement may be so great and so prolonged that you and your son may wish to discuss the possibility of cosmetic surgery with your doctor. Usually, however, the condition resolves itself within a year.

Girls

The process of sexual maturation in girls usually starts about two years before the first menstrual period. The first sign of puberty typically occurs at age 9½ for white girls and age 8 for African-American girls. For many girls development starts earlier and is still normal. Signs of pubertal development before the age of 6½ require an evaluation by your doctor. Although the ages at which changes occur vary widely, their sequence is pretty much the same. Breast development begins at age 11 and is followed shortly by the appearance of pubic hair. Hair under the arms begins to increase at age 12. Between ages 12 and 13, a growth spurt results in a noticeable gain in height. Menstrual periods begin around age 12½. Over this entire period, there is some change in the way fat is distributed over the body, particularly in the hips and buttocks. Widening of the hips, as well as development of the uterus, vagina, and external genital organs, also occurs. Problems associated with menstrual irregularity and vaginal bleeding are discussed on page 532.

Variation in breast development is often a cause for concern. Breasts usually begin to develop one or two years before the first period, but this may not occur for as long as two years after the onset of menstruation. Often one breast develops more rapidly than the other so that temporarily the breasts may be of unequal size. Frequently, this is of concern and embarrassing, but most breasts reach nearly equal size by age 18. Sometimes there is worry that the areola (the dark area surrounding the nipples) may be elevated too much or not enough. In about 75% of women the areola is elevated, while in the remainder it is not; both patterns are perfectly normal. Inverted nipples are not a problem and need no treatment other than reassurance.

Finally, there are the inevitable concerns over the size of the breasts. We are happy to note the decline in emphasis on huge breasts as the ultimate sign of sex appeal. Still, every girl worries about being flat-chested, while virtually none of them will be. Mostly, the problem is one of timing. Remember that the breasts may not begin to develop until two years after menstruation has begun and may not be fully developed until age 19 or 20. It is perfectly possible for one 13- or 14-year-old to have undergone very little or no breast development while most of her friends will be well into breast development and some will essentially have completed their development. Your reassurance is a most important factor in easing your daughter's anxiety. Efforts at stimulating breast development artificially have no place in the developing child or mature woman. Oral or topical estrogens are potent drugs that should not be used to promote cosmetic changes.

Excessively large breasts are also of concern. Emotional support from parents as well as physical support from a good brassiere is important during adolescence. Surgical procedures are available for cosmetic reduction, and this procedure can be considered by older adolescents and their parents.

Development

NEW SKILLS

Watching your child grow in size is only part of the fun. Development of coordination is even more exciting, and infants quickly acquire a variety of skills. They acquire muscle abilities, which give them mobility and allow them to explore. They develop extremely fine motor abilities, which enable them to accomplish complex tasks such as playing a musical instrument. They develop the ability to convey anger, disappointment, or love by use of facial expressions. Language develops in order to express needs. These needs are simple at first, but gradually these language abilities become more sophisticated until children are finally able to use words to express feelings.

Social skills provide early rewards for children. Smiling begins very early, often in the delivery room. Soon this smiling becomes more specific; infants are soon able to recognize their parents and recognize that parents are exhibiting warm feelings toward them. Gradually, children learn how to play games with others. Soon afterward, they learn how to share things and ultimately they develop the capacity to love. Learning is one of the more complicated skills, and progresses from primitive basic memory to abstraction, moral judgment, and creativity.

SEQUENCE OF SKILLS

The development of skills progresses in an orderly rather than a haphazard fashion. Pediatric and psychological research has focused on this sequence and has been helpful in establishing the expected order of events and the limits of "normal." Attempts to correlate the age at which something happens with future capabilities have not been very successful. This sequence is, in fact, a logical building of increasingly complicated skills. As an example, let us look at the sequence required for a child to use a pencil.

The first stage in the sequence is reflex control. You may remember that your infant kept his or her hands clasped in a fist for most of the first few months of life. All infants have what is known as a **grasp reflex**. They will grab and hold on to any object placed in their palm, such as a finger or a pencil. Doctors may demonstrate the strength of an infant by allowing him or her to clasp each of the doctor's index fingers, and then lifting the child off the table. The infant demonstrates his or her strength admirably, but this is not a sophisticated use of the hand; there is no control over this reflex, and anything in the palm will be clutched.

As the nervous system develops, the grasp reflex is overcome, usually by about three months. The infant is now ready to use the hands in a different fashion. By about four months, the infant's hands are held open and can hold onto objects placed in them. Although vision is good, coordination is not sophisticated enough to reach for an object and retrieve it. The infant will usually overshoot the object.

At five months, retrieval abilities get better. The process known as raking is begun; objects are brought closer by use of the entire hand. At nine months or so, the infant achieves control of thumb and forefinger and can bring them together so as to pick up tiny objects. Soon, small objects, such as raisins, pennies, and pills, are picked up, looked at, and ultimately placed in the infant's greatest exploratorium, the mouth. Development of this finger-to-thumb pinching motion is unique to humans; unfortunately, this ability also places the infant at risk for stuffing nose, mouth, and ear full of "goodies and baddies."

By now, the infant is able to let go of items voluntarily. The ability to release items in a projectile fashion—better known as throwing your food on the floor—begins around the child's first birthday.

After a year, things progress rapidly. Hands that were first able to release toys voluntarily only a few months earlier are now gentle enough to release one block precisely on top of another. By age two, the budding engineer can create a tower of six stories (or blocks) high. Like any engineer pleased with a building, our growing child feels the need to leave records of building successes, and the ability to draw develops. At first, the child draws with crayon in fist, but before the third year begins, the crayon is held as the adult holds it. Pencils are somewhat trickier than crayons, but this skill is also soon mastered.

Finally, the apprentice learning skills needs help from the master craftsman. A child cannot develop artistic skills without paintbrush and paper. A child needs raw materials with which to practice. Encouragement is also required. Children are social beings; a smile or kind word lets the child know you are pleased with the progress on the product.

This evolutionary sequence is illustrative of most other processes in child development, like walking. The general rules are:

- Involuntary reflexes must be overcome in order for a child to have voluntary control.
- There is a logical progression; cruder movements always precede finer movements.
- Coordination of several senses is necessary for skill development. For example, the child relies on sight, depth perception, and balance to build a tower of blocks.
- A skill can develop only when the nervous system is mature enough. Building a tower of blocks is impossible for a six-month-old, no matter how much exposure to blocks the child has had.

INDIVIDUALITY

The growth of every individual is unique; this is the most important point of this chapter. Even identical twins brought up in similar fashion grow differently and ultimately do different things with their lives. Children are all very different and will demonstrate their unique tastes, even for music, at an early age.

The infant is an active participant in creating his or her own environment. Babies quickly learn how to manipulate their parents. A child soon learns that jabbering will, as if by magic, produce parents. Parents sometimes substitute a favorite teddy bear when the child cries or jabbers, rather than spend time with the child. Or a parent may comfort the child vocally when physically occupied with other tasks. The infant quickly decides whether these responses are sufficient. For a time, they may suffice, and then the game begins again as the infant discovers a new way of drawing attention. Quieter infants may use smiling or eye contact, rather than voice, to make their demands known. In short, your infant will actively influence your behavior.

Toilet Training

Toilet training is an important developmental milestone both for toddlers and their parents. The skills required for bowel and bladder control are similar. Because these skills are acquired at slightly different ages, we will be discussing them separately. These skills are the culmination of a long series of accomplishments, and parents play an important role by recognizing, reacting to, and rewarding each accomplishment.

BOWEL CONTROL Young infants have a reflex known as the **gastro-colic reflex**. About 20 minutes after eating, the infant will have a reflex bowel movement. This reflex, like other early reflexes, will gradually diminish, usually by 12 to 15 months of age. Willful control of bowel elimination can only develop after the child has overcome this reflex.

Many parents recognize this happening and will begin toilet training by capitalizing on the reflex. We have all heard stories of children being toilet trained by the age of six months. These children were, of course, not really toilet trained, but were passing stools in a regular pattern determined by the gastro-colic reflex. Some parents have confused the natural disappearance of this reflex with a conscious decision by their 12-month-old to be uncooperative. This is far from the truth; a 12-month-old is still much too young to toilet train. Other skills must first develop and mature.

The child must first be able to sense that a bowel movement is occurring; most children acquire this knowledge after one year of age. (Parents can often recognize that their child is producing stool by facial expressions or other gestures.) Initiation of bowel control is often begun at this time and is most often begun unconsciously. You will often mention "BM" or some other term when the child is having a bowel movement. The association between act and language will have begun. But the final act of this complex learning task will not be accomplished for another year or two.

After the child is able to detect the sensation of a bowel movement, he or she becomes able to sense the presence of fecal material in the rectum before it passes. Following this, the child begins to achieve some muscular control over the passage of the stool. Newborn infants have no muscular control at the anus; this muscular coordination develops at different ages in different children. Some children develop it early in the second year of life; others not until the end of the third year. The rare child with a neurological condition may develop this control much later or not at all.

The length of time that a child can use muscular control to withhold the stool gradually increases. In the middle of the second year, the child may be able to tell you that a bowel movement is on the way, but the movement will probably have occurred before you reach the toilet.

Children should not be "rushed" to the toilet. Frantic activity can be a frightening experience for some children, who might associate normal body functions with dirt and disgrace. Of course, many households have such a high level of activity that rushing may be normal. In short, the speed with which you take your child should be similar to the speed with which you do other things with your child.

The Potty

Once your child begins to inform you of an impending bowel movement, you can begin taking the child to the toilet. Most children will be able to do this when they are two or three years old. Introducing your child to the toilet ("potty") should be a relaxed affair. Initially, there is no need to even remove the child's pants. A word of caution about the first approach being a "bare-bottom" one: The child's potty is often on the floor where cold tends to concentrate. Sitting on a cold seat can be a shocker to anyone, and especially so to a two-year-old. Merely associating the bowel movement with the potty is enough at first; staying too long should be avoided. The energy level of two-year-olds is such that they will not wish to remain on the potty too long anyway.

Potties that sit on the floor generally seem more secure than toilet seats placed way up high on that big toilet. It is also easier to push with feet on the ground. These portable potties can easily be taken on trips, and they become the individual possession of the child. Initially, when the child produces a stool, it should be allowed to remain in the potty for a while. Your child will be pleased with the product and may be disappointed when it is taken away. One friend, confronted with a shocked two-year-old noticing her missing bowel movement, invented the "B.M. birdy" who needed the stools. His daughter was satisfied with the explanation and loved to talk about the birdy. Remember also that the roar of the toilet can be frightening: Seeing this roaring monster consume a stool can upset and frighten a two-year-old; it may thrill a three-year-old.

A major accomplishment during the long process of development is the ability to put things off, or the ability to delay pleasure. Learning to postpone urges can take months or years. (Indeed, the postponement of some urges remains difficult for many adults.) Having a bowel movement is biologically a pleasurable experience; a child learns to delay this pleasure in order to receive something equally pleasurable—a parent's reward. The child is beginning to make decisions about the pleasures that social behavior can bring. The child of two is becoming more social and appreciates the social interaction with mom and dad. Smiles and praises, not overdone, should accompany potty sitting.

Praises can be more laudatory when a bowel movement is produced, but they should also accompany non-productive sittings. You should not keep your child sitting on the potty until a bowel movement is produced. Children take time to learn, and there will be many dry runs. Persisting will frustrate both you and your child. When your child is tired of sitting, time is up.

The second year is a period of negativism. The child will frequently say "no." This negativism is an important developmental milestone for the child, for he or she is developing an individual personality that is not a part of mother or father. The drive toward independence will eventually motivate the two-and-one-half- or three-year-old child to control bowel eliminations. However, the early part of this struggle for independence may mean a struggle over who will make decisions about bowel movements. It does not pay to engage in battles about bowel movements. Reward, or disengagement, is the key. Punishment will only intensify the opposition and prolong the struggle.

BLADDER TRAINING

Bladder control in children follows the same developmental sequence as bowel control, and lags behind bowel control in most children. At birth, infants urinate by reflex when their bladder is stretched to a certain point. As they get older, they can hold larger and larger amounts of urine.

By 18 months, children can sense when they are urinating. The muscular ability to hold urine in the bladder is usually acquired between the ages of two-and-one-half and three-and-one-half years. Usually by the age of two-and-one-half years, the child is well on the way to learning bowel control, and bladder control is often learned soon afterward. Many toddlers are very much concerned about being in control and learn quickly to hold on to their urine. They are training themselves by holding their urine longer so that they may spend more time at play or gain the social rewards offered by their parents.

Because calling mom or dad for assistance in going to the potty is a handy way of getting attention, parents can expect some dry runs. Accidents should be expected because a child at this age can become easily pre-occupied or forget.

Daytime bladder control is achieved by age three in 85% of children and by age four in over 95% of children. Occasional accidents will occur. Often, older children with the "giggles" will have accidents even in school. Children will usually learn quicker when their older brothers or sisters are around to help and demonstrate.

BEDWETTING

Nighttime dryness takes longer to achieve. Nearly 20% of five-year-olds are still subject to frequent nighttime wettings, and more than 10% of six-year-olds still wet the bed. On the other hand, some three-year-olds, if taken for a late-night trip to the bathroom, will make it through the night just fine. Expect an occasional accident, but avoid any sort of punishment. Children should not be kept in overnight diapers until they are 100% dry; these do not encourage the child to develop his or her own control. (See **Bedwetting**, page 315.)

Because bladder control, like bowel control, is the culmination of a maturational sequence of sensory, muscular, learning, and social development, all these areas will be investigated by a competent medical professional. Medication may occasionally be necessary. Surgery is virtually never needed. Most therapy focuses on the social aspects of changing this behavior. Frequently, improvement occurs when the child makes the decision that a dry bed is important, such as when visiting with a friend or relative.

Relying on a drug to increase urinary retention through increased muscular control, or an alarm to modify behavior through negative reinforcement, is a narrow approach to a complex phenomenon. These methods have a role, but only as part of a therapeutic process involving coordination of efforts with child, parents, and professionals. Don't buy the devices advertised with coupons on the back of magazines.

What is Normal Growth and Development?

Is my child normal? Why is Johnny so short? Why does George wet the bed still? He is almost eight! Why didn't my second child start to walk as early as my first? Our society seems pre-occupied with the prediction of the future success of children, creating a natural environment for parents to worry. The spirit of competitiveness lives in all of us. We worry unnecessarily about such things as whether Suzie will walk before the little girl across the street does.

Individuality   While children are constantly developing in all areas, certain areas can develop more rapidly because of the individual needs of the child. As an example, consider a child growing up in a large family. Dinner with many people is an experience that most of us have only at Thanksgiving or at large picnics. It is fun but chaotic. The relative chaos tends to produce two types of children, the quick and the hungry. In order to survive, the youngest child in a large family must develop quick hands. Some of these children will develop hand skills at an early age. But later in childhood, the child from the smaller family will catch up. Both are normal.

The opposite seems to occur in the development of walking ability. With many other feet belonging to larger bodies running about, the small child in a large family is likely to be knocked over frequently. It is not uncommon to see late walking in children from large families. These children, however, do have the need to explore and discover new things, and they develop other means of locomotion. Some are clever enough to get their older brothers and sisters to carry, push, or pull them about. Others develop a pattern of extremely rapid creeping or scooting. Children are constantly developing and adapting to their environment. They are adapting in the way that is best for them and *not* according to printed schedules in textbooks.

Variability   An important consideration in interpreting the limits of "normal" is the concept of **variability**. Variability describes the outside limits of age at which a given skill should develop. For example, most children begin to smile by about three to four weeks of age, but some children begin on the first day of life, and other children don't begin until they are seven to eight weeks of age. We say that the average child smiles at one month, but the variability is either a month earlier or a month later.

As tasks become more and more complicated, the variability always increases. Most children sit at about 6 months, but the variability is between 4 and 8 months. Most children walk at one year, but some begin as early as 9 months of age, and still others may not walk until 16 months or even later. Verbal ability has even greater variability. Some children say three words (other than mama and dada) before a year of age; others may be almost two before this is accomplished. Beethoven was said to be composing and playing at the age of three, but adults ten times that age have difficulties with these tasks. Variability limits are far more meaningful than single "normal" values in interpreting whether a given milestone has occurred on schedule or not. In other words, a child who does not walk at the average one year of age is still normal if he or she is walking at 16 months. However, the child would be considered abnormal if he or she is not walking by the age of three years.

Normality in one area of development does not guarantee normality in all others, and abnormality in one area does not signify abnormality in others. Consider a child who, for whatever reason, has acquired an injury that affects muscle (motor) development. This child may lag considerably in sitting, walking, and the more sophisticated locomotion skills, while acquiring language, social, and other skills at the appropriate age.

Intelligence

A common misconception is that a child's intelligence can be predicted by observing how rapidly he or she develops in the first year of life. Except for rare cases of severe retardation, development rates offer very little help in predicting ultimate intelligence, let alone whether or not that person will use intelligence creatively, productively, or not at all.

Experts can only put a fuzzy border around what is normal and what is abnormal. Nobody can say what your child will be like in the future. A child who is developing slowly at first may develop rapidly later on. Other children may develop very rapidly initially and then slow down. Most usual are periods of alternating rapid, slow, and average development in response to changes in season, changes in family composition, and other changes not yet determined.

Additional Reading

T. Berry Brazelton, M.D., *Infants and Mothers: Differences in Development.* New York: Delacorte, 1983.

T. Berry Brazelton, M.D., *Toddlers and Parents: A Declaration of Independence.* New York: Delacorte, 1989.

T. Berry Brazelton, M.D., *On Becoming a Family: The Growth of Attachment.* New York: Delacorte Press, 1981.

Selma Fraiberg, *The Magic Years.* New York: Macmillan, 1981.

Penelope Leach, *Your Baby and Child: From Birth to Age Five.* New York: Knopf, 1978 and *Babyhood*, Second Edition. New York: Knopf, 1989.

CHAPTER 5

Personality and Behavior

T he complexity of human personality development cannot be discussed in a single volume, let alone in a single chapter. Freud, Piaget, Erickson, Sears, Maccoby, and hundreds of others have devoted their entire lives and thousands of volumes to expanding our knowledge in this area. In this chapter, we provide a framework for personality development that can help you understand better some of the common behavioral problems that you may encounter.

Infancy: Birth to One Year

Each infant is born with an individual temperament. Theories that a child's personality depends entirely on environmental factors are now felt to be untrue; infants are individuals from the very start. Some come out active and screaming and remain at a high level of activity throughout life. Other children are very mellow at birth and continue with this personality trait.

We now know that infants exert a strong influence over their parents' behavior toward them. An active child will quickly learn how to attract his or her parents' attention. These active children with their high levels of demands will force parents to interact with them more often and will often encourage parents to provide them with more play objects. Quieter children may show the greatest pleasure while being held or merely looking at a parent. Quiet activity pleases this type of child the most, and parents, sensing this pleasure, respond with the level of activity that the child enjoys the most. Often children will interact differently with mom or dad.

When infants have quiet times, they are listening, learning smells, judging distances, and developing intellectually. During active times, they are moving and discovering limbs, exercising muscles, and developing coordination.

The newborn quickly learns how to obtain food. Babies are born with reflexes that enable them to suck vigorously and locate the nipple on the breast by rooting, but they must learn how to make the milk appear. Crying most often serves this function. Parents influence this behavior by modifying feeding schedules to their own preferences. Development progresses quickly, and infants develop an interest in their environment. They soon learn that toys can be manipulated by kicking and reaching; hence, they begin to control their surroundings.

Infants can be troublesome before the terrible twos. The following are descriptions of one-year-olds given by a *majority* of mothers in a recent study by Emily Szymowski and Robert Chamberlin: "attention-seeking, stubborn, clingy, temperamental, restless, noisy, and fidgety." However, the three most characteristic descriptions were "cheerful, curious, and friendly."

Within the first few weeks of life, children are already becoming social and beginning to smile. Smiling is clearly an expression of joy. From birth to six months, infants will often smile at anything and all persons that please them. By the age of six months, many children can recognize the faces of their principal caregivers. Faces other than those may bring anxiety, terror, and screaming. Children raised in households with many people may not experience as much anxiety.

CRYING

Crying is a normal activity of all infants and serves many functions. In a newborn infant, it helps the lungs adjust from the fluid-filled amniotic sac to an air-filled world. Infants cry in response to their needs. They cry when they are hungry, wet, cold, or, on occasion, in pain. This crying is a way of communicating with their parents and is effective because the necessary response—the parent to alleviate the need—is usually produced. Parents quickly become adept at distinguishing different types of cries: the cry of hunger, pain, boredom, or anger.

Some crying is not due to a specific need of the infant. With this type of crying, neither hunger, wetness, nor pain is responsible. The infant is merely fussy. Holding and rocking does not always relieve this type of crying, and parents need not be concerned about this fussiness.

Several years ago Dr. T. Berry Brazelton did a study on healthy infants that revealed that two to three hours of fussiness per day in the first few months of life was to be expected. Some infants were fussy for as much as four hours a day. The crying increased gradually, becoming most prolonged by six weeks of age, and declined thereafter. By three to four months, most infants were fussy less than 1½ hours daily. Most of the crying seemed to occur between the hours of 6 P.M. and 10 P.M. Unfortunately, this crying occurs at a time in the day when parents are becoming tired and the irritability is harder to tolerate. (For further discussion, see **Colic**, page 516.) In the young infant, sleeping on the stomach can cut down on the amount of fussiness, prolong sleep, and produce more regular sleeping patterns. However, crying is often independent of sleeping position.

Crying may also occur as a result of being frightened by a loud noise or a sudden movement. Other infants may cry in response to lights being turned on or being turned off.

As the infant becomes older, crying becomes less frequent. Children can delay a need for gratification longer and longer. Older infants, however, begin to miss their parents and will cry when they are lonely. By six months of age, the child recognizes his or her parents and consequently recognizes their absence. Some people recommend that the infant be held if loneliness is the cause of the crying. Others warn that cuddling the child after crying will spoil the child. We believe in a commonsense approach. A few whimpers when a child wakes up at night and is lonely need not be attended to. Frequently, playthings in the crib will serve to comfort the child. If the whimpering persists, often the voice of the parent is enough to quiet the child. If crying becomes intense and prolonged, the child should be held and comforted.

Beyond one year of age, crying is often a product of the child's frustration at not being able to control his or her surroundings. This is, of course, a healthy sign of personality development but can be disruptive to the family. It is better tolerated if it is better understood.

As children become older, crying can be triggered by emotions. It is helpful for parents to teach a child to be able to say that he or she is angry or frustrated rather than to continuously demonstrate this feeling by crying. Sadness and separation are also frequently accompanied by crying.

Crying can, of course, also be a sign of pain, especially in a young child who cannot verbalize the presence of a headache, stomachache, earache, or sore throat. If illness is causing the pain that is producing crying, other symptoms will usually be present as well.

SLEEPING

An infant's typical sleeping patterns can be extraordinarily fatiguing for parents. Frustration and futility can set in when you learn that a friend's infant has always slept from 8 P.M. to 8 A.M., or if you read an article which buoyantly claims that by three, four, or six months your child should be sleeping through the night.

Individual sleep patterns and requirements vary tremendously. The average infant sleeps approximately 16 hours a day, with some infants sleeping as little as 10 hours and others sleeping as many as 22 or even 23 hours. As children get older, the sleep requirements gradually decline. By one year of age, children sleep only about 14 hours. By four years of age, they sleep an average of 12 hours and by age 8 an average of 9, but as few as 6 and as many as 13 hours a day.

Newborn infants have as many as five or six sleep cycles throughout the day. By one year of age, most children have just two sleep cycles, the afternoon nap and evening sleep. Most children outgrow their need for an afternoon nap by the time they are five years of age, but even many adults choose to take an afternoon nap whenever possible.

While the length of children's sleep cycles increases markedly in the first year, a substantial number wake in the middle of the night all during the first two years. One interesting study that recorded the sleep of nine-month-olds revealed that two-thirds wake during the night while only 22% of their parents reported the night-waking. Thus, in a sense, "sleeping through the night" is a phenomenon that involves both children's sleep and parents' sleep. While night-waking can exhaust parents, slight solace may be found in knowing that this pattern is normal for infants and they will thrive. As infants become older, you should not encourage prolonged wakefulness with food, holding or other forms of attention, but merely attend to the child's immediate comfort and safety. As infants enter their second year, a pre-bed ritual can decrease anxieties over separation and facilitate a child's going to sleep.

Eventually, it is important for children to learn to fall asleep on their own. Most children awake several times during the night. If they are used to falling asleep in a parent's arms, they will expect these arms at 3:00 A.M. It is best to work toward this goal by having a pleasant nighttime ritual and then placing your child in the crib—awake. Do this either at naptime or evening bedtime. If your infant protests, you can listen to the protest for about three minutes (use a watch or you'll be back in 20 seconds). Then return, comfort your child with your voice or a touch, as appropriate, and leave shortly thereafter. You can now leave for up to four minutes. This process can be repeated with gradual time lengthening, and it is far preferable to deciding one night to let your baby "cry it out." Most infants learn how to put themselves to sleep within a week using this technique. Relapses will occur, and routines must be altered to accommodate illness and other events.

Toddlers

Toddlers are moving from a period when they have been totally dependent on their parents to a time when they are trying to control themselves and their universe. They begin to discover their individuality and try to determine how much power they have as individuals. This is an age of daring for toddlers; it is an age of exploration. They will climb ladders far too high, attempt to run much too far away, and attempt to eat dangerous things. Parents of children in this age group must set limits without overprotecting the child. The child must learn what is too high, what is too hot, what is too sharp.

Children are very negative during this period. Because of the child's extreme inquisitiveness, parents find they often must say "no." The child, on the other hand, is driven by a burning desire to explore and to control anyone who would interfere with these explorations.

The constant "no, no, no" of the child is meant to ensure the autonomy of this explorer. Children will often say "no" when in fact they have every intention of doing what the parents ask. Again, this is their way of telling you who is in charge. They will say "no" to the parent and really mean, "No, I'm not going to do it because you want me to do it, but I will do it because I feel like it and because I am in charge here." Parents would do very well to study some of the tactics of diplomats. Every diplomat knows how important it is for the opponent to be able to save face. Parents will do well to allow the child to protest "no" while at the same time making sure that the child does what is expected.

Like the "no" syndrome, temper tantrums in toddlers are often a reaction to a world they cannot control. It is their way of reacting to parents whom they perceive as trying to interfere with their autonomy. The tantrums are a sign of their frustration. They are frustrated not only with parents and the world but also at their inability to communicate their frustration. As children become more verbal and are more able to express their own anger, temper tantrums become far less frequent. A parent can be most helpful by encouraging children to communicate their feelings directly rather than using tantrums or other aggressive acts. Try not to let tantrums become a successful tactic for the infant to obtain a goal.

Despite the first steps toward independence, the child still requires a tremendous amount of holding and touching, which are important for personality development in this age group.

It is not easy being a parent of a two-year-old. For that matter, it's not easy being a two-year-old. This period often creates the greatest amount of anxiety for parents and it has been dubbed the "terrible twos." The "terrible twos" may not seem so terrible if the parents have some basic understanding of what is going on. The task of a toddler's parents is to aid the child in becoming independent. This requires not only tremendous affection and more patience than at any other stage of development but also a thorough understanding of what the child is going through. We recommend Selma Fraiberg's excellent book *The Magic Years* (New York: Macmillan, 1981) for a better understanding of this period, as well as T. Berry Brazelton's *Toddlers and Parents* (New York: Dell, 1989).

SLEEPING

Many of the sleep problems of childhood are based on the fact that children's sleep patterns are different from those of adults. Many parents would prefer that their children would go to bed at 7:30 or 8:00 in the evening and awaken shortly after the parents arise. These wishes do not, of course, correspond to the needs of all children. We feel that some sleep problems can be avoided by careful planning. Children get to sleep most easily when there is a pre-established, predictable routine. If children are highly wound up because of exercise or vigorous play, it will be difficult for them to go to bed; ordering a child into bed at such a time is futile. A good wind-down routine consists of a nighttime bath, brushing of teeth, and a nighttime story when in bed. These periods serve to calm the child down and to provide a time for parents and children to enjoy each other.

Young children often do best with night-lights in their room, and these should be provided when necessary. You can also expect your toddler to spend some time in the room talking or playing after you have left; it is unnecessary to interrupt this type of activity. Some children may begin whimpering after the parents have left their bedroom for the night, and common sense as well as the age of the child should dictate your approach to this problem. Further discussion is provided in the section on crying on page 88.

It is common for children to waken between 5:00 or 6:00 A.M. Here again, younger infants may prefer to spend time in bed babbling and playing with toys, and toys can be provided in the bed for this period. Toddlers generally rush off to their parents' room upon awakening. Depending on the hour, some parents prefer to bring the toddler in bed with them in order for them to get an extra hour of sleep. We feel this approach is entirely a matter of parental choice. Sometimes three- and four-year-olds will go through periods of trying to sleep in their parents' bed. While there is, of course, no harm in permitting this activity occasionally, we do not encourage children to sleep with their parents for prolonged periods of time.

Fears

Fear of the dark and nightmares are so common that we should probably not even consider them as problems. The best approach to fear of the dark is the use of a night-light. Nightmares occur commonly in children of pre-school age, and generally they can be handled with a few moments of cuddling and reassurance. Rarely, children experience something called night terrors, which are quite distinct from nightmares. Night terrors occur in a different stage of sleep than nightmares do, and the child is generally hysterical and cannot be comforted. Persistent night terrors will require medical help.

A frequent cause of sleep disturbances is the use of medications. This is particularly true with antihistamines and decongestants, which can interfere with the child's normal sleep pattern. If your child is having sleep problems and is simultaneously taking a medication, you should consult your doctor about the possibility that the problem is linked to the medication.

Another cause for concern is the use of a milk bottle to get children to sleep. Children who are put to bed with milk bottles are subject to increased numbers of dental problems and may have increased numbers of earaches. Children can be given a bottle of milk before being put to bed, but the milk should be drunk in the upright position and the milk should not be allowed to remain in contact with the teeth for a long period of time.

THUMB SUCKING AND PACIFIERS

All newborn infants are born with a sucking reflex. Because of this reflex, they will suck on a fist, finger, nipple, or anything else that comes in contact with their mouth. The sucking is, of course, necessary for feeding; hence, the infant soon associates sucking with a feeling of satisfaction and security. As the child grows older, it is normal to suck on fingers, pacifiers, or favorite objects. Often the child will hold a favorite object in one hand and suck the thumb of the other, or will suck a thumb and insert a corner of a blanket or teddy bear's ear in the mouth as well.

Sucking activities increase when the child is anxious, tired, hungry, or stressed. The sucking of objects, in addition to thumb sucking, most frequently occurs between the ages of one and two-and-one-half. There is no need for concern about these sucking activities. They are perfectly normal and demonstrate resourcefulness in finding a way to deal with stress and anxiety.

Several words of caution: **Pacifiers** should not be sweetened because this can encourage development of dental cavities. Pacifiers should be of hard rubber and should have a sufficiently large "skirt" to protect the infant from suffocating on the pacifier; several deaths have been reported from a pacifier with a small skirt lodging inside the infant's mouth. These pacifiers have skirts less than 4 centimeters (1½ inches) in diameter and have a clown, bear, horse, or sailor's head. Some liquid-filled pacifiers have been found to be contaminated with bacteria. Pacifiers with strings attached should have the strings removed because strangulation can occur.

Children will generally outgrow **thumb sucking** by the age of five. Parents should not make an issue over thumb sucking because this will heighten anxiety and hence increase the need for an object that can pacify. Persistent thumb sucking beyond age five should be discussed during a regular medical visit.

TEMPER TANTRUMS

Temper tantrums are common in children between 15 months and 4 years of age, and tantrums almost invariably arise when there is conflict between parent and child. The pattern of the tantrums is familiar to many parents of toddlers. A child is asked to do something—put down a toy, come in for a nap—and returns a few "nos" in response. As the child realizes he or she is not going to be indulged, an outburst begins. There may be kicking, crying, shouting, rolling on the floor, fist banging, shouting, and spitting. As often as not, these displays are put on in public. If a tantrum occurs in public, you will probably feel embarrassed and want to beat a hasty retreat with your child. If the tantrum is at home, you are also tempted to pursue an evasive course of action.

A good approach to temper tantrums is to ignore them as much as is humanly possible. It is especially important that your child not get his or her own way after a tantrum. Punishing the child briefly and then indulging the child ensures repetition of the tantrums. You are then essentially teaching your child that temper tantrums are effective in getting what is desired. Giving in even occasionally will prolong the persistence of these outbursts.

The best approach to temper tantrums is to try to prevent them. Understanding of a child's personality development, consistency in discipline, and a commonsense approach as to what demands and restrictions are reasonable for your child are all helpful. As children grow older, parents should teach them to verbalize, rather than demonstrate, their feelings. It is better for a child to say, "That makes me mad" or "I'm frustrated" than to display anger through physical acts of violence. It will also be easier for you to deal rationally with verbal, rather than physical, protests.

BREATH-HOLDING SPELLS

The factors that may precipitate a temper tantrum—conflict, frustration, anger, a contest of wills—are also responsible for breath-holding spells in some children. Breath-holding spells also occur as a result of pain in some children. The spells are most common around the first year of life but occur anywhere from a few months to five years.

Usually a child will begin to cry and suddenly will hold his or her breath at the end of a cry. After a few seconds, the child becomes blue and then relaxes and recovers. In moderate cases, the child prolongs the breath-holding period even longer, becomes blue, and then temporarily becomes limp and unconscious. As soon as the child becomes unconscious, the reflex system that controls breathing quickly resumes the normal breathing pattern. It is this reflex system that the child has overcome in the process of the outburst.

On rare occasions, a breath-holding spell can lead to a **seizure**. In seizures that are due to other causes, it is unusual for a crying episode and blueness to precede the seizure; in febrile seizures and in epilepsy, the blueness follows the seizure.

The approach to breath-holding spells is exactly the same as the approach to temper tantrums. Although these episodes are extremely frightening, damage is quite rare. Breath-holding spells almost invariably resolve by the age of five.

Discipline

MOTHER OF TWO:
A parent's most important job is not to make a child happy but to develop the child's character.

Children require many things from parents. Tremendous emphasis has been placed on the successful ingredients of healthy child development, including good parental and child nutrition, safety, and immunizations. However, greater challenges come from the emotional requirements of child rearing. Few parents need lessons in how to love their child, the most important emotional ingredient. However, a major dilemma occurs in knowing how best to discipline a child so that in the end we will have a happy, healthy adult with self-discipline. Am I too strict? Is my spouse too lenient? Will I spoil him? Will she learn to limit herself if I impose too many limits? These are all legitimate concerns of parents.

The most important concept is that discipline means giving knowledge and skills. It is very different from punishment, which is the consequence of an unacceptable behavior. A consistent, positive approach to discipline should minimize the need for punishment. Appropriate discipline techniques cannot be learned in five minutes, and unfortunately many of today's parents grew up without appropriate experience and witnessing considerable corporal punishment. Decades of research has taught us that neither a permissive attitude nor an authoritarian "do it my way because I said so" works as well as consistent limit-setting with explanations. We encourage your continued learning about these techniques.

There are some important principles to discipline:

- Young infants need a safe environment—entirely the parent's job.
- Discipline begins when infants become mobile.
- Discipline should be geared to age-appropriate learning.
- Catch the child being good. Constantly correcting mistakes is not enough.
- Do the right thing yourself. Nobody is as important to your child as you, and they learn by watching you.
- Praise and hug liberally *after* the discipline discussion. Be sure your child knows you dislike a behavior, not him or her.
- Children don't need to be hurt to learn. Parents don't need to hurt to teach.

AGE-APPROPRIATE DISCIPLINE

Infants

For infants, safety is of the utmost concern. Infants are responsive to a sharp "no" or "hot." They will learn your displeasure when your firm voice is accompanied by holding their hands if they scratch or pull.

Older infants and toddlers need:

- Structured environments that minimize the risk of ruining vases or expensive equipment

- A firm voice in explanation
- Redirection toward acceptable playthings
- To be ignored occasionally: Don't reward attention-seeking behavior (e.g., tantrums), but do limit problem behaviors (e.g., biting, hitting).
- Praise when they're doing a good job

Pre-schoolers

Pre-schoolers need:

- Parents with unbelievable patience, fortitude, and stamina
- Clear and consistent rules and expectations repeated and repeated
- Time to get ready: "In five minutes you need to put away your toys and wash for dinner."
- Acknowledgment with explanation: "I know you want to stay at Grandma's, but it's time we went home so we can all go to sleep."
- Removal now with talk later. A child hitting another needs quiet physical isolation followed with an explanation: "You know we have two important rules. You can't hurt anybody by hitting. And you can't hurt their feelings."
- Temporary time out: Removing a child to a corner for a few minutes can help defuse a situation, give a message, and teach about consequences.
- Praise when they're doing a good job
- Praise when they've learned something: "I'm really glad you and your cousin have been playing so nicely this week."

School-Age Children

School-age children need all of the preceding plus the following:

- Opportunities to explain
- Opportunities to express themselves
- Opportunities to choose: "I understand you really want to watch the end of this program. However, if we allow this additional 30 minutes of TV tonight, it will mean no more television this weekend." "If you do not make the effort to clean your room today, you will have to live in a dirty room and I will not make the effort to take you to your friend's party tomorrow."

- Opportunities to solve problems: "You seem to have a hard time finishing your homework at night. Why don't you work on a list of suggestions and then we'll go over them together?"

Adolescents

Adolescents will continue the limit-testing that they started as infants. Even the best discipline system will be tested during adolescence. In addition to the preceding, adolescents need:

- Discussions on the long-term consequences of today's behaviors
- Limit-setting that is agreed on rather than imposed arbitrarily (although parents must ultimately establish the limit boundary)
- Limits that increase according to their maturity and ability to choose wisely: "You can stay out till 10:30 this year, 11:00 next year, and after that I will evaluate how you use your own judgment to decide."

PUNISHMENT

While we oppose physical punishment of children, we recognize that parents in the heat of the moment, may occasionally resort to behaviors to which they were exposed and spank a child. Parents who were harshly treated physically as children need to be aware that they may resort to this damaging behavior and should seriously consider alternatives. It is also important that parents be aware that while all experts agree on a positive approach to discipline, there is disagreement on physical punishment. The American Academy of Pediatrics unequivocally opposes corporal punishment in schools and asserts that in the family "punishment or restriction, when necessary, must be *immediate* and *not physically harmful* to the child." Other experts say you should never have to hurt a child either physically or emotionally to get your point across. Saf Lerman writes, "Parents need to use methods that will make childhood a good experience for their children and parenting a satisfying experience for themselves. When children are being hit and parents are hitting, the process is not beneficial to anyone." Still other experts and authors tolerate one swat on the bottom with an open hand, but decry anything more.

Nevertheless, all parenting experts agree that there are better alternatives to physical punishment. Taking away privileges is very effective. It is important to remember such alternatives even at the emotional moment when you discover that your child has misbehaved. Take the situation of a mother who returns home one winter night to learn that her five-year-old twins had summoned the police by dialing

911 while Dad was attempting to cook dinner. (Not a true emergency!) After the parents explained why this behavior is unacceptable, one twin agreed not to do it again. The other was less remorseful, arguing that nobody answered his phone call; only his twin sister had gotten through. The mother strongly felt that the children's behavior warranted a punishment, but what? The parents' solution was to take away all of the Christmas presents given the previous week, with the understanding that they would have to be earned back. Thus the parents effectively taught a dramatic lesson about a serious behavior without using physical punishment.

If you find yourself shaking or repeatedly hitting your child, you should attend a course that teaches effective discipline techniques in your community. Most communities have helpful child abuse "hotlines." Your physician can also be a useful resource.

The Pre-School Years

The use of language is perhaps the most dramatic development at this age. Children progress in a few years from the use of two-word phrases to the ability to tell stories and describe fantasies. They begin grasping concepts such as size, numbers, orientation in space, and time.

Play begins to occupy a great deal of their time. Play in this age group accomplishes many things. Children learn many fine motor skills and concepts through play. In the early pre-school years, children will most often be seen playing independently and exploring their toys. If placed with other children, they will not interact. As children become older, they begin to play together, although the sharing of toys is not always accomplished easily. Besides aiding intellectual and social growth, play is also a way in which three- and four-year-olds, who seem to have enough energy to run for 36 hours every day, can divert this energy into activities that are less socially distracting to parents. Fantasy friends and stories are common at this age. Fantasy at this age is not fibbing; it is a way of learning.

This is an age of further emotional development. Parents can provide a tremendous service by beginning to teach children to discuss their feelings. Three- and four-year-olds are sometimes able to tell their parents that they are grumpy or angry or sad. When children can talk about these feelings, it is easier for parents to help them through hard parts of the day. Inability to recognize one's own feelings can lead to problems in later life; we encourage parents to assist their children in developing a sense of their own feelings. This is also an age of sexual exploration, which we discuss later in this chapter. Lying is also common at this age and in the early school years. Children of this age are not trying to deceive; they often have difficulty drawing a sharp boundary between reality and fantasy.

Many children attend day-care centers or nursery schools at this age. The activities there aid in socialization and help prepare for the activities of formal schooling. Children who do not go to day-care or nursery schools still hear a great deal about school from other children or older siblings. Most children are therefore socially ready for schooling at the time that they enter.

SHORT ATTENTION SPAN (Hyperactivity)

All children are different, and there are as many different temperaments as there are children. Some have temperaments that are considered by many parents to be easy. They are social, affectionate, react to new situations with curiosity, and often considered "even-tempered." Others may be shy, more difficult to warm up, or cautious of new situations; these traits may be viewed as stubbornness. Infants may be innately intense or difficult to comfort.

Some children seen to be unable to focus on any activity for more than a few moments. Often these children with short attention spans seem to go from one activity to the next with barely enough time to take a breath before changing focus. Many such children also seem to have boundless energy and are in constant motion and viewed as "hyperactive." A child's ability to focus is a gradual developmental process that progressively lengthens with age. However, a child with a very abbreviated attention span can be particularly frustrating for parents if a child cannot pay attention for at least 20 to 25 minutes by age five. There is also a potential for difficulty in kindergarten.

Most parents will find it reassuring to learn that their active toddlers, pre-schoolers, and early school-age children will gradually develop the ability to sit still and pay attention. However, this can be fairly frustrating to deal with while waiting for maturation to occur. In older children, for whom this problem can seriously interfere with both academic and social achievement in school, we recommend formal evaluation as soon as this problem is reported.

For younger children, there are some specific techniques that can be helpful.

- Don't blame yourself. Children are born with different temperaments, and the very active child is one of the many varieties. In fact, this temperamental style adapts well in many cultures as well as in many adults' careers. It can, however, seriously interfere with parental mental health as well as a child's progress in school.

- Remember that routines are important for all children and even more so for children with short attention spans.

- These children often seem to have boundless energy and are literally built for physical activity. By all means, follow their lead and allow them ample opportunity to "burn off" this energy. However, many of these children have difficulties making transition from one activity to another. Therefore, avoid strenuous physical activity before attempting to put them to bed.

- Remember that these children often have difficulty focusing on one thing at a time. This can result in overload if they are allowed to wander from task to task continuously in an unstructured environment. Therefore, having periods during the day when they are in a playpen or a quiet area of the room with a limited number of choices can be very helpful. Many of these children even ask to be put in a playpen when they sense their circuits are about to go into overload.

- Remember that with their increased activity and energy level these children seem to be more apt to get into trouble. As with all children, limits need to be set, and discipline administered in a firm and consistent manner. Negative behavior should be confronted immediately, in a firm tone of voice, and without physical punishment. The child should know certain behavior is unacceptable. After that message is conveyed, it is also important to follow up a short time thereafter with the fact that you still have affection for your child, but no biting, hitting, running, kicking, throwing, and the like are tolerable.

- Use common sense in considering which social situations to enter with your child. While you can almost make it through a quiet dinner in a romantic restaurant with some three-years-olds (although we're not sure why you would want to do this), very active three-year-olds will make your life miserable in these situations. Fast-food restaurants are fine for these children, particularly those that have accompanying play areas that cater to the active child. Supermarkets are also a high-risk zone. We are not suggesting you and your child remain socially isolated, just that you consider carefully what time of day your individual child is best able to deal with highly stimulating situations.

- You can help by spending time with your child alone doing things such as reading or storytelling. It is also important to have time away from older siblings who always seem to have tendencies toward exciting younger brothers and sisters.

- Active children drain energy from the most patient parents. Not only is "time out" in the playpen good for children, it is especially important for you to find time to do things for yourself.

The Early School Years (6 to 11)

During the school years, children make the transition from being members of a family to being members of both a family and society. A teacher becomes an extremely important figure in the child's life. Playmates, however, remain the most significant persons in a schoolchild's society.

FRIENDSHIPS

This is a period of organized games, which tend to stress competitiveness as well as cooperativeness. Psychology texts often stress that this period is one of sexual identity, where 8- to 11-year-old boys associate chiefly with boys, and similarly aged girls associate chiefly with girls. With new societal attitudes toward the dichotomy of sexual roles, we expect some of this polarization to diminish. Indeed, the degree of polarization between boys and girls varies greatly, depending on the circumstances of the moment. For example, ball games may require full participation of girls and boys in order to provide the necessary number of players as well as the necessary equipment. In addition, jumprope, jacks, potsy (hopscotch), and other games are frequently played by both sexes. Even in single-sex schools it is important that children learn both the value and the means of cooperating on meaningful projects with members of the opposite sex.

Because of the importance of friends in this age range, parents must be sensitive to their children's needs if a move is planned. A promise of a bigger house, a room of your own, or even a swimming pool is negligible when compared with the loss of friends. Fortunately, new friends are made, but the parents must be supportive during this period.

ETHICAL CODES

Children in the school years are also developing their own code of justice, their own code of what is right and wrong. Moral judgment begins at about the age of 6 or 7, when children begin looking at the circumstances surrounding an event in order to decide whether the event was right or wrong. Piaget has written extensively on this subject in *The Moral Judgment of the Child* (New York: Free Press, 1965). The moral codes of 7- and 8-year-olds come chiefly from their parents and other members of the adult world. Between the ages of 8 and 12, children begin to talk more with friends and evolve their own code of justice.

Stealing is frequently a problem in the 8 to 12 age group. Some of this stealing is merely the child's testing of boundaries in order to determine what proper limits are, and most children who steal during these years have a sense of guilt. Occasionally, children may steal as a group act or be forced to steal as an initiation rite. While it is a parent's role to maintain the limits demanded by society and parents should define the limits clearly, they should not overreact to such actions. Stealing by a child is a profound disappointment to the parents but does not often herald the beginning of a life of crime. Let your child know firmly, but with love, that certain behavior is not acceptable. Your child is essentially saying, "Here's what society says, but is society right? How do I really feel about this action? The only way I can really tell how I feel is to test it." The ability to question society's values is important. The right to question should be encouraged. Mature reaction to the discovery of stealing by your child can be a learning experience for both parent and child, can establish the limits of acceptable behavior, and can teach the child the relationship between actions and consequences.

SCHOOL

Schooling accomplishes many things for our children. Children acquire academic skills and learn about the nature of our universe. Besides knowledge, schools foster a child's self-esteem, code of behavior, and ability to function within society. Students discover how to work with other children and begin to appreciate what will be required of them. Often, especially in large classes, it is easy for children to be denied individual attention. For a teacher to deal with 30 children, a tremendous amount of order and uniformity of action is a necessity. However, socialization can be detrimental to the extent that it denies an individual the ability to think or act differently; this is clearly undesirable to mature development, and creativity can be stifled. Children require an atmosphere that allows learning for all at the same time, yet encourages each child's development as a unique individual.

All school systems are under perpetual attack for fostering either too much competition or not enough. We are enthusiastic about the development of alternative schools. We we anxious, however, about schools that see their *only* role as developing individuality or creativity. Preparing children either for the rugged individualism necessary in a pioneer society or for cooperativeness that calls for complete submission to the needs of society can potentially produce children who are historical or cultural misfits. Children must learn a combination of competition, cooperation, and creativity from their parents and from schools. The parent must understand the school program and complement it so that the child is prepared for a world that is not always fair and not always good.

Adolescents

FATHER OF TWO:
It's like trying to steer a battleship with the rudder from a dinghy.

EXPECTATIONS "The trouble with today's youth is ..." How many times have you heard that expression? How many times have you heard the same tired clichés: "They are irresponsible, hung up on sex, never show any respect, only interested in drugs, cars, and so on"? Youths and adolescents have suffered from countless dehumanizing generalizations and oversimplifications. By lumping all teenagers together, we essentially deny youth the freedom to be individuals by *expecting* certain adverse behavior. When this expected behavior occurs, the media leap upon it, and we all nod our heads knowingly and feed our own prejudices.

It is interesting to see how the media portrays adolescents. For instance, in the 1950s, adolescents were portrayed as angry young rebels. Marlon Brando in *The Wild Ones* and James Dean in *Rebel Without a Cause* rose to fame in roles portraying youth at odds with the world. When the youth of that era controlled the media, they portrayed the black-leather-jacket-wearing youth, such as "the Fonz," as polite, well-groomed, respectful superheroes. The good old days of today were the terrible times of yesterday.

Some authors have portrayed adolescents as the conscience of our country. Others view youths as passive and merely reflecting the common value system of our society. The truth lies in between.

Studies of adolescents have revealed several interesting facts. Most adolescents have attitudes and values that closely reflect the attitudes and values of society as a whole. Their political orientation is usually that of their parents. The majority of adolescents approve of their parents' attitudes toward discipline, although they, of course, would not be expected to enjoy being subject to the discipline. Most adolescents feel their parents understand them, and most feel that the communication lines to their parents are open. There is, of course, a wide spectrum here because many adolescents are at odds with their parents. No individual can be expected to agree with a parent, spouse, or friends all of the time. Disagreements will produce tension, and because of their unpleasant nature, will often be remembered. A thousand productive acts are often forgotten in the wake of one heated argument.

INDEPENDENCE

Adolescence is often a difficult period because of the many complex tasks required of young people and their parents. The adolescent is defining his or her identity while rapidly moving toward legal, economic, and psychological independence. **Legal independence** is coming at an earlier age than ever. The ancient custom of legal majority at the age of 21 is nearly extinct, and most states now recognize the age of 18. Youths are recognized to be at the age of majority for making choices about health care at an even younger age in many states. Emancipated minors or youths living by independent means away from home are also recognized as having achieved the age of majority.

Economic independence means that young people must learn to plan for the future, an ability that develops during adolescence. This planning includes assessing the educational tasks that must be accomplished to reach economic goals. Work-study plans are available in many communities. Vocational counseling is very important at this age. Goals that are unrealistic or unclear are difficult to achieve and can lead to lifelong frustration. Many adolescents do not have clear-cut goals. Some do not develop goals until they are in their 20s. Parents can offer support but statements of what adolescents "should" do are not helpful. Maturation is complex and occurs at different ages.

Psychological independence is, by far, the most important and most difficult part of adolescence. This is a period in which individuals develop abstract thinking and begin to develop a theoretical framework by which they will live. This involves developing a way to deal with the contradictory values of society, as well as developing ethical and moral values for the adolescent's personal life. During this time, our children first begin to recognize the contradictory values with which we all must deal on an everyday basis. The gray zone between black and white is a supremely disturbing discovery. Contradictions about war, pollution, exploitations, and class differences are difficult enough for anyone, and even worse if you have been raised by parents and teachers to believe in peace, love of fellow humans, and equality. How do the idealistic teachings of parents and teachers measure up to the reality of how the world is truly constructed? This contradiction is extremely difficult for young people to resolve.

Adolescents also come to realize that they must soon move away and separate from their parents. Although many adolescents talk enthusiastically about leaving home, this is a major source of anxiety. At the same time, adolescents are developing their own personal value system. Most of these developments have something to do with how they relate to other people. Friendships are extremely intense, and very often these friendships are with members of the same sex. Best friends often become inseparable in early adolescence. Younger adolescents will sometimes deal with their sexuality by homosexual encounters. This can be frightening to adolescents and even more so to their parents. However, it is a normal developmental phase for many adolescents. Younger adolescents are vulnerable to sexually exploitative homosexual or heterosexual acts. These acts are far more significant than the transient homosexual episodes more commonly encountered.

Finally, the task of learning how to develop meaningful inter-personal relationships must be accomplished. Learning how to deal with sexuality is only one aspect of developing meaningful relationships. The sexual drives that adolescents experience cannot be denied; they provide a learning experience, requiring the youth to decide on a set of actions that are consistent with his or her own evolving system of values.

The availability of methods of contraception has become a double-edged sword. It has allowed many adolescents to test their sexuality in a mature and responsible way, but it often leaves some young people without an "excuse" for not engaging in sexual activity. Adolescents need support from parents and sometimes from professionals in sorting out the path of action that is best for them. Good sexual education programs stress the importance of mature choice, as well as the mechanics of contraception. It is important to stress the maturity of most adolescents toward sexuality. The vast majority of sexually active adolescents have one partner. The level of promiscuity among adolescents seems to be less than that of adults.

CHOOSING MEDICAL CARE

One of the best ways in which parents can assist adolescents is to encourage them to make a personal choice for their medical care. We feel that the best way to encourage good health through life is to encourage responsible decision making about health and illness. Pediatricians and family doctors are often viewed by adolescents as people that their parents have chosen for them, and this lack of control in choosing their doctor can discourage the development of mature medical behavior. An important issue in the adolescent's right to select his or her own health professional is the right to privacy and confidentiality. Although very few doctors would violate an adolescent's confidentiality, the fear that a doctor might do so may interfere with the seeking of necessary medical services.

The development of the free-clinic movement, primarily serving youths, is a testimony to the desire of adolescents to seek medical care in confidence. Because no bills are sent home by these clinics, confidentiality is closely maintained. Most adolescents with good pathways of communication to their parents will wish to discuss their medical concerns with the parents. But they should be allowed to decide when. Responsible approaches to medical problems will develop more quickly in youths who are treated as mature individuals capable of rational choices. We encourage parents to begin fostering independent medical decision making in young children and to allow adolescents their individual choice of medical provider. We feel the best care can be given by the doctor who has been seeing the adolescent throughout the years. We therefore encourage parents to permit a "letting go" of the old parent–child–doctor relationship and consent to confidential communication between doctor and adolescent.

How can parents best support their children through this intense and often emotional process of adolescence? It's not easy. You are actively involved in gradually granting more independence, yet you are often uncertain about how much independence is appropriate. Your children may be becoming very independent in one area, while still showing youthful tendencies in another. For their part, youths often become anxious about their own independence. They look forward to becoming an adult on the one hand, and yet many of the freedoms of childhood are also appealing. You can help by providing support and by helping them to acquire the best information possible in matters such as health, education, employment, and sexuality.

What should be done when problems arise? Many people adopt the attitude that problems of adolescence are temporary and will disappear with time. However, the percentage of adolescents who are emotionally impaired corresponds rather precisely to the number of adults who are similarly impaired. While many of the minor turmoils of adolescence give way to the different but nonetheless tumultuous problems of adulthood, the severe ones do not go away by themselves. Severe emotional impairment in adolescents or adults is a problem requiring help from doctors, social workers, or counselors. The 6-year-old who cannot function in school, the 16-year-old in trouble, and the 36-year-old who cannot hold down a job all need help.

PROBLEM BEHAVIORS

While adolescents are less likely to be troubled by the illnesses that so often brought them to doctors' offices in their earlier years, new troubles emerge. Unfortunately, some problems are worse for the adolescents of today than they were for their parents. Only half of teenage deaths in the 1950s were due to accidents, homicides, and suicides; now these three causes represent 80% of the mortality for adolescents. Increasingly, researchers have identified three problem behaviors as being responsible for more than half of the serious illnesses and deaths of teenagers:

- Substance abuse

- Motor (and recreational) vehicle use and misuse
- Premature sexual behavior

There are clear-cut associations among these behaviors. The relationship between alcohol and injury, as in a driving accident, is well known. Substance use is also linked to earlier initiation of sexual activity. For more advice on sexual behavior among adolescents, refer to page 539.

When parents notice a teenager smoking, drinking, or abusing drugs, there is a clear-cut problem. That's why teenagers rarely do these things in front of their parents. But other signs may indicate a hidden problem with alcohol or other illicit substances. Families who have experienced alcoholism should be especially alert. If you see any of the following signs, start the process of getting your teenager in a program to discourage participation in problem behaviors:

- Decline in school performance
- Lack of motivation in school
- School absenteeism
- School discipline problems
- Increased behavior problems
- Loss of interest in previous hobbies
- Lack of interest in family activities
- Changing groups of friends (especially to groups involved in high-risk behaviors)
- Friends with criminal histories or behavior
- Neglect of appearance, relative to peers
- Excessive concern with money
- Changes in mood
- Depressive symptoms (see page 326)
- Running away from home
- Driving while intoxicated

Some of these signs, such as changing interests and appearances, are normal for adolescence. It is the degree to which adolescents change that might tip you off. Get to know your child's peers, and use your judgment about your child.

It is often difficult, and sometimes seems impossible, to change a teenager's problem behavior. However, your physician may be helpful in steering your adolescent to an effective community program. Schools are also becoming more involved in preventing problem behaviors. Most communities have groups for dealing with substance abuse geared specifically to teenagers; one such group is Alateen, associated with Alcoholics Anonymous. Parents can help by controlling their own behavior: eliminating smoking and excess consumption of alcohol!

Parents must ensure their teenagers receive appropriate health care, especially when their behavior indicates that they are taking bad risks. Usually the first step is to arrange a regular health appraisal by the child's physician. Not all doctors are equally well prepared to confront adolescents with risky behaviors and identify a way to deal with the problem. Most physicians will probably not treat the problem alone, however; they will help you to make sure that your adolescent is enrolled in an appropriate program to help him or her solve the problem.

Sexuality

Like all other aspects of human development, sexual development begins at birth and continues in a logical progressive order throughout life. This process includes the development of gender roles as well as the development of sexuality.

The concept of gender roles is formed early and re-inforced continuously throughout life. A child's gender identity begins in the nursery when pink or blue cards are used to identify girls or boys. Little boys are handled in a much different manner than little girls are, even in the first few weeks of life. A close friend of ours, firmly committed to sexual equality in raising children of both sexes, began calling her newborn boy "little tiger," a term she had never used for her daughter. This subtle and not-so-subtle discrimination of gender roles that begins in the nursery is re-inforced by families and schools and continues throughout an individual's life.

INFANTS

A child's sexuality begins within the first few moments of life. Achieving pleasure through genital stimulation is only one small part. Affection and fondling of the newborn infant is a natural thing, affording pleasure to both parent and child; this is probably the first sexual interaction.

Parents often notice that newborn boys have erections. These erections are controlled through the nervous system and can occur spontaneously or in response to a number of different stimuli. There is no reason to believe that infant boys do not feel pleasure during these periods. Certainly by 1 year of age, both boys and girls are able to reach their genitalia and notice that rubbing will produce a pleasurable sensation. At this age, most parents recognize that their children are merely exploring their bodies and are not concerned about genital manipulation. The infant is more concerned with other stimuli. In the first year of life, the infant derives more pleasure through the oral gratification of feeding and sucking.

TODDLERS

As the infant moves into toddlerhood, pleasure continues to be derived from many other areas, including interactions with parents, food, toys, and eventually the ability to control bowel movements. However, when a child is 3 and 4 years of age, there is often renewed interest in genital manipulation, and masturbation occurs in both sexes. Masturbation to orgasm is common in this age group. Parents often have questions about how to deal with masturbation in young children. Certainly, this natural experience is not to be condemned in children. However, parents may wish to talk to their children about when and where it is appropriate to masturbate. Teaching children how to control their sexual impulses should not be regarded any differently than teaching them not to talk loudly in church.

At this age, it is often common for children to barge into their parents' bedrooms during sexual activity. During these episodes complex explanations are not needed—just a reassurance that the activity was playful and that any noises did not indicate anyone was being hurt.

AGES 3 TO 6

This is the age of questioning about how children are born. Explanations appropriate to the level of the child's understanding are important. Again, detailed anatomical dissertations certainly are not what the child wants. Answering a single question at a time is better than a confusing full discussion. Children between the ages of 3 and 6 have a great deal of curiosity about their genitalia and are frequently anxious to compare theirs to those of their parents and to those of children of the opposite sex. Questions should be answered openly and frankly. Detailed explanation of precise anatomical differences is not what is required. Little girls concerned about why they do not have a penis should be told that they have a clitoris instead and may be told that they have a vagina in addition. Focus on what the girl *has* instead of why she does not have something that the boy does have. It is more satisfactory to have something else than to have something missing.

These are also the years in which children like to examine the genitalia of their playmates, especially those of the opposite sex; these activities sometimes lead to conflict with parents of neighboring children. Although you will probably want to acknowledge to neighbors that you are sorry that they are upset, there is no need to apologize for the activity of your child. This is especially true in the presence of the child, for the child may begin to think that he or she has done something terrible. Convey to your children *your* feelings on appropriate behavior, as opposed to what Mr. and Mrs. Jones feel is appropriate behavior.

As children reach the age of 5 or 6, they may play "house" or "doctor" with children of the opposite sex. Often these games are played where the children may be discovered. Children wonder whether this type of activity is right or wrong. Often they have guilt feelings and need reassurance from their parents that they will not be rejected. An acceptable parental attitude is again to acknowledge the activity without encouraging it in inappropriate social situations and without punishing the child for it.

AGES 6 TO 12   Between the ages of 6 and 12, children are occupied chiefly with tasks of learning, developing, and making friendships. These activities take precedence over all others. Genital manipulation does persist throughout this stage although it is less visible and important than at other ages. Although parents may feel that the scouts and school projects are molding their child's development, this is also a period when children learn a great deal about sexual relationships. At this age, children have a certain "street knowledge" about the mechanical aspects of sexual intercourse. All schoolchildren have their repertoire of "dirty" jokes. In rural areas, children often see farm animals and dogs copulating. We doubt that anyone learns very much from the birds or the bees.

Children at this age are also astute observers of parental as well as male–female interaction in general. The most significant impression a child will form about how men and women interact is by observing his or her parents. Parents must provide as good a model in this area of development as they do in others. It is important for children to witness affection between parents. Naturally, parental battles will occur; when they do, you should explain to your children that they are not responsible. Children can adjust to a spectrum of emotions between their parents. They are apt to suffer, however, if they are witness *only* to fights or hostile interactions.

When children are between the ages of 10 and 12, most parents attempt a discussion of "the facts of life" with them. Parents should recognize that their children's education in these matters has been ongoing for many years. The "facts of life" are often presented differently for girls and boys. Girls tend to be taught about the biology of ovulation and menstruation and the importance of preventing pregnancy. If a girl's mother has been having menstrual difficulties, often the daughter will be apprehensive about her coming periods. Very often children form early impressions about the pain of menstruation. In young children, blood is a sign of pain. Therefore, finding a tampon soaked with blood or witnessing their mother changing a blood-filled tampon or napkin may suggest to the child that mother is being hurt. Explain to children what is going on.

Boys, on the other hand, often receive a discussion about how to acceptably channel their developing drives toward sexual intercourse. In other words, boys are told about masturbation, whereas girls are not. We do not feel that discussions that deny the pleasurable parts of sexual activity are helpful. It's important to acknowledge that sexual activity, including masturbation or intercourse, can convey a great amount of pleasure. However, you will probably also want to discuss the personal, emotional, and ethical aspects of sexual activity. Emphasize the long-term benefits that mature interpersonal interaction and sexuality can afford and discuss how much more pleasurable sexual activity can be if accompanied by affection, intimacy, and love.

ADOLESCENCE

Recent studies have pointed out that the number of adolescent males having sexual intercourse has changed very little over the past 50 years, although more adolescent women are having intercourse now than 50 years ago. The number of women actively having sexual experiences is approaching that of their male counterparts. This has created, however, new stresses for many women. In the past, women were made to feel guilty if they had a premarital sexual experience. Today, however, many women feel that rather than being pressured into celibacy, they are being pressured into sexual activity. Both men and women at this age need support from parents in dealing with these complex sexual issues.

The emergence of the AIDS epidemic is beginning to change many adolescents' views on sexuality. The risk of sexual activity has always been present. The negative consequences have become far more dramatic and for some will influence their decision making. Besides AIDS, sexually transmitted diseases such as syphillis, gonorrhea, herpes, *Chlamydia,* human papilloma virus (genital warts), trichomonas, lymphogranuloma venereum, pelvic inflammatory disease, lice, scabies, monilia, hepatitis, chancroid remain prevalent and, in some cases, are on the rise. Dispite these dangers, biology, personal goals, risk taking, role models, and social pressures may all weigh into a teenager's decision to have sex. According to a 1991 survey by the Centers for Disease Control, 54% of high school students report having had intercourse.

Adolescents need information, but they also need to be able to discuss their feelings with adults who will understand the difficult decisions they must make and who will encourage responsible actions on their part—actions that can mean the difference between continued maturation into adulthood and a life prematurely encumbered with a child or prematurely terminated with AIDS. Teens need to be able to turn to their parents for guidance through this difficult period.

For more discussion of adolescent sexuality, how to talk to your teenager about this difficult subject, and the pros and cons of contraception, see Chapter Q, page 539.

Summary

You are the cornerstone of your child's personality development. You are required to provide constant love and yet set limits. Children need discipline, but it is important to make clear that a specific behavior, and not the child, is being chastised. Parents should teach their children to be sensitive to and to respond to feelings. They can help children to talk about their feelings, to be able to say "I am frustrated," "I am sad," "I am depressed." Sadness, depression, joy, and affection are emotions of young children as well as adults. And it is important to state in capital letters that WE ALL MAKE MISTAKES. Making mistakes is part of being human. We are all faced with uncertainty many times. There is no certain approach to all problems and we must all make decisions in the face of uncertainty. We all do the best we can, we all do things differently, our children all grow up differently, and that's the way things will always be.

Additional Reading

American Academy of Pediatrics, *Caring for Your Adolescent: Ages 12 to 21.* Elk Grove Village, IL: The Academy, 1991.

Mary S. Calderone and Eric W. Johnson, *The Family Book About Sexuality.* New York: HarperCollins, 1990.

Louise Bates Ames, and Frances Ilg,
Your One-Year-Old: The Fun Loving Fussy (1982);
Your Two-Year-Old: Terrible or Tender (1976);
Your Three-Year-Old: Friend or Enemy (1976);
Your Four-Year-Old: Wild and Wonderful (1976);
Your Five-Year-Old: Sunny and Serene (1979);
Your Six-Year-Old: Loving and Defiant (1980).
New York: Dell.

Ruth Bell, and Others, *Changing Bodies, Changing Lives: A Book for Teens on Sex and Relationships.* New York: Random House, 1980.

T. Berry Brazelton, M.D., *Infants and Mothers.* New York: Dell, 1969.

T. Berry Brazelton, M.D., *Toddlers and Parents: A Declaration of Independence.* New York: Dell, 1974.

T. Berry Brazelton, M.D., *Touchpoints: Your Child's Emotional and Behavioral Development.* Reading, Mass.: Addison-Wesley, 1992.

Richard Ferber, *Solve Your Child's Sleep Problems.* New York: Fireside, 1985.

Selma H. Fraiberg, *The Magic Years.* New York: Macmillan, 1981.

CHAPTER 6

Today's Family

*T*here is no need to mourn the death of the American family. The American family continues to survive but is changing as our society is changing. As many functions of the family evolved, so did our views of the purpose of the family. Clearly, children are no longer produced in abundance in order to assure economic viability as required in an agrarian era.

Many of us carry around images of the typical traditional family with mom and dad, two kids, and their dog gathered around the fireplace, reading. There is clearly a contrast between this image and the reality of today's family as it has evolved over the past century. In fact, although a snapshot of many families might capture the above ideal, a motion picture taken through the years would reveal the continuing changes in almost every family. For example, it may come as a surprise that in a classic study of first-grade children done nearly 20 years ago, only 34% lived in households with a mother and a father and without other adults. In contrast, over 36% lived with only their mothers, 1% with their fathers, 2% with their grandparents, and the remaining 27% in households with their mothers and other adult relatives. Today, there are a growing number of children (about 500,000) living in foster homes. Adoption, which dates back at least to the time of Moses, is also a process in transition as more "open" adoptions become prevalent.

We cannot deny that divorce is a major issue affecting children. While the divorce rate has finally dropped after a 20-year rise, it is not a phenomenon invented during the 1970s. The U.S. government began keeping tabs on divorces in 1867 (0.3 per 1000 population then, compared to an all-time high of 5.3 per 1000 in 1981) and has witnessed periodic surges in rates as a response to societal change. For example, immediately after World War II, the rate climbed to over 4 per 1000 population (or 18 per 1000 married women), a figure that was not again achieved until 1973. Today, more than 5 million children are living with a divorced parent. This is about 9% of children under 18 years old. Millions of others live with parents who are separated. It is estimated that over a third of today's children will experience the absence of a parent from the home. Today, divorced parents often share child-care responsibility, known as co-parenting. Many parents with children re-marry, forming blended families.

Another major change in how children now live is the renewed surge of women working outside the home. Although there have always been "working" mothers, the demand for day care today is unprecedented. In our highly mobile society, Grandma may not live close enough to help out everyday, and even though Aunt Gigi lives around the corner, she may also be at work. Due to these factors, there is a growing national focus on "latchkey children."

In a nation that is still predominantly only a generation or two away from immigrant grandparents, we have grown up with the stresses of leaving relatives and roots behind and with losses through separation and death. The average child born today is expected to move nearly 13 times by age 18. All children and parents experience major stressful life events as part of growing up; coping with these stresses can be both painful and maturing.

In the rest of this chapter, we focus on some of the more difficult challenges facing families today:

- Adopted children
- Child care
- Parents divorcing and/or remarrying
- Common parental feelings, and finding help for them

These challenges are not new. In fact, some of them are as old as families themselves. We try to give advice and point the way to the best solutions for today's families.

Adopted Children

FATHER OF THREE, AFTER NUMBER TWO:
We decided to adopt an infant after medical problems interfered with our having a second biological child. When our social worker from the state adoption agency cautioned we might react differently to this second child, something she called "differential bonding," I was a bit surprised because I regarded myself as a competent and caring father who wanted and would love this new child equally. However, several weeks after we brought Michael home, my wife pointed out that our three-and-one-half-year-old was being more affectionate than I was with the baby. The attachment process really was different. I'm glad Theresa confronted me with what was going on because it seemed to speed up the process, but I'm really grateful to the social worker who anticipated and informed us about this potential problem.

MOTHER OF TWO:
Kristy first learned she was adopted when she was three and one-half. Even though her older sister was also adopted and we thought we had given her just the right amount of information, she become very angry with us. Maybe it was because so many of the other mothers at nursery school were pregnant. Then, one of the other children came to school with his newly adopted baby brother. Suddenly Kristy's attitude changed completely. I think this really helped her understand that this was another way babies come into families.

MOTHER OF ONE:
When I first brought Daniel home, I was very anxious that all evidence of his true biological mother be hidden. I even got into a battle with the hospital about the name on his medical record. My attitude has really changed in the last eight years, and now I tell him as much as he wants to know about his parents.

FATHER OF ONE:
Now that the baby is here, how do I finally feel about his being a different race? I don't know how I feel. I'm too exhausted from changing diapers.

FATHER OF FOUR:
My biologic son is white, my black (adopted) son is white, my other black son is black, and my Asian daughter is also black. She was even elected vice president of the African-American group at college.

FIVE-YEAR-OLD ADOPTED GIRL:
My mom stole me from my real mom. She said she was just going to borrow me for a short time in the hospital but never brought me back. She lied.

SIX-YEAR-OLD ADOPTED BOY, RESPONDING TO FIVE-YEAR-OLD SISTER'S COMMENT THAT SHE WAS TALLER AND STRONGER:
Yeah, well I'm older, I'm faster, and I have two moms instead of one.

We are going through a period of increasing enlightenment about adoption. In the past, many adopted children had information concealed from them by well-intentioned adoptive parents. This resulted in confusion, frustration, and anger. Lies and secrecy have never been a sound model for child rearing. Fortunately, new societal attitudes about adoption are resulting in less pressure on parents to protect their adopted children with secrets.

OPEN ADOPTION

In part as a result of the growing fertility issues confronted by baby boomers delaying childbearing, there has been a growing focus on adoption issues. While surrogate parenting has been receiving extraordinary media coverage, a quiet revolution has been going on. Private adoption has been rapidly increasing, accounting for up to half the adoptions in some areas. Private adoptions are open. In other words, biological and adoptive parents are aware of each other, know names, and often meet. In **closed adoptions**, agencies are aware of all parties, but confidentiality is characteristic. Private adoptions have grown in popularity for several reasons. The wait is considerably shorter for potential adoptive parents. Many mothers who will be relinquishing their infants have a desire to meet or even know who will be caring for their child.

Private adoption is not baby buying. State laws strictly forbid providing monetary expenses to the mother except for medical and living expenses incurred to sustain the pregnancy. Some lawyers charge considerable amounts for legal services surrounding adoption; many charge their customary hourly fees. Lawyers are helpful but not necessary for adopting a child, however. Many parents are successful on their own in adopting infants through networks including churches, doctors, friends, and even newspaper ads. State agencies also must interview and approve prospective adoptive parents before children can be adopted under state law.

This new openness is helping foster the realization that adoptive children will always have two sets of parents: their parents (the psychological, or adoptive family) and their biological parents. Biological parents are part of a child's ancestry, and we all have curiosity about our ancestors. There are certainly many unique issues about raising adoptive children: they may look remarkably different and be temperamentally different from their parents (but so are many biological children). By and large, they confront the same medical and social problems as most children. We are not contemplating an edition of *Taking Care of Your Adopted Child.* There are some unique psychological features confronting adoptive families, for which there have evolved a number of excellent books, newsletters, and support groups. Some of the central issues include telling children they were adopted and helping children understand and sort out their feelings about being adopted.

WHAT TO TELL YOUR CHILD

Although there are no foolproof recipes for dealing with these issues, we do offer the following practical approaches. When telling children about where they actually came from, always:

- Tell the truth.
- Don't tell too much.
- Tell children things they are capable of understanding.

All children have natural curiosity about where babies come from and in particular about themselves. However, no three-year-old really wants to hear about sperms, eggs, fertilization, implantation, gestation, labor, delivery, and so forth. Similarly, adopted three-year-olds need not be subjected to the entire saga of the adoption process. Both three-year-olds comprehend on a very concrete level. They are looking for simple answers such as "babies come from women's bellies" or "babies come from hospitals." Children under six cannot truly integrate or understand any notion such as having two mothers (or fathers).

These terms should not even be applied to the biological parents of a young child. If pressed by a young child with a direct question such as, "Was I inside your belly?" you can follow rule 1 (tell the truth) by explaining, "You came from a woman's belly and then daddy and I took you home from the hospital. I'm really glad I brought you home, and I'm your mommy. Would you like to see (pictures, clothing) from then?" Your child now knows a number of things and still retains a firm sense of who mom and dad are.

At around five or six, children can distinguish that they came from "a woman's belly" and they have a mother, whereas most of their friends came from "their mother's belly." This is therefore an age when you can focus on the fact that children can come into families in many different ways. The word "adopted" can be used, but don't make an issue of it. The use of the adjectives such as "special," "chosen," "unique," or "adopted" or terms such as our "special gift" aren't really necessary. Too many adjectives make children feel differences and can be seen as problematic by biological children at home who may be adjectiveless by comparison. All children are special. Use a wide array of adjectives.

At this age, children seldom want to know why this process occurred; but by age eight, children certainly want to know reasons, and they begin making moral judgments. Ultimately, the question will arise about why the other woman gave away her child. At this time, facts, feelings, and the future are important. Information can be given: for example, "She was too young and it would have been very difficult to be your mother [employ the example of the teenage baby-sitter if appropriate]. She wanted you to have a very good mom and dad. With the help of some other people, we all decided that we would be your mom and dad." Supply as much additional information as requested. Also take this time to let your child know that it must have been a difficult but caring act. This is also an important juncture at which to emphasize that you are mom and dad for keeps and will not be planning another turnover.

When the chips are down and offspring feel that cleaning up their room one more time is intolerable, parents are apt to hear, "You don't love me! I'm leaving here to become someone else's little boy." For a typical child, this might be followed by, "I'm going to live with grandma." For many adopted children, the phrase might become, "I'm going to live with my real parents." It is natural to think of what might have been, and few adopted children persevere with these fantasies.

Adolescence is fraught with many issues of establishing a sense of identity. It is inevitable for adoption to play a role in this process. Again, tell as much as you know. Background information and pictures are helpful in establishing this additional sense of background. Individual decisions should be made about circumstances surrounding any meetings between adolescents and their biological parents.

Child Care

The day-care concept is not new. It dates back to 1854 when New York City provided day nursery facilities for unwed mothers required to work as wet nurses. By the end of the 19th century, Maria Montessori began teaching three-, four-, and five-year-olds in such day-care centers, beginning a movement in pre-school education that is still popular today.

As the times changed, so did the appeal of and interest in day care. The economic and defense necessities of World War II sparked a re-birth of day care because many young women took the desk and factory positions formerly occupied by men who were at war. Today, women are working due to financial need and by choice. The result is that two-thirds of American households have children regularly cared for by someone other than the core family of parents and siblings. Over 25% of children below school age are enrolled in a formally licensed day-care center. Although it is a common practice to refer to planned, half-day programs for three- to five-year-olds as nursery school, and programs for younger children lasting part or all day as day care (or infant day care), we henceforth use the term "day care" to discuss all these programs.

YOUR CHILD'S NEEDS

Adequate facilities can be found for children of all ages. Infants and toddlers require more attention and should have at least one staff person for every four children. By the age of three, most children have moved beyond the period of independent or parallel play and are ready to begin playing with other children. Besides enjoying the company of others, three- and four-year-olds usually have a fair language ability as well as some control of their bowels and bladders. This is a good age to start children in day-care centers where there are fewer staff members.

Before deciding on a particular day-care center, parents should assess the needs of their child:

- Is the child able to communicate needs to others?
- Is the child toilet trained?
- Does the child have any feeding difficulties?
- What are the child's sleeping patterns?

BENEFITS AND RISKS

The major concern when contemplating day care is how the experience will affect your child. Within this concern are two separate questions: Will day care adversely affect my child's health? What impact will it have on development including intellectual, social, and emotional maturation?

There is no doubt that children in day care have greater opportunities for exposure to a multitude of germs. The greater the number of children your child is exposed to, the more likely he or she is to acquire an illness. Here the attention to proper control of infections by the individual facilities is important.

Children in day-care centers learn at a rapid rate. They also learn to cope with separation from their parents by spending part of the day away from home. To help ease the shock of this separation, parents should introduce their children to the day-care center gradually. Spending a few hours a day there and gradually decreasing this time often helps. A second child usually has much less trouble if an older brother or sister is already there. Many parents participate regularly by helping out and teaching at the day-care center; this lessens the burden of separation. Children who learn to accept this separation from their parents in day care and nursery school will generally make a smooth transition to the full-day separation that the first grade demands.

A number of studies now indicate that, in general, the separation experience has not interfered with children's attachment to their parents. In the realm of intellectual development, day care seems to enhance the learning of socio-economically disadvantaged children. Socially, children in day care become socialized at an earlier age than their peers who stay at home. The nature of socialization has been found to differ across different cultures and, by implication, in different day-care centers.

Ultimately, all children return home. Thus, the most important factor is likely to be you and your efforts to discuss with your child what happened during the day, re-inforcing the experiences that are consonant with your values and explaining any differences or objections to any that are not.

CHOOSING A PROGRAM

While research on day care may be interesting, what we all need to do is make sure the day-care center our child attends is the best available to meet his or her needs. Few parents would require letters of reference for a grandparent who is to provide day care or expect the grandparent to change some family habits that have bothered the parent for a number of years. But in reviewing day-care centers, the following considerations are generally applicable: types, cost, environment, program philosophy, and health policies.

Types

The basic arrangements include care in your own home by either a relative, a live-in employee, or a home baby-sitter; care in another's home, also known as family home care; or care in a group type of care center. Facilities range from a neighbor's home to an elaborate institution. Licensing requirements vary from state to state; always ask whether the facility is licensed.

Costs

Costs may vary considerably. For example, live-in help, depending on accommodations, hours worked, age, nationality, and the market rate in your area, can range from $300 to $1000 monthly, plus room and board. The most expensive option for a single child might be a home baby-sitter; but if there are two children at home, the cost may not be that much more. For example, in our area, family home care costs about $250 to $350 per month; group centers run $275 to $450 per month; and home babysitters about $800 per month for a 50-hour week.

Environment

Location is obviously important. Today's family often has four or more individuals needing to be in remotely different places in town between 8:00 and 8:30 A.M. Is a nursery school really worth the added chaos of driving across town?

Consider the facilities that the day-care center provides:

- Do they look safe and well cared for?
- Do the facilities offer children ample opportunities for physical activities and exploration? Are there places where they can play independently?
- Are climbing apparatuses at a reasonable height and surrounded by sand?
- Are the equipment and toys appropriate for the two-year-olds as well as the four-year-olds?
- Does the center allow for an afternoon nap, which most pre-schoolers need?
- Are meals and snacks suitable, nutritious, and reasonably familiar to your child?

Staff Philosophy

Most important in day care are the staff—their experience, philosophy, diversity, compatibility, and, of course, number. The supervisory staff should be accustomed to children of your child's age. At day care your child should experience a balance between play with other children and interaction with adults. Most important, the staff members should be willing to sit down with you and discuss the individual needs of your child.

Determine whether the staff's attitudes are consonant with yours. Are activities carefully structured for groups, or designed to accommodate individual preferences and development? For instance, can your child keep a favorite blanket or other plaything? Ask the staff how they would handle your child if he or she misbehaved. Discussing discipline problems in advance can avoid conflict later on. Some parents find out much too late that staff members have completely different attitudes about child rearing from their own.

The most useful way to evaluate a day-care center is to observe children's behavior there. That will give you a good idea on how it is run. Observe as many activities as possible: group activities, individual instruction, group quiet time, and mealtime (for social interaction, decorum, and nutritional value). Ask about the curriculum and ability to meet needs of individual students. Finally, talk about the role of parents as participants in the child's care. After choosing a day-care program, plan on frequent conversations with the teachers to ensure that your child's needs are being met.

Health Policies

Determine what health policies are in effect as well as procedures for handling emergencies. Find out about toilet-training procedures. Make sure there are adequate hand-washing facilities for children. In general, children in diapers should have changing areas kept away from the older children.

Facilities should keep the child's personal doctor's number on file and require a screening exam and up-to-date immunizations. Ask about outbreaks of infections in the previous year. Many infections, such as chicken pox and group A strep are quite common. Make sure that the center has a system to notify you during such an outbreak. Day-care centers should be notified and in turn notify you about outbreaks of the following:

- Campylobacter gastroenteritis
- Chicken pox
- Giardia infections
- Lice
- Type B *Hemophilus influenza* infection
- Hepatitis
- Measles and German measles (rubella)
- Meningococcus
- Mumps
- Salmonella
- Scabies
- Shigella
- Group A strep infection
- Tuberculosis

For more information on this topic, read *The Working Parents' Guide to Child Care* by Bryna Siegel-Gorelick (Boston: Little, Brown, 1983) or *Mother Care/Other Care* by Sandra Scarr (New York: Basic Books, 1985).

Divorce and Remarriage

The types of crises families encounter can be classified as (1) status shift, for example, sudden or prolonged loss of income; (2) loss, for example, divorce, death; (3) addition—new family or stepfamily members; and (4) dysfunction or demoralization—drug abuse, sexual abuse. In this section we discuss the crises of divorce and new spouses.

FATHER IN A NEW FAMILY:
At the end of the argument, Annie really blew up and told me I had to get out of her house. For a few seconds, I felt I had to leave but then realized that this new house was my house too, and that I had just paid for it. Somehow, with all her kids and her furniture, I sort of believed I was living in her home and not ours.

DIVORCED MOTHER:
It was really hard on [my son] George when he heard his father was going to have a baby with his new wife. At first I thought that George was sad because he was going to have to give up the position he had held for so many years as the baby in the family. But then I realized that he was in utter terror that this child was literally going to take his father away from him and that he was about to lose his father.

CHILD, AGE 10:
I like changing from house to house because it's nice being in different backgrounds every now and then.

CHILD, AGE 11:
There are times when going to different houses can be trouble because at one house you can have most of your things but not the other. Spending a single night at one parent's house is hard because there really is no reason for it. You go there and only eat dinner, do homework, and go to sleep. When morning comes, you must get up, eat fast, and go to where you are going, but I like spending single nights.

CHILD, AGE 11:
Having two different fathers is not a great deal of stress or nothing to feel sad about unless you like one family more but see less often, but I think of it as just having two normal fathers. Besides I should be glad that I have two.

DIVORCED FATHER:
The kids seem to be doing fine. We talk about things periodically, but the urge to talk is more mine than theirs. They listen tolerantly to my suggestions of their feeling anger and confusion toward me. The other night, [my son] Tom responded to an overture with: "Is this going to be one of those serious talks?" So I'll try to lighten up a bit.

In general, divorce represents a loss to children. While the loss is most obviously the amount of time the child will spend with one or both parents, there is often an accompanying loss in financial status. A mother working part-time may now need to return to work full-time; new child-care arrangements may develop.

In addition, it is common for children to believe they were responsible for the divorce. Younger children often regress and display more infantile behavior, including bedwetting and night-waking, and may resume or begin tantrums. Older children often have difficulties in school and may withdraw from friends, complain frequently of pain, show signs of depression, or engage in new risk-taking or anti-social behavior.

There is no way to shield your child from the pain and loss that divorce represents. You can help your child cope by answering all questions and anticipating ones they may be afraid to ask. If true, you should let your child know:

- That he or she did not cause the divorce
- That each parent will continue to love him or her
- The way he or she will be able to interact with each parent
- Where he or she will live, and when
- Where he or she will go to school
- What will not change (friends, pets, and so on)
- What will change, in addition to the loss of parents functioning as a couple
- That he or she should not attempt to patch things up and that those efforts have been attempted unsuccessfully

Almost all parents have some help in getting through divorce, usually from friends and sometimes through professionals. Often children's peers are not as capable of lending support. You are the primary help for your child, but this is one time you shouldn't hesitate to use other resources, such as books or, most importantly, professional counseling. Your doctor is the best place to start; he or she may refer you to a colleague with greater skills in this area.

As the above quotes indicate, the forming of new families can also provide opportunities for trying times. The vast majority of parents who divorce will re-marry. Stepparents have had bad press since the days of Cinderella, although there was some truth in the fairy tale—adolescents do experience far greater difficulties in adjusting to stepparents than younger children. The typical divorce occurs less than seven years after marriage, making younger children most liable to experience the separation and reconstitution of a family. When courtship between two divorced parents ends in marriage, a major transition in roles occurs. It is important at this stage to clearly decide on the roles and authority of the stepfather versus the biological father and the stepmother versus the biological mother. While a new family member at home often evokes jealousy, children also may perceive the added security or attention that may develop. The tendency for men to marry younger women continues. If dad has custody of older children and stepmom is somewhat younger, authority problems are common. A successful transition will not occur overnight.

Give it time and, if stress is not declining, don't hesitate to seek help. Occasionally, the entire family (including former spouses) may need to gather to negotiate a working system. Helpful books are *Divorce and Your Child* by Sonja Goldstein and Albert Solnit (New Haven: Yale University Press, 1984) and *Surviving the Breakup: How Children and Parents Cope with Divorce* by Judith Wallerstein and Joan Kelley (New York: Basic Books, 1982).

Parents' Feelings

An ancient wise man once said: "If you're alive, you've got troubles." Having children proves the point; along with an enormous amount of pleasure comes trouble. This section is directed at helping you as a parent anticipate and recognize trouble and know when to seek outside professional help.

In the same way we talk about child development, we can think of parental development. Each milestone achieved by the child implies a response on the part of the parent, which is also developmental. These transitions require that a family system re-adjust to accommodate the maturational needs of the child. Sometimes the child sets the pace; if he is learning to move around, parents will respond by "baby-proofing" their home. In other situations, the parents may encourage a child's development by shifting him or her from crib to bed. Just as there are wide variations in normal child development, there is an equally large variability in the normal response of parents. The job of the parent is to be responsive to and supportive of a child's development, while simultaneously monitoring their own development as parents and marital partners. So, we'll start at the beginning.

THE ARRIVAL

You are two mature, caring adults, anticipating the arrival of a baby. Your conversation is making a slow shift from politics and daily activities to the advantages of disposable diapers versus a diaper service. You have considered how to re-organize your house to accommodate a child and may have spent vast sums of money or time on decorating a nursery. You are contemplating when to allow grandparents and friends to come to help out. The big day arrives. Now you two are three. You arrive home. Your fantasies of the first few weeks—being treated like a king and queen, nourished by healthy meals, napping when baby naps, and having a clean house—are compromised by reality. There's always a crowd of relatives ooing and gooing, and it seems as if you only see your baby when he or she is crying or in the middle of the night; you have had to feed a cast of thousands; your house is irrevocably messy and chaotic.

If you are feeling out of control, don't be surprised. The reality of parenthood is almost never what you expected it to be. However, there are some things you *can* do to help yourself make the difficult transition into parenthood.

- Set firm ground rules for visiting friends, relatives, and colleagues.
- Don't expect to be able to maintain your household at the same standards you had before the baby.
- In the first few weeks, try to use whatever opportunities you have to catch up on your sleep—when the baby is napping or when friends and family can help out with meals or baby-sitting.
- Remember that you are still adults who have adult interests and needs. As soon as it is feasible, leave the baby with a friend or relatives—even for an hour—and go out together for a walk or drive.

Once the excitement of the first few weeks passes, you will need to make new adjustments. This can be an especially difficult time for a new mother, especially if the father must spend most of his time away at work. Confusion, crying, and a sense of isolation at this time is normal. You may be distressed that none of your pre-pregnancy clothes fit or that most of your friends without children are busy with work or other activities. You may find it hard to locate a baby-sitter with whom you feel comfortable or to figure out how to leave your baby—especially if you are nursing—in the care of someone else.

Some helpful advice to follow during your first few months is to look for opportunities to lessen your sense of exhaustion and frustration. Try to evaluate your priorities. While housecleaning, writing thank-you notes, and cooking may seem like obligations you feel compelled to fulfill, a nap, an exercise class, a playgroup with other mothers, or just some free time to take a walk in solitude may be more beneficial. Scheduling time for yourself will enable you to face the demands of baby care with more energy. Although you may want to continue with your life exactly as it was before, try to feel comfortable with making positive changes in your schedule that acknowledge the presence of a new person in your household.

Fathers

For a father, the delight in parenthood may be tempered by feelings of exclusion and jealousy. If you are not the primary caregiver, it is normal to feel that getting to know your baby is taking you longer. Because your wife may be pre-occupied with nursing, diapering, and other tasks, you may experience a period of adjustment to her new role as well. Try to work out methods that will make sharing child care satisfying for both of you. Volunteering to help with household chores, as well as offering to spend time alone with your baby, will aid in establishing new ties of intimacy within the family unit. A special task for a father is to support and encourage the marital relationship. A new baby can sap the energy of both parents; a father can be enormously helpful in reminding a mother that she is a woman and wife as well.

Talking about feelings and searching for new solutions together is the first development test for parents. By the time your baby is four months old, you should both have the sense that things are settling down. Ideally, parents should be able to enjoy each other's company alone for an evening without their child.

If feelings of being overwhelmed persist, talk to your pediatrician. For the first few months of your baby's life, you will see your pediatrician frequently. Part of his or her job is to help you integrate your new baby into your lives. Because pediatricians have observed many parents and have dealt with the common problems this transition time creates, they can offer suggestions on how to overcome the hurdles of the first few months. Other resources include support groups, playgroups, and many fine books and articles.

SEPARATION

From the first time you leave your child with a sitter to the first time your adolescent goes out alone at night, separation between parent and child is always difficult. It can frequently be traumatic, but generally separation is harder on the parent than the child and a necessary experience for growth.

Some early issues of separation are weaning, leaving your child with a sitter, and letting a child cry himself or herself to sleep. The guideline for all of these experiences is to try to be gradual, loving, and consistent. Try to sort out whether the problems surrounding the issue are yours or your child's. Is your 15-month-old's reluctance to give up nursing for comfort a reflection of your own ambivalence to forgo this intimate and special relationship? If you must call or check up on your child when you go out, are you really fearful about your baby-sitter's competency, or are you worried that your child's independence is a form of rejection?

Bedtime

If bedtime is a major battle or your child ends up sleeping in your bed more often than not, you are not providing your child with the opportunity to learn how to comfort himself or herself. Often, a stuffed animal or favorite blanket can be an effective object to facilitate a transition from dependence on parents to self-sufficiency. Sometimes, it is helpful to buy more than one "lovey" and rotate; if one is lost or worn out, you have a well-used replacement. If, by six or seven months your child cannot sleep alone in his or her crib—with or without a favorite object or night-light—or if separation situations continue to cause tears and struggles, you may need some guidance on how to help yourself and your child to accomplish the developmental task of separation.

FRUSTRATION

All parents become frustrated and angry with their children. These are natural human feelings. Although we all accept anger as a normal part of everyday living, there are many ways of dealing with anger that are unacceptable.

There are many indications that children are often the ones to suffer most, often physically, when family relationships are strained and anger flares. Abuse of children is becoming an increasing societal problem.

We learn how to handle our emotions from our own parents. It is not uncommon for us to encounter three-year-olds who will throw a puzzle to the floor and shout, "Oh, damn!" Our temperament clearly has its foundation in the temperament of our parents. It is highly likely that if you remember your parents being impatient with you or having difficulty controlling their temper, some of this may have worn off on you.

At the end of a long and hectic day, we all deserve to sit down to a peaceful dinner. But more often than not, dinner with a one-year-old child will be punctuated with shouts, tossed spoons, and thrown food. On the wrong day at the wrong time, this can be unnerving. Becoming exasperated and angry is likely. However, it is important to remember that your one-year-old is behaving exactly as a one-year-old should. If it were a ten-year-old throwing food and utensils, punishment, such as being sent from the table, might be in order. However, it is inappropriate to punish a one-year-old who is exhibiting appropriate behavior. What would you do in this situation? Would you start shaking your child violently? If you felt you were going to explode, who could you turn to, who could you talk to in order to prevent this explosion?

Asking for Help

The problem of extreme anger with accompanying potential abuse can be dealt with most easily once people accept that it is a problem of being human and therefore not a problem to be ashamed of. However, it is difficult for most of us to admit that we have a problem that may require help. Asking for help is difficult. Often the first person to turn to is your mate. Neighbors can sometimes help. While it is often difficult for us to pry deeply into or expose extensively personal problems, a true friendship should give you license to talk openly with friends about your or their temper problem. Your doctor is another possible contact. If there is no close person to talk with, it is possible to call the parental stress hotlines that many communities have organized. The presence of so many hotlines is an admission by most communities that we all become angry, and it is important for us to keep this anger from harming our children.

The solution to a complex problem is seldom simple. While most communities have services available to help parents and children involved in severe abuse, services geared to preventing abuse are only in the earliest developmental stages. Many childbirth preparation classes now spend time focusing on preparing expectant couples and individuals for parenthood. Other courses deal specifically with the problems of becoming a parent.

Very often we do not know what to expect when we become parents. Parenting involves more than just adding a family member. It involves the re-adjustment of relationships between the existing family members. Some parents will find it easier than others to re-adjust. However, none will find it easy. The time that parents once had to unwind at the end of a day or to talk out problems between themselves may no longer exist once a child is added to the family. And yet, this time is important.

We should all be working to make resources available to our neighbors in need of help. If you feel that programs in your community are inadequate, you should contact your doctor, county department of social services, or members of the local American Academy of Pediatrics in order to inquire about the development of such programs.

Getting Help   BY ANNE S. BERGMAN, M.S.W., DR.P.H.

Many pediatricians and family doctors now receive considerable training in caring for mental health problems. If you and your pediatrician are stumped about a problem with your child or your family, you may wish to seek a mental health consultant. Usually, your pediatrician will recommend someone with whom he or she has worked in the past; however, you as parents can help in the selection of such a person. This chapter will give some information about the training of mental health professionals and the way to find the best person to help.

First, let us calm your fears: a referral to a mental health worker does not mean that you are "crazy" or that you are doing something terribly wrong. Moreover, although people find it easy to talk about going to see their doctor, there is a much greater reluctance to admit that one has seen a counselor. In fact, you would be surprised at how many of your friends, relatives, or co-workers have seen a mental health professional at some point in their lives. Mental health professionals do not read minds, nor do they make judgments. Their job is to help you understand why a certain problem may have developed and what you, as parents, can do to help your child overcome a problem. Very few people are forced to seek mental health consultation, and for a mental health worker to help you, you must actively want to be helped. The light-bulb joke says it all. *Question:* How many mental health workers does it take to change a light bulb? *Answer:* Only one, but the light bulb has to *want* to be changed.

There are a growing number of pediatricians who take one or two years of additional training to become skilled in behavioral pediatric techniques, but the majority of help is offered by special mental health workers. There are three primary disciplines in which these workers are licensed and trained to provide services.

Social workers comprise the largest group of mental health workers in the United States. Social workers have two years of training after the completion of college, at which time they are awarded a Master's Degree in Social Work (M.S.W.). As is true in many disciplines, there are areas of specialization in social work; some social workers work in welfare agencies, some in hospitals, and some in specific mental health clinics. In most states, social workers are licensed to practice mental health care; this usually implies that they have had several years of experience and have passed a state-administered exam certifying their competence. If your child has a medical problem that has an emotional component, a good way to find a social worker is to call your local major hospital and ask if the staff includes social workers who work with children with problems similar to those which your child is experiencing. An advantage to you as a family is that social workers often work in agencies that have a sliding scale of fees; this means that the cost of evaluation and treatment will be determined by your income. A potential disadvantage is that some insurance policies do not cover care by social workers. In most communities, there is a Family Service Agency that provides mental health services on a sliding scale of fees. These agencies are set up to meet local community mental health needs and are staffed by a variety of well-trained mental health workers.

Clinical psychologists are trained at a doctoral level (Ph.D.). They provide evaluation and counseling and are often trained to do psychological or educational testing. These tests alone are not sufficient to help you understand or care for your child, but they can offer evaluations of intelligence, personality, and psychopathology in terms of comparisons with other individuals. Thus, a good mental health worker will use testing to enhance his or her understanding of you or your child; testing, along with meeting with you to talk about the problem, is a good beginning. If you or your child's school or doctor think that your child is having learning problems, testing is a well-accepted way to measure how your child compares to others. While questions have been raised about cultural bias in some of the commonly used tests, a good clinician will keep the test results in perspective as he or she learns more about you and your child.

Remember, the point of these tests, as with all professional care, is to offer your child the best chance to develop to his or her full potential. Psychologists often work in mental health centers or in the school system. You can also call your local County Psychological Association to find the name of a psychologist experienced in working with children and their families.

Most insurance companies will pay part of what a psychologist charges; it is always helpful to know in advance what your mental health professional charges and whether it will be covered by your insurance policy. Often educational testing will be paid for by your school.

Psychiatrists are licensed medical doctors (M.D.) who have been specially trained in neurology and psychiatry. Some psychiatrists have done three years of training (after they receive their M.D.) in adult psychiatry, and a much smaller number do two more years of training in child psychiatry. There are only approximately 3000 board certified child psychiatrists in the United States; however, for certain problems it may well be worth your while to find one. Often sorting out the difference between neurological and psychological symptoms can be complicated. Because the treatment of psychological problems can be varied, from medication (an M.D. is the only professional licensed to prescribe medication for your child), to psychotherapy, to behavioral interventions, having a good diagnostic evaluation is crucial to selection of the correct treatment. If your child is experiencing significant depression, or problems in conduct or attention, or the pediatrician suspects a relatively uncommon disorder, you may wish to seek out consultation from a child psychiatrist. There are many professionals competent to perform psychotherapy; there are few who integrate both organic and psychological factors. If you feel that your child would benefit from medical and psychological expertise, a child psychiatrist is highly skilled to provide this evaluation. You can find a child psychiatrist by asking for a recommendation from your doctor, calling a local medical school, looking in the Yellow Pages, or by calling the American Academy of Child Psychiatry at (202) 966-7300.

A few words on what to expect in a mental health evaluation. Keep in mind that a mental health professional needs to know as much about you and your child as possible to give you accurate information about the problem and its potential solution. For parents, this may involve joint interviews about you, your relationship, and stresses and strains in your marriage. A family history will be asked for, including information about your family of origin and your upbringing. The mental health worker is not only interested in facts but also about feelings. The mental health worker will be sensitive to your hesitation and/or discomfort in talking about your personal life, but it is often the very things that we are most afraid of that are important to talk about. Remember, the mental health worker is there to help you, and the more open you are, the better able he or she is to share his or her knowledge and offer you understanding.

A psychological evaluation of a child is somewhat different than that of an adult. As we all know, children express their feelings in many ways, not all of them verbal. It is possible to evaluate the psychological development of even an infant, and often the tools used are those with which a child is most familiar. "Play therapy" is exactly what it is called; in the process of play, a child can communicate many feelings and ideas that he or she may be unable to communicate verbally. Don't be surprised if your mental health professional has a toy box, puppets, or a sand tray in his or her office. These are very helpful in making children feel at home in the office and allowing them to express themselves in a usual fashion. You *should* prepare your child to see a mental health professional. It is always helpful for a child to know that his or her mom and dad are also seeking help. You may wish to acknowledge that you are concerned because your child appears sad or worried; tell them, if they are able to understand, that you are taking them to see someone who understands and helps children. Beyond that, once you have brought your child to see someone, it is up to the mental health professional to establish a relationship with your child. If you have concerns about how to prepare your child to see someone, ask the professional for suggestions about what to tell your child. If you are genuinely interested in helping your child, you will find that a good mental health professional can help you to understand the complex relationships that exist in families and the many good qualities that can be used to help your child be a happy one.

In seeking the best person for you and your child, the selection of the discipline of the individual may be less important than his or her experience with the type of problem your child has, their licensure, and how much you can afford to pay. It is helpful to ask other professionals, including your pediatrician, what kind of reputation the person has in the community. You should feel comfortable asking the person you select how he or she intends to evaluate the situation, how many visits it will take, and what his or her experience in the past has been. And remember, psychological care is a joint venture; your anxieties and fears are as important for the therapist to know about as the strength you and your child can bring to bear upon the problem.

Some situations can be resolved in a short period of several months; some are more complicated and can take a year or more. It is more difficult to measure progress in psychotherapy than in medicine; in most cases, an immediate response is unlikely. If you are willing to work on a problem, giving it time and thought, one can increase one's understanding of why a problem may have developed and devise strategies to improve the situation. Finally, remember that although therapy requires work and may at times be stressful, the ultimate goal is to make you and your child feel and function better. If this is not happening, you may wish to re-evaluate the therapy.

CHAPTER 7

School Days

S chool is a critical factor in the growth and development of your child. Not only does it teach a child the necessary academic skills for functioning in an adult world, but it also attempts to teach social skills. To help your child get the most out of this experience, you need to understand and complement the experiences that your child has in school. In this chapter, we briefly discuss some of the common concerns that parents have about the academic and social aspects of school.

School Readiness

Children mature at different rates; hence, there is no particular age at which it is best for all children to begin school. Some children are ready at age four; others may not be ready until age seven or later. Having a child begin school too early may lead to frustration and early school failure and may hamper the child's future school experiences.

Many things are important for success in school. For example, good general health is required. Children with serious diseases will be under medical supervision long before school begins. Knowing that your child can see and hear adequately are important prerequisites for entry. In addition, most states require that immunizations be up to date for entering children.

Social skills are important for doing well in school; sitting in one place all day is not easy. Most children should be able to play well with other children, separate easily from parents, get dressed alone (except shoe tying), have daytime bowel and bladder control, and enjoy games.

Language ability is, of course, very important. Children with severe speech defects will require additional help.

Interest in learning is also an advantage. It helps if the child likes books, is curious about things, knows colors, can repeat a few numbers, can hold a crayon, and can draw a square. Many children will demonstrate ability to grasp complex concepts before they have entered school. They may know about the difference between summer and winter and understand concepts such as "over" and "under." They may even be able to tell you the color of grass and the sky without looking at them directly. Children also must be able to follow a series of commands such as "Go to the closet, select a puzzle, and return with it to your seat."

If your child is able to do many of the above tasks, there should be no problem in beginning school. However, if you are concerned about your child's inability to profit from school because of any one of a number of factors, you should consider a conference with school or medical professionals before enrolling your child.

School Problems

School offers children more than book learning. This is a time for children to begin to experience a world different from their home. In school, children have the opportunity to become social beings; relationships with peers offer a child an opportunity to feel both uniqueness and commonalities with other children. School offers children a variety of experiences that encourage a sense of self-esteem. They may excel in sports or academics; they may discover a preference for some experiences over others; they may learn that certain things are required, while others are voluntary. Because the repertoire of a child grows dramatically in school, it is a particularly important area for parents to monitor.

Schools also offer the parent an opportunity for the next true crisis: the first parent–teacher conference. Nothing strikes more terror in the heart of well-intentioned parents than the thought of someone else judging their child. What if all those adorable characteristics, like climbing on furniture at home, are enough to make the child *persona non grata* at school? Never fear. There's hardly a more understanding and supportive group than schoolteachers. Regardless of what your child does, they've seen worse. However, if they have concerns, it's important to consider them seriously. Schoolteachers are good observers of children; often they can see things in our children that we can't, or don't want to.

The identity your child forms in school is an important part of the development of his or her own personality and style. Problems in school can be trivial or indications of potentially greater troubles down the line. Get to know your child's school, teachers, and classmates. Take seriously the concerns of your child's teachers. An adversarial stance with your child's school is not in your child's best interest. If your child is substantially different at school than at home, find out why. An isolated child with no friends is not a happy child.

How can a parent tell if a child is having problems in school? Obviously, listening to what the school personnel tell you is important. It is equally important to listen to your child. The reams of drawings, tests, and other tidbits he or she brings home are invaluable products of a child's work at school. Spending a few minutes each day asking your child about school is as vital to your child as your own work is to you. If your child appears excessively anxious about an exam, fearful of relationships with his or her peers, or persistently reluctant to go to school, you should make an attempt to understand why.

Children with difficulties in school may complain of vague physical symptoms on school days or have difficulties falling asleep at night. It is often very hard for a parent to discern whether a child is really ill or has something else on his or her mind. A good rule is to keep a child home if he or she has a documented temperature, and if the child stays home, he or she must stay in bed all day. If complaints such as tummy ache or headache persist, first check your own behavior. Do you tend to experience pain when you are upset? If mommy or daddy often has a headache, children learn that this is a way to either gain more attention or avoid unpleasant tasks.

While all children have normal complaints about school and schoolwork ("What did you do at school today?" "Nothing, it was boring," or "How's your friend Ricky?" "Ricky has a club, and he won't let me join"), some problems can be symptomatic of stress that your child is experiencing. If talking to your child and his or her teachers does not resolve the problem, speak to your pediatrician. It may be helpful to have an outside observer evaluate your child.

Failing in School

Children learn at their own rate. The best learning is an active process involving an exchange between the teacher and the student. This one-on-one type of learning experience can best be done by parents. The active exchange that takes place when an adult is reading to a child and the child is questioning is an extraordinary learning experience; this experience is not matched by the slickest educational television program. While we applaud some of the excellent television educational programs, they are a supplement to and not a substitute for other types of learning. They are a passive technique and not an active exchange. The feelings that children acquire when interacting aid their emotional development as well as their mental development.

Many problems can impede a child's learning; some are so complex that it is impossible to deal with them here. **Dyslexia** is a term that means inability to read. It is not a single diagnosis; there are literally hundreds of reasons why children may be unable to read. There are just as many reasons why learning in other areas may be impaired. What follows is a list of a few factors that may interfere with your child's learning. If you suspect any of the following, a professional should be consulted.

- **Visual problems**. Vision testing is simple in children over four years of age. Strabismus or "lazy eye" can be detected in children over the age of two years.
- **Hearing problems**. Children must not only be able to hear sounds, they must also be able to identify fine differences in sounds such as "p" and "b." Hearing and complex language disabilities should be suspected if there is excessive nonsense verbalization after 18

months, if the child is not talking at all by age two, if the child began talking and then stopped, if the child is not using sentences at all by age three, or if there is no verbal communication of the child's wants.

- **Coordination problems**. These can, of course, interfere with performance in school. Much of the time in the first grade is devoted to writing and drawing. Children who develop coordination skills slightly later are at somewhat of a disadvantage. This, however, may be merely a maturational lag and is not necessarily something to be concerned about.

- **Visual motor ability**. The child's ability to see an object and then copy it involves coordination of eyes and hands. Inability to copy designs may be a sign that visual motor ability is lacking.

- **Auditory perception**. Hearing sound accurately is not enough. The child hearing the word "boat" must be able to picture a boat in his or her mind and recall what a boat looks like and what a boat does. Furthermore, the child must be able to respond to describe the images that have been evoked.

- **Attentional problem**. Some children have difficulties concentrating on the task at hand. They may become easily distracted. It may be difficult for them to concentrate if there is a distracting sound or sight going on. Other children may dwell on a task for an unusually long period of time.

- **Maturational lag**. Children mature differently. Not all six-year-olds are capable of learning in the same way. Some need a few months more before they are developmentally *capable* of learning normally. This delay does not mean that these children will always be behind. It is similar to the case of the child who does not walk until the age of 18 months. Nobody can tell this child from the child who walked at age 9 months when they are both five years old.

- Children from homes where the **primary language is not English** are at a disadvantage when beginning an English-language school. These children generally catch up very quickly, but ostracism in the first few months of school (by either classmates or teachers) can seriously hamper a child's confidence and development. Bilingual education programs in areas where there are numbers of students who speak Spanish or other languages in the home have helped many students. Bilingual ability is extremely advantageous for the child where the primary language at home is not that of the overall society. These programs are also excellent opportunities for English-speaking students.

- Children who **change schools frequently** may encounter school problems. Some of this may be due to differences in curriculum, but much of it is due to the child's need to re-adjust after losing friends and familiar surroundings.

- Special types of **seizure disorders,** known as "petit mal," generally cause short periods in which the child does not respond to the environment. The periods may last only two to three seconds, but children may have up to several hundreds of these in an hour. During these periods, the child is unaware of the surroundings and will not be learning. This is a very unusual reason for a learning problem but can be treated effectively, which is why we mention it here.

- Children with **physical defects** are often teased by classmates. While the deformity may not interfere with the child's ability to learn, the feeling of being an outsider certainly does.

- **Lack of sleep** can also interfere with the child's learning ability. Many children now have television sets in their rooms, a practice we deplore, and so stay up quite late at night; inadequate sleep can hamper their learning abilities. In addition, children who are often tired in the morning should be suspected of the "reading the book with the flashlight under the blanket" syndrome.

- **Physical size** may also interfere with learning. Surprisingly, this generally affects taller children. A tall seven-year-old who looks like a nine-year-old will be expected to act like a nine-year-old. The discrepancy between the adult expectations and the child's ability creates the problem for the child. Remember, a child should be judged by his or her chronological and developmental age and not by how old he or she looks.

- **Hyperactivity** is an extremely subjective complaint. Often one parent will think a child overly active, and the other will think not. Hyperactivity is often first noticed by a teacher and brought to the attention of parents. Some extremely curious children who are learning at a rapid rate appear to be overactive. Other children's overactivity and/or inability to focus interfere with learning. It is important to focus on the quality of the child's activity and not just the quantity. If hyperactivity is noticed at school and not at home, a learning problem should be suspected. We all are bored by lectures that we do not understand. Children who are having difficulty learning become uninterested in what is going on in the classroom and consequently find something else to do. Some children have difficulty adjusting to the different standards of behavior at school than at home. Children who speak little English in an English-only classroom can be expected to find something to do other than sit in a seat listening to talk they do not understand. Finally, many medications, including the antihistamines in common cold preparations, can cause hyperactivity responses. See page 312 for further discussion.

INVESTIGATING LEARNING PROBLEMS

The child with serious learning or school problems needs thorough evaluation; these problems are best handled early. The number of teenagers who are unable to read is distressingly high and is partially a reflection of the failure to intervene appropriately at an early age. Although some children have school problems because their nervous system development is temporarily behind their age group, many children have causes other than maturational lag. Parents should be told that their child will "grow out of this phase" only after a thorough evaluation. The following is what we feel constitutes an adequate evaluation for a learning problem.

The **medical history** will focus on the pregnancy, labor, and delivery. Early health and early functioning, such as feeding, activity levels, and behavioral problems, will be discussed.

The child's **early development** will be assessed. This will include questions about the child's coordination, language development, and social development.

A **school history** will include questions about day-care, pre-school, kindergarten, and school failures and successes. If a doctor is performing the evaluation, he or she may request a copy of the teacher's reports as well as any achievement or psychological testing performed at school. An evaluation of a school problem cannot be conducted by a doctor without cooperation from the school. The doctor may spend considerable time discussing the child's behavior and how certain types of behavior are rewarded and punished.

Finally, a doctor will perform an extensive **physical examination**, an extensive neurological examination, and a developmental examination. He or she will be checking for the impediments described in the previous section. Depending on the results of the history and physical examination, additional tests and psychological testing may be advised. Treatment of a school or learning problem necessitates cooperation between doctor, both parents, and the school.

Intelligence Testing

Intelligence testing is more than 70 years old and was originally designed to predict school failures. Instruments of prediction, whether they be crystal balls or intelligence tests, are never completely accurate. Group testing, most often done in schools, is much less accurate than individual testing. Group tests depend greatly on the child's ability to read in order to assess intelligence. A bright child who may be lagging in reading abilities will perform poorly on these tests and will be falsely labeled as having a low overall intelligence.

Individually administered tests are far more accurate. They usually test both a child's verbal and performance abilities. It is easier to take into account a child's poor performance due to a cold, sleepiness, or non-application if an individual is working with only one child.

Verbal skills tested include vocabulary, general information, understanding, and arithmetic. Performance tests evaluate picture completions, block designs, and mechanical skills. Separate verbal and performance scores, along with a total score, are obtained.

IQ

The term "IQ" stands for intelligence quotient. In most tests, a number, such as 105, is the result. This number is derived by dividing the child's mental equivalent age by the child's actual age. If a five-year-old scored as well as the average six-year-old, the fraction 6/5 would be formed and multiplied by 100 to give the score 120. A five-year-old scoring as well as the average five-year-old would have a score of 100.

Tests of younger children are far less predictive than tests of older children. Group tests are often grossly inaccurate, and all tests are inappropriate if they do not match the child's previous experiences. Black urban children have different experiences from the children of a Mexican-American migrant farm worker, who in turn have little in common with children living in a New York suburb.

Parents should be aware of the frailties of intelligence tests. Your child's IQ is not an indelible number that will be carried throughout life to grant or deny access to schools, jobs, and success. While there are some who would use the IQ for this purpose, the most responsible use of IQ tests is to help design an educational program most suitable for your child. Children who do poorly on group IQ testing often do not do well in the standard group teaching of our school systems. They may be fully capable of learning but often require a different teaching approach. A child who does poorly on a group IQ test should have an individually administered intelligence test. Many low IQ's disappear quickly when this is done. Remember, a group IQ test measures how interested your child happens to be in taking that test at that particular time. It does not measure accurately your child's intelligence or potential success.

Avoiding School

Tom Sawyer and Huckleberry Finn led a glamorous life by avoiding school. There is scarcely a woman or man alive today who has not thought of repeating the adventures of these two folk heroes. Almost everyone at some time or another has considered staying home from school or has actually played hooky for a day or two. These feelings are especially common after a child has returned from vacation or recovered from an illness. However, while thinking about avoiding school may be considered innocent, actually avoiding school is another matter.

Most children who avoid school state that they like school and that they actually want to go. However, a morning headache, a morning stomachache, morning weakness, morning nausea, or other symptoms conspire to keep them from going to school. There is always an excuse; it is a question of threshold. Most children have a runny nose much of the time; as we have stressed earlier, runny noses and colds are really not illnesses in children. A minor cold is not a reason for a child to stay home.

Avoiding school is really a symptom. The cause is not very often a terrible teacher, although this is the reason often given. Sometimes a teaching program that is suitable for most children may not meet the needs of an individual child. More often, the child may be experiencing a learning problem, a problem with friends, a physical problem such as being too small or too tall, or a problem at home. Often parents feel guilty during periods of stress at home and want their children to stay home so that they can demonstrate that they still care about them.

School avoidance is a complex problem with multiple roots. It must not be treated lightly. Children who are missing more than 5% of their school days because of problems not accompanied by a fever or chronic disease should be considered to be avoiding school. A program to return the child to school will involve the cooperation of a doctor, the teacher, and both parents.

Childhood Athletics

Children have been running, throwing, climbing, and swimming for thousands of years; there are tremendous benefits to a child from athletics. Children learn new skills and how to control these skills. The exercise that is part of athletics is important for conditioning; adults who follow a regular exercise program are less likely to succumb to a cardiovascular catastrophe. Children and adults who are in good physical condition generally feel good about themselves.

Learning how to play a sport is a learning exercise that is as useful to a child's development as other learning experiences are. There are beneficial socializing aspects. The competition that children impose on themselves is also potentially beneficial. Most of us can fondly remember that some of the best moments of our lives occurred during pick-up games of stickball or basketball or hide-and-seek. Athletics for all children of all abilities should be encouraged. Unfortunately, there are many aspects of athletics that are sorely in need of improvement. Here is the way we see it.

First, most of the sports glorified in our culture and, consequently, of great importance to children are sports in which adults have little opportunity to participate. These include football, baseball, basketball, hockey, and others requiring a number of players. Sports such as running, skating, biking, rowing, and tennis can be enjoyed on a lifelong basis. We favor programs that emphasize sports with long-range benefits; we want healthy children to become healthy, fit adults.

Second, the sports that receive the most glory are often the most violent and therefore the most hazardous to a child's health. We prefer sports emphasizing speed, skill, and coordination to those emphasizing violence and mayhem. The growth of soccer in recent years has been a refreshing addition to team sports.

Third, competitive sports for adults should be clearly separated from competitive sports for children. The sets of rules that govern professional athletics should not be the model for childhood athletics. Children often have a difficult time dealing with failure. When children interpret their parent's love as being dependent on their winning a game, a psychologically dangerous situation exists. We applaud the many organized athletic programs that do not keep score for children less than eight years old.

Fourth, children need to develop in many areas. To the extent that competitive sports or any other single skill precludes their intellectual, social, and emotional development, that sport or skill is to be condemned. A full day of practice for a mature adult making a conscious decision is a different situation than a 12-year-old's practicing for the same amount of time because of pressure from school or parent or coach. We do not believe in pressuring any child to practice any sport for long hours daily.

Fifth, school athletic programs spend large sums of money on inter-scholastic football, basketball, and baseball. The most money is spent on the best athletes, that is, on the children who need it least. Funds for girls' sports may be smaller than funds for certain boys' teams. Physical education, as all education, should recognize the principle of equity.

We support organized sports, but we do not feel that the competitive drive to be number one should be the underlying basis for a child's participation. There are over 25 million children now participating in organized athletic activities. They can all be winners, but they can't all get trophies. Your child must make the decision about the extent of his or her participation and you should be supportive of your child's decision.

IS YOUR CHILD READY FOR A SPORT?

You should ask yourself the following questions:

- Does my child have the **coordination** for the sport? The hand-and-eye coordination necessary for some sports such as tennis and baseball requires an older child.

- Is my child the **proper size** for the sport? There have been several 5'6" college all-American basketball players, but no 100-pound college football tackles. Children mature at different rates and ages. Children who enter puberty late can be physically injured if matched with opponents of greater weight. Forcing children to compete with younger opponents of the same weight can hurt their ego almost as badly. Varsity "lightweight" programs should be encouraged in the schools.

- Is my child playing at the appropriate **skill level**? Most organized sports now group children by age. However, while older players may be excluded from younger teams, younger players are often permitted on older teams. Even an athletic child will encounter difficulties playing against older, stronger children.

- Is my child in proper **condition** for the sport? Size is not everything. A child wishing to join a team in mid-year may have the size and coordination but not the stamina to play successfully.

- Does my child have any **medical conditions** that might be limiting? Many schools have rules that prohibit children with certain problems from playing contact sports (football, basketball, wrestling, ice hockey, lacrosse, boxing, soccer, rugby). Often these children can participate in non-contact sports. Children with HIV infection can participate in all sports programs, although they are discouraged from wrestling because of the high likelihood of bleeding while in close contact with another person.

The following conditions often disqualify children from contact sports, although a doctor or the courts may sometimes permit a child with some of these problems to participate:

- Brain concussion
- Head injury with residual skull defects
- Absence of an eye
- Detached retina
- Glaucoma
- Lung infection (tuberculosis, pneumonia)
- Certain heart rhythm disturbances
- Severe heart defects
- Severe or recent heart inflammation
- Certain types of undescended testes
- Missing kidney
- Bone infection
- Hemorrhagic blood disease

Temporary conditions that often disqualify students from formal sports competition include:

- Active infections
- Perforated eardrum
- Hepatitis or enlarged liver or spleen
- Healing fracture
- Injured growth plate of bone
- Pregnancy

Conditions that may disqualify certain students from some sports, and for which individual decisions are most appropriate, include:

- Physical immaturity
- Diabetes
- Severe visual handicap
- Hearing loss
- Asthma
- High blood pressure
- Absent testicle
- Seizure disorders (epilepsy)

Children with HIV infection can participate in all sports but are discouraged from wrestling because of the likelihood of bleeding and potential transmission of the virus.

SPORTS SAFETY

Children should be taught safety as a basic skill continually; this is the best insurance against serious injury. They should also learn to warm up and stretch appropriately for their sport. They must heed pain as a warning sign of an injury and allow proper healing time. Finally, they must consider the implications of a minor illness on their sports participation; an earache may ground a swimmer, and a swollen knee put a runner on the bench, whereas an archer could participate with these problems but not with a nail or eye infection.

Supervising adults should be prepared to deal with common emergencies. In inter-scholastic competition, where risks of injury are high, professional help should be nearby. The following are some often forgotten problems:

- **Contact lenses.** We do not feel that contact lenses are necessary for all those participating in high school sports. Safety lenses in steel frames with safety straps are often a less costly alternative. The safety record of soft contact lenses is excellent and recommended by many for older competitive athletes. The use of hard contact lenses is permissible in some circumstances.

- **Heat exhaustion and stroke.** Workouts should not restrict water from athletes. Loss of water and salt from the body through prolonged sweating can cause heat exhaustion. Under severe conditions, heat stroke can occur, resulting in rapidly elevated body temperatures, and can be fatal. Salt and water should be replaced.

- **Weight loss.** Losing weight to place into a certain weight category, as in wrestling, should be restricted to loss of fat. Weight reduction of more than two-and-one-half pounds per week should be discouraged. Weight loss by attempts at dehydration are dangerous and can compromise a child's strength. A sub-standard diet to maintain a weight category deprives a growing child of food necessary to grow normally.

Have fun. That's what sports are for. If taken too seriously, though, they may not be healthy. That's the important message.

PART III

The Healthy Child

CHAPTER 8

Working with Your Doctor

*I*n the normal course of growing up, your child will need various kinds of medical attention—from physical examinations and laboratory tests to encounters for behavioral issues to surgical operations. In this chapter, we discuss some of the encounters that your child will have with medical professionals and provide guidelines to help you choose a doctor or other health worker to meet the needs of your family.

Finding Someone to Care

MEDICAL PRACTITIONERS

A variety of professionals provide medical services to children and their parents:

- **Pediatricians** are doctors who have spent at least three years after medical school in formal training for the management of childhood and adolescent problems. The pediatrician is no longer a "baby doctor"; many treat a large number of adolescents. In fact, adolescent medicine has been a field developed almost exclusively by pediatricians.

- **Family practitioners** (general practitioners) are doctors who spend part of their formal medical training in the management of children's problems and who are trained to treat the whole family. Those who have become family practitioners recently have had training in the psychological as well as medical aspects of family interactions.

- **Internists** (specialists in internal medicine), although not trained in the care of children, are familiar with many diseases that occur not only in adults but in children as well and sometimes see children referred by general practitioners.

Several new types of practitioners are also capable of dealing competently with the common problems of children:

- **Nurse practitioners** have an R.N. degree and have attended a program (usually from six months to a year and a half in length) that gives them further training in the management of common medical problems. Some nurse practitioners spend their entire time with children; others, family nurse practitioners, spend part of their time with the problems of childhood.

- **Child health associates,** trained at a Colorado school, undertake a four-year training program after graduation from college. They have not attended medical school but have had more exposure to childhood problems than most doctors except pediatricians.

- **MEDEXs** and **physician's assistants** have graduated from a one- or two-year training course, enabling them to work alongside doctors in the management of common problems.

The most important factor in selection of a medical care provider is your confidence in the individual. Confidence develops with time and experience; be careful about first impressions. Initially, you may do best to ask friends for their opinion of local providers.

MARKS OF GOOD CARE

Look for the following attributes; they are hallmarks of good medical and pediatric care.

- Does the practitioner **listen** to you? He or she must perceive the same problems that you do.

- Does the practitioner encourage **questions**? Are your questions taken seriously?

- Do you receive satisfactory **answers** to your questions? The quick phrase "She'll grow out of it" may not always be enough explanation.

- Does the practitioner take an appropriate **medical history**? Be wary of practitioners who listen for only a few seconds before deciding on a course of action; actions taken on a partial story are often in error. Injuries and simple problems may not require many questions by the provider, but a complicated illness requires more.

- Does the practitioner do a careful **physical examination** before ordering laboratory tests? Some practitioners will hear that a child has fallen off a bicycle and order a whole series of X-rays before performing an examination; with some severe injuries, sending the child for X-ray examination can be dangerous. A good medical history and physical examination will suggest diagnoses to most competent practitioners. Laboratory tests should be used only when necessary to confirm the diagnosis. The test-oriented practitioner will not only waste your money; dangerous side reactions or radiation exposure may result from many tests.

- Does the practitioner try to solve the **underlying problems**, or does he or she merely make sure that there is no biological illness? Being told that your child does not have meningitis, sinusitis, or a seizure disorder may ease anxiety; but if your child is having headaches three times a week and is missing school because of it, the practitioner's job is not yet done. Your doctor's job is to assist in returning your child to an optimal level of functioning, not merely to monitor the return to normalcy of laboratory tests.

- Is your practitioner concerned about the child's **development** and about safety matters? About preventing illness or just treating it?

- Is there a **backup person** available when the practitioner is out of town on vacation or at educational meetings?

- Does the practitioner take throat and other cultures or does he or she give antibiotics or other medications carelessly?

**SECOND
OPINIONS**

Feel free to seek additional opinions where your child's health is concerned. However, remember that you can always find a number of different opinions about any given problem. Much has yet to be learned about illness, and there are many different ways of interpreting what is already known. Two explanations for the same phenomenon may sound totally different to you. We recommend a "second opinion" strongly if your purpose is to ensure the correctness of a serious decision. On the other hand, if you are merely trying to find agreement with your own opinion about how to manage a problem, you are playing a dangerous game. There may well be some practitioner who will agree with your interpretation of a given problem. But the real question is which solution is best for the child!

Finally, some advice from a mother:

The doctor is such a mixed blessing. You have this kid whom you know so intimately ... you know him as nobody else does. And then there is this guy who has all this medical knowledge who knows things about your kid that you don't know, and there can be a subtle something in this relationship between the mother and the physician that can undermine altogether her confidence in herself and her mother role ... and at the same time the support can be terrific. Someone else is there especially in the crisis times ... someone who takes part in the decision making. Pick the doctor with care and don't accept an autocrat.

Child Health-Supervision Visits

Child health-supervision visits have also been called well-baby or well-child checks. They represent opportunities for you to assure that all aspects of your child's health are going well. Consequently, biological, developmental, psychological, and social information will be exchanged between you and your doctor.

Initially, there will be considerable emphasis on detecting physical and biological abnormalities through frequent examinations and screening tests (such as for thyroid disease), as well as preventing disease with immunization. Your doctor will also be interested in you and your infant's adaptation.

As your child grows, there will be continued interest in physical problems; older children will be checked for vision and hearing and adolescents for sexual development. However, it is likely that more time will be devoted to guidance about injury prevention, parenting issues (such as discipline), and your own concerns about behavior, day care, schooling, sex education, and the like. The time devoted to these issues is just as important as the time devoted to looking for physical abnormalities during the first few visits. During your initial visit with your doctor, developing rapport and a relationship are of paramount importance.

The first visit generally occurs within the first two weeks after birth. By now, parents have come to know some of the unique characteristics of their baby and often have many questions. Besides this initial visit, at least three more visits in the first year and two in the second will be required. Immunizations will be given during these visits. There is nothing magical about well-baby visits. The purpose of the visits is to help you and your child, so schedule additional visits periodically when and if you have concerns.

CHILD HEALTH-SUPERVISION EXAMINATIONS

Well-baby visits provide an opportunity for parents to question the practitioner about *any* concerns including feeding, safety, learning, and so forth. Many parents have their first contact with their child's doctor before the baby is born. This visit is an excellent opportunity to discuss early concerns. The visits also allow a check on the growth, development, and behavior of young children.

MEASUREMENTS

Besides a check on development, the most important parts of the examination are the simple measurements of the infant's height, weight, and head circumference. Head measurements are important because there are several correctable problems in development of the skull and of the brain that may be detected during the first few months of life. One of these is an early fusion of the bones of the skull preventing further growth of the skull, known as **craniosynostosis**. Another is obstruction of the fluid system bathing the brain, known as **hydrocephalus**. Both of these problems are detectable by careful head-circumference measurements. Periodic measurements of height and weight should be carried out throughout childhood. These measurements are useful in detecting nutritional problems and in detecting metabolic or glandular growth disturbances.

Hips

Examination of the hips is especially important in infants. Detection of **dislocated hips** is difficult at birth, and this examination should be repeated after a month or so. Congenitally dislocated hips are very easily treated in the first few months merely by the use of extra-thick diapers. Failure to detect this condition until later may require surgery or casting.

Feet

The feet should also be examined carefully in newborn infants and during the first few months of life. Foot abnormalities such as **metatarsus adductus,** when detected early, can be corrected simply with a few weeks of casting. Metatarsus adductus causes an exaggerated curving of the outside of the child's foot. Children often have a certain degree of toeing-in, and this is normal until the age of two. Toeing-in may also result from the twisting of one of the bones in the lower leg or rotation of the thigh bone at the hip joint. Careful practitioners will examine the child's feet and legs and assure you of their proper development.

Spine

By the time your child reaches school age, your doctor should be checking for curvature of your child's back (**scoliosis**) by having the child perform a toe-touching maneuver.

Eyes

Periodic examination of your child's eyes is also necessary. Infants are capable of seeing at birth, and their eyes can follow you around the room. Infants begin focusing clearly at about six weeks of age, and at this time the wandering of their eyes starts diminishing. A wandering eye, also known as **strabismus** or a weak eye, is not normal after six months. You can check for strabismus by having the child focus on any bright object and alternately covering and uncovering one of the child's eyes. When the eye is uncovered, the child's eye should not move; any movement is a sign of a weakened eye muscle and should be checked out. Untreated strabismus can result in blindness of the weak eye. Three and four year olds should be tested during doctor visits.

Ears

Hearing disturbances can also interfere with the child's development. Newborns will startle after hearing a sound, and later turn to locate the sound. Although elaborate methods have been devised for testing the hearing of infants, the best way to assess hearing ability is to assess the ability to talk clearly. You don't learn to talk if you can't hear. Thus, more precise hearing can be most readily tested after 12 months of age. Parents are usually the first to suspect hearing problems. If you are concerned, be sure to have a satisfactory evaluation done.

Teeth

While adequate dental care requires regular visits to a dentist, as well as routine flossing and brushing, good doctors will also check the development of your child's teeth.

Blood Pressure

Blood pressure measurements are not hard to do in older children and should be done at least once.

Heart

A word about **heart murmurs:** The casual statement by a doctor that a child has a heart murmur is enough to strike terror into the heart of many parents; some perspective is needed. Certain heart murmurs are *normal* in children! A murmur is just the sound of blood rushing through the heart. In fact, during a fever, during excitement, or after exercise, a heart murmur can be heard in almost any child; this does *not* mean heart disease. Such murmurs occur in the part of the heart cycle known as systole, and are frequently called "innocent" murmurs, "benign" murmurs, or "functional" murmurs. They will usually disappear as the child gets older, although many adults have these murmurs as well. Don't worry about them. If your doctor hears a murmur in the diastolic part of the cardiac cycle, or a particularly loud murmur, or if the child is growing poorly or is blue (cyanotic) or short of breath, then further evaluation will be required and your doctor will explain this situation to you. If you are in doubt, ask the doctor if it is an innocent murmur. We have seen serious problems in family interactions develop from a misunderstanding of the statement that a child has a murmur.

Genitals

Genital examinations are also important. Boys should be checked for undescended testicles and foreskin problems. Girls should be checked for problems with the external genitals. We feel that regular external examination of girls will make the first internal (pelvic) exam seem somewhat less frightening.

There are many other aspects of well-child examinations; almost anything can be noted for the first time at such an examination. Most important problems, however, are noted at home first.

Your Doctor, Your Child's Doctor

"Children should be seen and not heard" is a familiar expression that few would accept as a policy for dealing with children. Unfortunately, too many children are treated in precisely this fashion in the doctor's office. Doctors may seem too busy to talk with children. Yet most doctors caring for children, especially pediatricians, chose their professional specialty because they genuinely like children.

Often, doctors may not be able to initiate a dialogue because of a child's anxiety at being at the doctor's office. The fear of shots is universal among children. We feel strongly that if children are to develop competence about dealing with their health, illness, and the medical profession, they can profit from the experience of communicating during an office visit. Once children are school-age, and assuming, of course, they are not too ill, they should begin expressing their own concerns to doctors, as well as participating in discussions of any plans that directly involve them.

You can help. Answer as best you can any questions your child has before a visit to the doctor. Encourage the child to speak up and ask questions. You may need to introduce an opportunity for your child to ask the doctor a burning question. Children often do not respond to a doctor's questions because (1) the panic of the moment has made them forget their age, their sibling's name, or whatever information the doctor is requesting and (2) they do not understand the question. Again, you can be helpful as an initial intermediary in assisting your child in achieving these important communication skills.

During the visit, your child can participate in the examination process as well as in decisions about follow-up care. Discuss, for example, who will be responsible for remembering medicine or who will decide about when to return to school. Ultimately, you must decide when your child should begin to have time in confidence with the doctor and opportunities to telephone directly. We recommend gradually transferring responsibility so that by the middle-school years your child spends some private time with the doctor, and by adolescence the vast majority of time is between your doctor and child.

The First Pelvic Examination

While hearing and vision are the most important parts of routine childhood examinations, the pelvic examination in girls arouses much concern because of its social and sexual connotations. When should the first pelvic examination be performed? How often are such examinations necessary? Briefly, here are some of the considerations to be taken in reaching this decision.

Pelvic examinations are performed in adult women to detect problems, such as cancer, before they cause symptoms. These problems are so rare in childhood and adolescence that routine pelvic examinations are seldom necessary. There are two exceptions:

- If the child's mother received a drug called **diethylstilbestrol** during pregnancy: This drug was used fairly widely until 1971 to help prevent miscarriages and by some physicians after that date. Female children of such pregnancies have occasionally developed cancer of the vagina as early as age eight or nine. The present standard medical recommendation is for yearly pelvic examinations in these few children.

- Adolescent girls who have regular **sexual activity** at an early age should also begin regular pelvic examinations. Moderate–to–heavy sexual activity, especially with multiple partners, has been shown to increase the possibility of sexually transmitted diseases (STDs) and possibly cancer of the cervix (mouth of the womb).

We feel that the routine of a mature woman should be assumed when regular sexual activity begins. The opportunity to express fears and worries about sex is at least as important as the possibility of detecting disease. We suggest Pap smears every few years until age 25, and more regularly thereafter. If a Pap smear is totally normal, a five-year interval is not unreasonable. If minor abnormalities are noted on a Pap smear, do not exceed two-year intervals. And, if the Pap smear is suspicious of early cancer, follow the recommendations of your doctor closely. Pap smears do *not* detect STDs; a special culture is needed.

If a child requires a pelvic examination prior to adolescence, the procedure should be done by the family doctor or pediatrician with the mother present. Returning to the family doctor eliminates the necessity of an introduction to a strange doctor and a strange office. Only occasionally will a gynecologist be necessary.

During adolescence, the decision about which doctor to use is more difficult. Some adolescents view their pediatrician as a "baby doctor" and feel that they should go to a gynecologist, internist, or family doctor. Of course, many pediatricians spend a great deal of time practicing adolescent gynecology; but the choice should always be left to the young woman.

Whoever the doctor is, confidentiality is extremely important. The mother should *not* be present during the examination or during discussions between the doctor and the adolescent, unless requested by the adolescent. Do not demand that either the doctor or your daughter tell you exactly what happened; these are adult problems and conversations and must be treated as such.

You can help the most by explaining the procedure *before* you get to the doctor's office. Discuss the reasons for the examination and the purpose of each one of its steps. Make it clear that there may be some discomfort, but that there will be no cutting or severe pain. If the procedures are not understood, then fantasies and fears of damage to sexual organs may be aroused. Not only do these make the examination more difficult but they may also result in psychological harm.

The pelvic examination is usually performed by a doctor with a nurse in attendance. Amenities are important. The metal "stirrups" on which the heels rest can be wrapped, and the speculum can be warmed. A sheet will usually be placed as a drape over the knees so that the patient feels less exposed.

The full examination consists of inspection of the outside genital parts, palpation (feeling) of the internal organs, inspection of the inside of the vagina, and the taking of a Pap smear, not always in that order. There should be no douche for the preceding 24 hours because it may make diagnosis of some conditions impossible. Inspection of the outside parts includes the genital lips, the clitoris, and the anal opening. Rashes, sores, small growths, or other problems may be identified.

Palpation of internal organs includes the vagina, the mouth of the womb (cervix), the womb (uterus) itself, and the ovaries. Usually, this is performed with two gloved fingers inserted into the vagina. It can be uncomfortable, but it is not painful unless a serious infection is present. Much of the same information can be gained by palpating with a single gloved finger in the rectum, and this may be done in virginal patients. Or a "bi-manual" examination may be performed, with one finger in the rectum and one in the vagina—uncomfortable, and somewhat undignified, but not painful.

Internal inspection requires a light and a speculum. The speculum is metal or plastic, and gently spreads the walls of the vagina so that the inside may be seen. The speculum is not a clamp and will not pinch or hurt. It will be lubricated with water or a Vaseline-type lubricant so that insertion is easier. Speculums come in several sizes, and smaller ones will be used with young girls or virgins. The procedure should always be performed gently, with all actions explained in advance. The patient can assist by relaxing as much as possible; taking slow deep breaths or thinking of an enjoyable leisure activity may help.

The Pap smear is performed through the speculum, with a stick similar to a narrow tongue blade. It does not hurt. Some of the secretions are collected on the tip of the stick and examined under a microscope by a pathologist, who classifies the cells seen as normal or malignant. Sometimes cultures will be taken as well. This is a very similar painless procedure except that a cotton swab is used instead of a stick and the secretions are cultured so that any germs present may be identified.

In some parts of the country, women are performing their own pelvic examinations in self-help clinics. Medically, this practice seems to work out well, and it is a matter of personal preference.

The manner in which the examination is performed will help you judge the quality of care that you are receiving. The examiner who is abrupt, rough, or doesn't explain is insensitive to a fault. Pelvic examinations are medically essential in many situations, but the examiner who is not sensitive to the complex psychological and human dignity considerations that are involved cannot be giving good medical care.

Screening Laboratory Tests for Children

Eye tests and hearing tests are the most important screening examinations for your children. Some of the more common laboratory examinations are discussed below. Except for tuberculosis screening, we do not feel that these tests need to be done routinely for all children; they are needed only for particular children.

TUBERCULOSIS

Tuberculosis is still common in some parts of the country and is on the increase, particularly in low–income, inner–city areas. Nevertheless, it is no longer considered necessary to regularly screen all children for the disease. Instead, infants and children should be tested if they are known to have been exposed to tuberculosis or to people at high risk for tuberculosis (such as people with AIDS). We recommend the skin test known as the Mantoux or PPD test; it is more accurate than the Tine test.

ANEMIA

One way to screen a child for the most common form of anemia, iron-deficiency anemia, is to ask what the child is eating. Children fed a nutritionally balanced diet are almost never iron-deficient. Children who have experienced a number of infections may become iron-deficient; although in the presence of a diet with adequate iron, this condition will spontaneously resolve after infection. We see iron-deficiency anemia most commonly in children about one year old whose diet consists mainly of milk. Milk is low in iron; by one year of age, the child should be eating lots of different things. Detection of most other types of anemia at an early age is usually suggested by a family history of spherocytosis, sickle-cell disease, and others. **Hemoglobin** is a measure of the amount of protein that carries oxygen in red blood cells. **Hematocrit** is the percentage of your blood that is composed of red blood cells. Some doctors routinely screen all infants, while many selectively screen their patients.

SICKLE-CELL

Sickle-cell anemia is a severe genetic blood disease occurring almost entirely in blacks. For a child to have **sickle-cell** (anemia) **disease**, a sickle gene must have come from each parent; two genes are required. Someone with a single sickle gene is a "carrier" and is referred to as having "sickle trait"; he or she does *not* have anemia. The two-gene disease causes anemia, infections, and other problems; the one-gene trait does not cause any problems and appears to protect the individual from malaria.

We believe that sickle-cell screening should be available to adults planning pregnancy who would like to know whether or not they are carriers of a single sickle-cell (trait) gene. Sickle-cell disease causes problems in children, and it would be most unusual for an adult to have sickle-cell disease without knowing it. For parents who both have sickle trait, fetal screening is available. Screening of newborns for sickle-cell disease is also available and routine in some states. Routine screening of children for sickle trait is *not* appropriate, because it is not a medical problem. However, all parents with known sickle-cell trait (carriers) should have their infants screened because this may result in early detection and treatment of children with sickle-cell disease; we repeat that this disease can occur only if *both* parents have sickle-cell trait. Unnecessary concerns raised by detection of the carrier (trait) individual in mass-screening programs have outweighed any benefits of these programs because there is no need to treat these individuals. Any screening program must have an associated educational component to ensure that anything learned from the program is correctly interpreted by the individual.

URINE

Controversy currently surrounds the question of routine urine examinations to detect bacterial infections that are not causing symptoms, especially in girls. Most studies suggest that treatment of these asymptomatic problems is of no benefit. In general, treatment of "diseases" that are not causing trouble often does more harm than good. For children who have had urinary tract infections in the past, a periodic analysis of the urine may well be in order. Use of the urine for other screening (for example, protein or sugar) is also unwarranted.

DIABETES

Juvenile diabetes mellitus is a different disease from adult-onset diabetes mellitus and cannot be detected by screening examinations. The first episode will bring the child to the doctor with increased urinary frequency, excessive thirst, nausea and vomiting, shortness of breath, or coma; prior to this time the screening tests are negative.

CHOLESTEROL

The blood cholesterol level and heart disease are linked, but an elevated cholesterol is only one of many factors that contribute to heart problems. Children should not be routinely screened for cholesterol. Childhood cholesterol levels correlate poorly with adult levels. Many children have been tyrannized by a value that is a poor predictor of an even at least 40 years away.

If one of the parents has had a heart attack before age 45, a cholesterol screen and a comprehensive approach to the family will be required. This will include discussions about exercise, cholesterol, diet, stress, and anxieties. We believe that attention to these factors, including a diet moderately low in saturated fats, is indicated for everyone, regardless of the cholesterol level. (See further discussion in Chapter 9.)

LEAD

Lead poisoning in children is a problem that has been improving slowly in the United States. The decline in the use of lead-based gasoline and the limitation of lead in paint sold after 1978 has been helpful. However, many children still have blood levels high enough to interfere with their intellectual growth and functioning. As recently as 1984, 17% of school children had elevated levels. Recent studies suggest levels previously thought to be safe may be problematic. As a result of these studies, in late 1991 the Centers for Disease Control and Prevention recommended that all children should be screened. Unfortunately, this would involve considerable expense, discomfort, anxiety about borderline levels, and extensive follow-up for many children with marginal elevation. Consequently, many physicians are not routinely screening all children. We support the CDC recommendation for screening high-risk groups. We also believe physicians should be familiar with local surveillance data and intelligently advise you about the value of screening for your child. The CDC priority groups for screening include children 6 months to 72 months who:

- Live in housing built before 1960
- Live in housing with remodeling going on
- Have siblings or friends with lead poisoning
- Are exposed to lead because of household members' jobs (metal workers, plumbers, auto repair, pottery, etc.) or hobbies
- Have environmental exposure by living near lead smelters, battery plants, or industries releasing lead

Screening for the Cause of Frequent Illness

All children contract a large number of colds (upper-respiratory infections) in the process of growing up. We do not like to consider these episodes as sickness or illness. Many parents are surprised that we would call a three-day episode of a temperature of 100°F, runny nose, and slight cough anything but an illness. The point that we are trying to make is that respiratory infections are a normal part of growing up and actually provide benefits to the child. We regard common respiratory infections in children as a type of immunization. Like other immunizations, viral infections may have side effects, such as fever, runny nose, and cough. And, like most immunizations, many viruses produce no symptoms at all and yet afford lifelong protection against the virus. Children usually fare better with a given viral respiratory illness than adults do.

In careful studies conducted at several university medical centers, infants and young children have been found to develop between six and nine viral infections per year. The older the child, the fewer the number of infections, but even first graders average about six viral episodes per year. We are not usually disturbed by a child who has nine viral infections a year unless these interfere with the child's daily routine or are causing slowing of growth or maturation. Two bouts of pneumonia or two bouts of skin abscesses are more significant medically than nine colds are. Certain types of pneumonia or other unusual findings may occur in children with AIDS, but frequent colds should not raise such anxieties. In a child who is having frequent illnesses, the question is not so much how frequent, but how serious. (For a further explanation, see the decision chart and discussion in **Frequent Illnesses**, page 339.)

Don't be afraid to send your child to school with the sniffles if he or she feels like going. For most illnesses, the period when the child is most contagious is the period *before* he or she gets sick. It is regrettable that children are often sent home from school at the first sign of a cold; the cold can serve to immunize the other children against that virus. Here's a good general rule:

- Fever above 101°F, home
- No fever, school

Hospitalization

Each year, 1 out of 18 children will leave the familiar surroundings of his or her home, the companionship of friends, and the support and care of parents to enter a frightening social system and institution—the hospital. In this foreign environment, the child is not only expected to recover from an illness but to demonstrate good citizenship as well. These expectations are, to say the least, a tall order for a sick six-year-old.

PSYCHOLOGICAL EFFECTS

Parents worry about the psychological consequences of a hospitalization as well as their child's prompt and full recovery from the problem necessitating the stay. Fortunately, most hospitalizations are brief. Long-term psychological problems have been found primarily in children who have experienced numerous prolonged hospital stays at an early age. During hospitalization, it is not uncommon for children to experience considerable anxiety about separation from parents, mutilation, and especially death. Being put to sleep can easily be construed as permanent. Some behaviors are common, such as increased crying and activity levels. Occasionally, a child will become depressed or have other psychological disturbances in the hospital.

Immediately after discharge it is quite common to see a child regress to immature behaviors such as thumb sucking, bed-wetting, baby talk, clinging, crying, increased dependency, and a decreased attention span. Sleep problems occur frequently, and some children become aggressive. Anxiety may be manifested through refusal to leave home or excessive fear of doctors or nurses.

In contrast, some children may have positive psychological responses to a hospitalization. This is particularly true of 8- to 11-year-olds who often gain self-esteem by making it through a procedure and often show accelerated learning.

It is clear that the way a child experiences a hospital stay and what a child learns from this experience can be influenced by both medical professionals and parents. Your child may participate in a variety of hospital-sponsored preparation procedures including puppet therapy, coloring books, or films designed to help children cope with stressful procedures. Most important, however, is the comfort, reassurance, and information you give your child.

COMMON CONCERNS

Step one is to make sure that your concerns have been addressed by your doctor and the staff. If you are overly anxious, your child will pick up on this and become anxious. The next step includes addressing the following concerns of children in various age groups.

- *Ages 0 to 3.* Allow child to handle, where possible, new and strange equipment. Bring favorite cuddlies or dolls. Assure child you will return.

- *Ages 4 to 7.* Bring a favorite toy. Encourage rehearsal of responses to procedure. Help child distinguish attainable goals. For instance, blood tests cannot be avoided, but can be made easier if children help by holding still or by choosing which arm they prefer. Assure child about separation and mutilation concerns. Reassure child the hospitalization is not a result of evil thoughts or a form of punishment. Explore any misconceptions. Some common ones include: All diseases are contagious; procedures that hurt (or don't hurt) aren't therapeutic; blood is a fixed quantity that can be permanently depleted.

- *Ages 8 to 12.* Explore concerns about how hospitalization will affect the child's status with peers. At this age, preventive measures may not be grasped. The relationship between mood and disease may not be understood.

- *Ages 13 to 18.* Explore the adolescent's concerns about his or her body image and independence. Adolescents have a tendency to deny or minimize the severity of illness. They also may desire more information than the doctor provides about the etiology and future consequences of their problem. In addition, many have real concerns about the financial impact of a hospitalization on their family.

Remember too that almost all children are interested in the immediate events that will affect them. Finding out when they must leave for X-ray can be quite important. Here again, you can be an important intermediary between your child and your doctor and increase the likelihood your child may learn from this experience.

Surgery

Any recommendation for surgery for your child should be thoroughly discussed with your family doctor. You should understand why the operation is being done and the possible implications and complications. Many procedures that required hospitalization several years ago can now be performed as day surgery in which a child is admitted to the hospital in the morning and discharged the same day (for example, hernia repair). In this section, we discuss some of the more common operations performed on children.

After circumcisions, tonsillectomies, and adenoidectomies are together the most frequently performed operations in the United States today. Because they are operations on separate organs and because they are done for different reasons, they will be discussed separately.

TONSILLECTOMY

The tonsils are located on both sides of the back of the throat and can be seen quite easily when your child's mouth is open. Tonsils consist of lymphoid tissue, which is useful in fighting infections. Because children between the ages of four and seven develop so many colds (upper-respiratory infections), the tonsils are largest in this age group. Large tonsils are natural; they are enlarged because they are busy fighting infection. Occasionally, the tonsils themselves may become infected but only when there is an infection of the entire throat.

Removing a child's tonsils does *not* reduce a child's chances of developing a sore throat. Nor does it reduce a child's chances of developing a cold. Nor does it reduce a child's chances of developing an ear infection. Then why are so many tonsillectomies done in this country? This is indeed a difficult question to answer because every carefully performed study in the past 45 years has demonstrated the futility of removing tonsils to eliminate sore throats or any of the consequences of sore throats. And children die unnecessarily each year as a result of tonsillectomies.

There are, of course, some good reasons for tonsillectomy. If the tonsils are so large that they are interfering with the child's breathing, the tonsils should be removed; however, this condition is exceedingly rare. When such airway blockage occurs, it is the heart and lungs that are put under stress, and the problem is detected by evaluation of the heart and lungs. Merely looking at the tonsils will not reveal obstruction to breathing; many normal tonsils nearly touch in the mid-line of the throat. Although we have never seen it occur, it is conceivable that very large tonsils may interfere with eating or swallowing; this may also require their removal. In addition, if a serious abscess occurs in one of the tonsils, it is likely that this abscess will recur; this may be prevented by a tonsillectomy.

ADENOIDECTOMY   Tonsillectomies have been traditionally accompanied by adenoidectomies. So long as a patient was going to be placed under anesthesia, the surgeon often decided to remove the adenoids as well. The adenoids, however, are located in a different region and have their own functions. Adenoids can create problems, and there are indications for removing adenoids. However, removal of adenoids when only a tonsillectomy is indicated will increase the risks of a bad result from the operation. In addition, removing adenoids can lead to ear problems in certain children.

The adenoids also contain lymphoid tissue. They cannot be seen by looking at the throat; they are in the back of the nose above the palate and near the opening of the eustachian tube that drains the middle-ear secretions into the nasal passages. Adenoids enlarge as they fight infection in the upper-respiratory tract. If they become too large, they can block the opening of the eustachian tube and increase the likelihood of an ear infection.

An adenoidectomy used to be considered for a child who had frequent bouts of ear infections. Today, ear tubes generally accomplish the same function, providing adequate air circulation in the middle ear. If the ear infections are caused by allergy, children generally do not profit from an adenoidectomy. Children who may benefit can be identified by X-ray studies, and all children who are being considered for an adenoidectomy, in our opinion, should have such studies done beforehand. Some children will actually be made worse by an adenoidectomy, and it is sometimes possible to identify these children before the operation.

Severe obstruction of breathing or interference with speech may also be indications for considering removal of the adenoids.

Again, if an adenoidectomy is indicated, there is no need to remove the tonsils during the operation.

EAR TUBES

Ear tubes (ventilation tubes) are small plastic tubes that are inserted through the child's eardrums. They are placed in children who have frequent bouts of earaches in order to provide adequate air flow in the middle-ear chamber. Aeration aids the healing process and prevents the middle-ear chamber from becoming sealed off; an ear infection generally does not develop unless the middle ear is blocked.

The placement of ear tubes is a very simple procedure usually done as day surgery after the summer swimming season is over. Indications for ear tubes include:

- Recurrent ear infections that have not responded to medical therapy, including prophylactic antibiotics
- Presence in the middle-ear chamber of material too thick to drain under normal conditions

Considerable controversy surrounds the indications for this procedure. Discuss the benefits, harms, costs, and alternatives thoroughly with your physician. The purpose is both to reduce infection and to ensure adequate hearing.

REPAIR OF UMBILICAL HERNIA

An umbilical hernia results from the failure of stomach muscles to grow together in the area surrounding the belly button. Umbilical hernias are very common in black children but do not pose any risk. Generally, the hernia disappears within the first few years of life, and most others slowly heal themselves by the time the child is of school age. Because these hernias are of no risk to the patient and because they usually heal by themselves, we do not advocate surgical repair except in rare cases where cosmetic improvement is psychologically important. In general, time is a better and safer healer than a scalpel.

UNDESCENDED TESTICLES

It is not uncommon to find that one or both of your boy's testicles are not in the scrotal sac. The testicles are often in the inguinal canal, located just above the scrotal sac. One of the most common reasons for the testicles to find their way up and *out* of the scrotal sac is the presence of an examiner's cold hands. Occasionally, the testicles may withdraw all the way into the abdominal cavity. (This is a feat that can also be accomplished voluntarily by some Japanese wrestlers.)

Testicles that are permanently located within the abdominal cavity should be brought down surgically. Doctors differ about their recommendations for the best time for operation, but the procedure should be done by five years of age.

It is of the utmost importance to determine whether the testicles can or cannot descend into the scrotal sac normally. The best way to decide this is to have your child sit with legs crossed (the Buddha or lotus position) and check for presence of the testicles within the scrotal sac in that position. We do not recommend hormonal treatment for bringing testicles into the scrotal sac; it usually doesn't help and may have side effects.

HYDROCELE OR HERNIA

Hydroceles and inguinal hernias both cause swelling of the scrotal sac in boys. It is difficult to tell the two apart, and a doctor's visit will be necessary.

A **hydrocele** is a fluid sac within the scrotal sac and is often present in newborn boys. The fluid does not create any problem for the child and will be re-absorbed in time. Some doctors prefer to remove the fluid from the scrotal sac with a needle. Nature will work just as well, though not as quickly.

An **inguinal hernia** will also cause a swelling in the scrotal sac. Hernias generally first appear *after* the newborn period and often following a vigorous bout of crying. Severe problems can result from inguinal hernias because a hernia is a small portion of the child's bowel that has found its way through the canal and into the scrotal sac. Unlike inguinal hernias in adults, which doctors are able to feel on examination, hernias in children can only be evaluated when the scrotal sac is swollen, and the child should be seen by a doctor at a time when the sac is swollen.

APPENDECTOMY

Appendicitis is discussed in **Abdominal Pain**, page 511.

CIRCUMCISION

Circumcision is discussed in Chapter 2, pages 41–42.

TRANSFUSIONS

Transfusions are seldom required except for specific severe diseases including severe jaundice or anemia of the newborn, destructive (hemolytic) anemia in infants and children, and to replace blood lost in trauma or surgery. While blood used in transfusions is evaluated to make sure it is compatible with your child's blood and free of severe virus infections, there has been concern over the risk of acquiring AIDS (Acquired Immune Deficiency Syndrome). Since the spring of 1985, all blood has been screened for antibodies to HIV, which cause AIDS. Due to the nature of the test, it was estimated in June 1989 that approximately 1 in 40,000 units of transfused blood has HIV that has gone undetected. Transfusion with blood from a family donor whose medical history is known can be reassuring but has not been shown to reduce this low risk even further.

CHAPTER 9

Staying Healthy

*T*his chapter presents, as simply and clearly as possible, the groundwork for a healthy life. Some of this information you already know, and much of it you have heard before. We realize that most of our readers will rely heavily on Part IV of this book, in order to respond promptly and appropriately to new medical problems and to save time and money. This chapter is less dramatic. But overall, the measures recommended here are the most important health investment that you can make.

We believe that the best approach to dealing with serious problems is to prevent them. Only 3% of the money we spend on health care is for prevention. We believe that greater national emphasis on disease prevention and health promotion, as well as on changes in personal behavior, will have enormous health benefits. This chapter discusses four major problem areas in which prevention is essential:

- Injuries
- Immunizations
- Dental care
- The diseases of adult life

Injuries

Injuries are the number-one threat to your child's health. More children are seriously injured or die as a result of accidents than the combined attacks of cancer, infectious diseases, and birth defects. In adolescents, injuries and deliberate violence (homicide and suicide) account for more than 70% of deaths.

HOUSEHOLD

Young children spend most of their time at home; this provides them with the greatest opportunity for getting into mischief. Playpens may be somewhat restrictive, but they can be an island of guaranteed safety when you must spend time on the phone or at the front door.

Remember that as children grow, the opportunities for accidents increase. You must keep one step ahead of your child at all times. Before you know it, crawlers become climbers, and they will be able to get up to that previously unreachable cabinet.

Older children should be taught safety responsibility. They should know how and when to dial 911 for emergency assistance. For burns they should know to immediately plunge the burn into cold water. If on fire, they should not run (which fans and increases the flames) but stop, drop, and roll on the floor. Teach them a plan for exiting the house during a fire.

Here is a checklist to help you evaluate the safety of your home. You should review this important list every year.

Throughout the House

- Keep electric cords in good repair. Put away extension cords when they are no longer needed.
- Cover electric sockets not in use with plastic safety shields.
- Build a guardrail around space heaters, Franklin stoves, and other hot items. Keep fireplaces adequately screened, and keep any combustible liquids used in fireplaces out of reach.
- Beware of poisonous plants, such as poinsettia leaves, daffodil bulbs, and castor beans.
- Keep long cords from venetian blinds, mobiles, and so on well out of a small child's reach.
- Lock windows above the ground floor, or protect them with gates.

- Block stairways with non-folding gates as soon as a child becomes mobile.
- Block tables with sharp corners, and clear them of objects that could fall and hurt small children.
- Do not put your child in a baby walker, because his or her enhanced mobility can lead to numerous dangers. Each year over 27,000 emergency room visits are due to baby walkers.
- Always strap babies into infant carriers, and place the carriers on solid, level surfaces. Never leave children unattended in a carrier.
- Check all fire escapes, balconies, and terraces for danger of falling.
- Install a smoke detector in the house, and keep its batteries charged.

Kitchen and Dining Room

- Lock the following materials out of reach: oven cleaners, drain cleaners, ant and rat poisons, insect sprays, furniture polish, bleach, lye, and other poisons.
- Turn pot handles away from the edges of the stove.
- Remove electric cords on appliances such as coffee makers from the sockets immediately after use.
- Keep electric irons away from children.
- Place a fire extinguisher near the stove, out of a child's reach.
- Keep small children away from dangling tablecloths when a hot meal is on the table.

Bathroom

- Dispose of razor blades safely.
- Set the hot water heater to around 120°F.
- Lock all medicines out of reach, including aspirin, iron tablets, and sedatives.
- Flush prescription drugs left over from previous illnesses down the toilet.
- Lock all bathroom cleansers, especially drain cleaners, out of reach.
- Keep the toilet lid down so infants cannot fall into the water.
- Never leave a child unsupervised in a bathtub.

Nursery
- *Never* leave an infant unattended on a changing table or in a crib.
- Don't leave an infant too young to turn over on his or her stomach.
- Do not put anything on a cord, such as a pacifier, around a baby's neck.
- Do not string toys across a crib, because the cord is a strangling hazard.
- Periodically check and tighten all connections on a crib. Cribs should have no more than 2½ inches between slats and between mattress and frame, so a baby's head cannot become wedged in those spaces. Remove tall knobs from corner posts, because they can entangle an infant climbing out of the crib.
- Take rattles, squeeze toys, and other small objects out of the crib when the baby sleeps.
- Do not use infant cushions, which have been banned by the Consumer Products Safety Commission; an infant could suffocate while lying face down on one of them.
- Install spring-loaded lid-support devices on toy chests so that the top cannot fall on a child's head.
- Do not leave infants in mesh-sided playpens or cribs with the sides down; the baby may roll into the narrow space between the mattress and the mesh.

Parents' Bedroom
- Keep cosmetics and perfumes out of reach.
- Keep mothballs and shoe polish out of reach.
- Dispose of all plastic bags (dry-cleaning and trash bags especially) so children have no chance of suffocating inside them.
- Never leave an infant unattended on a flat bed.

Children's Room
- Check toys for sharp edges.
- Check toys for small parts that can be easily removed and swallowed, including marbles.
- Discard all broken toys and pieces of balloons.
- Check all connections on a bunk bed. There should be no spaces more than 3½ inches between mattress and frame.
- Lock model glue away from toddlers.

Workroom

- Keep small children away from large buckets, including diaper pails.
- Store guns and bullets separately, and lock them away. Handguns are a frequent cause of accidental death in children. Their potential benefit is far outweighed by their real risk. Consider getting rid of any handgun kept in or around the house.
- Lock electrical tools and tools with sharp edges out of reach.
- Lock paints, solvents, automobile fluids (especially antifreeze, which tastes sweet), pesticides, fertilizers, and snail bait out of reach. Never store them in soda bottles or other tempting food containers.

Yard and Garage

- Never leave a child unattended in a stroller.
- Invest in a helmet for your child. Helmets reduce serious injuries in bicycling, skateboarding, skiing, surfing, and horseback riding.
- Keep children away from rotary lawn mowers, and make sure those mowers have protective shields.
- Remove the doors from old refrigerators or freezers in which children could play, and dispose of them properly.
- Build fences around swimming pools, and watch children playing in them.

Although the child spends the most time within the home, there can be an increased risk of accidents away from home. The surroundings are unfamiliar, and families without children or with older children often have hazards not present in homes with young children. This is one reason why we recommend carrying ipecac (page 234) in your car's glove compartment and leaving an additional supply at grandma's or regular baby-sitters' homes.

While it is important to make your child's environment as safe as possible, it is also important for the child to gain experience with his or her environment and to learn to make personal safety decisions. This requires a testing of the environment and some painful experiences. The adult world is not totally safe, and there are perils in both underprotection and overprotection. Common sense is the best guide.

AUTOMOBILES

Automobile accidents are the number-one killers of children. There is absolutely nothing more important for your child's health than protection in the automobile. Drive prudently and use seat belts or safety restraints. (Air bags supplement seat belts; they do not replace them.) The average infant or child seat costs considerably less than the average doctor's office visit, and can be used for years for many children and then passed on to a friend or relative. It is the number-one health bargain of all time.

Car Seats

In selecting an infant or child seat, be sure that you understand how it must be installed. Many require anchoring that will involve placing a bolt through the floor of the car. There are currently a number of acceptable seats available. Models change frequently, but be sure that the one you purchase has a label certifying it has been "dynamically crash tested." Check that the model fits in your car and that it is properly installed according to instructions. If a tether strap and anchor bolt are required, the seat will not protect your child without them. Also, seat belts that lock only on impact are inadequate for holding seats in place around sharp turns or minor bumps. Special clips are available for these seat belts to allow them to accommodate infant seats. Finally, check prices. They vary considerably. Many communities have rental programs, but we consider a child seat a vital purchase because your child will need one for four to five years.

There are three basic seat types:

- Infant car seats must face backward, are usually installed in the front seat, and are good up to ten months or about 20 pounds. These seats should be used with caution in cars having front passenger air bags. Injuries can occur if the air bag opens. To avoid this, rear-facing infant seats should be used in the back seat, or the front seat should be moved as far back as possible.

- We recommend the seats that convert from infant to child seats. Initially they are installed in the reclining position facing backward; as the child grows, they face forward. They are good from birth to age 5 (or 40 pounds). There are a number of models to choose from. Don't face the seat forward until the child is about 18 pounds. A padded chest protector is another benefit of most of these seats.

- Child car seats are for children over 15 pounds and face front. Seats that require anchoring are ineffective if installed improperly. It is also important to consider convenience. Some parents have found highly recommended seats so complicated to use that they were forced to buy a different seat. Effective infant seats are now available that do not require anchoring.

It is important to purchase your infant seat *before* you bring your baby home from the hospital. Make the first ride a safe ride. But remember to make it a comfortable ride for baby by bringing along towels or blankets to prop your child. Also remember that as long as your child weighs less than about 18 pounds, most infant carriers should be facing the rear of the car. Consult the specific suggestions about your model from the manufacturer.

Remember, even the best seat will not protect your child if it is not used. Be sure you understand proper usage before buying any restraining seat.

Other Measures

Harnesses, seat belts, and shoulder straps can be used for older children. Children over 40 pounds can use seat belts. Children under four-and-one-half feet tall should not use shoulder straps since they can cause neck injuries. Place the strap behind the child. The best way to teach a child to use a seat belt is to use one yourself; the child will quickly learn that when you enter a car, you automatically put on a seat belt.

An extra word of caution must be added about leaving small children in cars during shopping trips. Families who would never leave pills or solvents in reach of children at home often leave them unattended in cars with shopping bags full of these same dangerous products. Although safety caps on both pills and solvents have decreased the risk, children display remarkable ingenuity in removing these "childproof" caps. In addition, remember that with a car's windows rolled up it can take only *moments* before children can suffocate or die of heat exhaustion. Leaving a child alone is also an invitation for abduction. NEVER LEAVE SMALL CHILDREN ALONE IN A CAR!

Finally, parents must become advocates in their communities for full implementation and enforcement of laws designed to protect children, including enforcing speed limits, minimum drinking age, and license suspension for drunk driving.

BICYCLES

A bicycle, for a child, is a friend, companion, horse, and source of transportation and exercise. Bicycles have become a means of transportation and exercise for many adults as well. It is important that a child learn the basic principles of bicycle safety. Here again, the principles that a child learns with a bicycle will provide lifelong benefits for general and automotive safety as well.

There are many excellent books about bicycle safety that can be purchased for children. However, parents' lessons are the ones that are best remembered, and books should be used in order to supplement, not substitute for, parents' teaching. You will want to teach your children the following:

- Always wear a **bicycle helmet**!
- How fast can the bicycle be ridden safely? How fast is too fast going down a hill? How quickly can the bicycle stop when the brakes are applied? Remember, bicycles traveling at faster speeds require a much greater distance in order to stop.
- Can the child interpret and obey traffic signs and signals?
- Is the bicycle ready to ride? Children should be taught to check the handlebars, the tires, and the brakes before beginning a ride.
- Do children ride on streets where it is safe to do so? If they ride on the sidewalk, are they careful of other people? Are they careful about riding near automobiles where doors may suddenly open? Are they careful when riding around small children and animals?

Children should not be allowed to ride their bicycles in the street until they are about eight years of age and can demonstrate good control of their bike and knowledge of the rules of the road. The exact age depends on your child's maturity and skill and local traffic patterns. Riding in the street around sundown adds additional hazards. Depending on the traffic flow in your neighborhood, street riding may need to be deferred or it may never be safe.

As parents, you will be familiar with the local hazards of your neighborhood. Teaching a child how to balance a bicycle is not enough. You must also teach the child to maneuver the bicycle safely. And you teach a child to stop at stop signs by doing so yourself.

Even with good safety preparation, bicycle accidents will occur. "Spider" bicycles with small front wheels and big handlebars are notorious for children flying over the handlebars and hitting their heads. We discourage the purchase of such bicycles. First bicycles should be simple and sturdy; the flashy accessories won't last long anyway. Many parents move the child to sophisticated multi-speed bicycles far too early. Only after the child has learned to care for a simple bike and is ready for a permanent adult-size bicycle do we think that this investment should be made.

Adult bicyclists who carry children on their bicycles should use a rear-mounted child seat with a safety strap and side panels. Wheel covers will prevent foot-in-spoke injuries. Again, always have your child wear a helmet. Purchase a helmet when you buy the bicycle. For less than the cost of a single office visit, you can protect your child from serious injury, and you may save your child's life.

WATER SAFETY

Children and water are a natural combination. The beach, swimming pool, bathtub, and lake offer more opportunities for fun than all the toys ever invented. However, water tragedies are all too common.

Young children must be watched carefully near water. It is easy for parents to fall asleep while lying on a beach or at poolside, and fatal consequences for children have followed even short lapses in supervision. Drownings also occur frequently in bathtubs. Use your common sense in deciding when your child is old enough to be left alone in the tub.

Waterproofing

More and more classes are being offered in infant swimming for children as young as six months of age. Although these classes do not actually teach infants how to swim, they do teach them to kick sufficiently to get to the surface. This will help if a child falls into the far side of the pool with the parent watching. The infant can reach the surface and kick long enough for the parent to reach the child. However, organized programs for children under age three are not recommended by the American Academy of Pediatrics. Excess water swallowing in this age group has led to seizures, and children are at an increased risk of infection. Also, a false and fatal sense of security may develop.

The concept of water-proofing your child is only a relative one. You *must* remain alert and observant. Also be sure to check out your child's swimming classes. Young infants with overzealous instructors have been documented to swallow so much water that seizures have resulted.

Swimming Lessons

It is important for all children to learn how to swim. The local YMCA, the local Red Cross, and many other organizations provide these services either free or at minimal cost. If you yourself are a competent swimmer, you will enjoy the time spent in teaching your child how to swim. Water safety is as important as bicycle or automotive safety. Children should be taught *never* to swim by themselves and never to go too far from the shore without at least one experienced adult swimmer. Children should *never* swim in irrigation canals or other fast-moving water. Finally, children should never dive head first except in designated pool areas.

First Aid for Drowning

We prefer to talk about prevention. Yet it is important to know what to do if you are first on the scene at a drowning. The techniques for resuscitation of a drowned child cannot be learned by reading a few pages in a book. The Red Cross, in all parts of the country, offers an eight-hour multi-media safety course that we recommend highly to all individuals interested in being able to deal with emergency situations themselves.

DOGS

Children and their parents are too often unaware of what constitutes good pet safety. It is natural for small children to be afraid of dogs. Even medium-sized dogs tower over toddlers and weigh as much as seven- or eight-year-olds. It is the parent's responsibility to teach a child how to act around dogs, even if the family does not own one, because all children come in frequent contact with dogs. Every year, thousands of children are bitten, sometimes severely, by dogs. Very few of these dog bites are by vicious dogs or dogs with rabies. Most often the bite comes from the friendly old dog next door, Phred, who hadn't done anything more energetic in the past few years than wag his tail while lying on his back. The friendliest dogs bite children. It is not because the dog is having a grumpy day. Almost invariably it is because the dog has been threatened by the child in some fashion.

Here are some hints to help prevent bites:

■ Explain to children that they should not squeal in loud, high squeaky voices around dogs. Dogs have sensitive ears to high tones and loud squeaky noises will often terrify dogs, who then react in order to protect themselves.

- Teach children not to make sudden moves around dogs. These may be interpreted as attacks and dogs may react to defend themselves. At other times, a dog will respond to a quick move in a playful fashion. Dogs, when playing with each other, can be seen to nip at each other in fun. (Parents and children do not consider dog nipping as much fun. This is a cultural difference between dogs and humans.)

- Children should not wave sticks or throw stones around dogs. Again, the dog may feel an attack is imminent or may try to retrieve the stick out of the hands of the child.

- Children should be cautious about running behind lying or sitting dogs. Most dogs have their tails stepped on too frequently and may react to protect them.

- Children should be taught how to approach a dog. The child should approach the dog slowly, allow the dog to sniff his or her hand for a few seconds while talking calmly, and then proceed to pet the dog gently.

Finally, let's protect the puppies too. Small children often love puppies, and because the child is bigger, most parents consider this a safe relationship. However, children of two, three, and even four often delight in swinging a puppy by the tail and throwing it. This type of puppy abuse may produce a very nervous or even vicious older dog. Remember also that because dogs grow more quickly than children do, the abused puppy may become a vicious adult dog in several short months, and the child will suffer.

In general, we feel that five should be considered a minimal age for a child to get his or her own puppy. Children should be old enough to demonstrate responsibility in keeping up with the routines of feeding, grooming, and cleaning up after their pets before being given the responsibility of caring for them. Parents remember the joys of their own childhood pet and tend to rush into getting pets for their own children too early. This can convert a joyous situation into a troublesome one.

Missing Children   BY MAUREEN SHANNON, M.S., F.N.P., C.N.M.

Unfortunately, worry about a child being abducted is a valid anxiety of parents. Newspaper headlines and magazine cover stories are indeed terrifying. They proclaim 1 million children are reported missing every year and 100,000 of these children are never heard from again. It is important to realize that only a very few children are victims of stranger abduction each year. Each year there are 4600 abductions by nonfamily members reported to police and 300 where the children were gone for a prolonged period or murdered. While this is disturbing, your child is at far greater risk of being injured by an automobile while walking to school. The overwhelming majority of missing children (more than 90%) either are kidnapped by parents in custody disputes or are runaway teenagers.

Runaways

Many children have threatened to run away in the past and leave home with special clothes, items, or friends that are important to them. When these items are missing, foul play is seldom at work. Runaways are a special concern to the medical care system with services now being established in many communities specifically to serve runaways.

Preventing Abduction

The success of law-enforcement agencies in recovering young children abducted by strangers is related to how quickly they are informed about a child's absence. It is also related to how organized and experienced these officers are when it comes to these types of emergencies. The more successful agencies have often developed a special bureau which has its own officers solely assigned to deal with missing children situations. There are also national organizations such as the "National Center for Missing and Exploited Children" (1–800–843–5678) that are organized to help locate missing children.

The most important response to child abduction is knowing methods to prevent it. While it is impossible for any parent to make sure his or her child is abduction-proof, there are things that you can do and teach your children. These may reduce slightly the chances of your having to confront this tragedy. The first step is to acknowledge that there definitely is a problem, a very serious problem, that is occurring nationally. It crosses all socio-economic groups and can happen to your child in your neighborhood. Educate your child in an age-appropriate manner about the possibility of this situation and set firm guidelines (rules) for the child regarding interaction with persons when you are not present: with whom, when, and where it is acceptable.

The following are some suggestions:

- Set up the strict rule that your child never speaks with strangers regardless of the situation. Give your child permission to be "impolite" by not responding to adults not known to them.

- Tell your child *never* to approach a car when a driver has requested directions.

- Although it may be difficult to keep track of an active toddler or young child, it is particularly important not to let him or her out of your sight in department stores and shopping malls. Allowing a child to get "lost in the crowd" may invite real danger. If, after a few minutes, you cannot locate your child, notify store authorities *immediately*. Prompt action may be lifesaving.

- Never leave a child of any age alone in a car. This is double jeopardy. Heat exhaustion and abduction are real risks.

- Teach your child that he or she has certain rights—not to be touched, not to be asked to keep secrets, not to answer strangers. Your child should be told to tell you of any adult asking the child to keep a secret. Warn your children that adults with friendly puppies or dogs are not safer than other strangers.

- Never allow your child to wear a sweatshirt or T-shirt or carry any item with his or her name imprinted. This is a common way for abductors to achieve familiarity.

- Warn of adults who claim they are police but are not in uniform or patrol cars. Even uniformed police seldom talk with children unless they are creating a nuisance or taking risks.

- If your child is picked up from school by neighbors in carpools, caution the child never to go with anyone who is not a regular driver. Individuals have been known to learn parents' names and can easily con young children into a ride. If your work situation is such that you must ask friends or neighbors to pick up your child, make sure that you establish some system with your child, such as a code word that must be provided by this neighbor before the child will travel with him or her. Generally, children who are older than seven years of age are capable of this sophisticated screening process. For younger children, it is best to inform them never to get in a car with anyone who is not a family member or neighbor that you have approved.

- Older children (those over eight) can use the buddy system to walk to and from school, church, or sporting events. Make sure your child does so with a group of children having at least one other child his or her age. If your child is under eight, insist that an older sibling or adult accompany the child to his or her destination. The same rule should apply to the buddy or group system that applies to your family's basic ground rules regarding your child's interaction with strangers.

- When setting up carpools with the parents of your children's friends, make sure you meet them and feel comfortable about their carpooling your children. If you are uncomfortable with a particular person, you should not use this carpool and check out your feelings with other parents. If your child must use public transportation, teach him or her to wait for the bus at bus stops used frequently by other people and/or bus stops that are in front of stores, schools, or near the home of a person whom you trust. In this way, if someone does bother your child, he or she can run into the store or school for refuge. Also advise your child not to stand on the curb right next to the street. A person could drive up in a car and pull the child into it without much trouble. The child should stand away from the curb to prevent this easy access to a possible abductor.

- Teach your children the police emergency number and have them use this number if they are ever being followed or harassed by someone and cannot reach you by phone. Most public phones do not require a coin in order to dial 911.

- Instruct your children always to call you or another trusted person if they need to stop anywhere on their journey home.

- Report any suspicious people you notice on your neighborhood school grounds or playgrounds to the police. It is far more reasonable to report such persons to the police than to assume they are harmless or that it is simply your imagination.

■ Have your children memorize, as soon as they are able, your home and business phone numbers and the phone number of a trusted friend. Always have a backup location to which your child can go if, for some reason, you are detained and cannot be home to meet him or her.

If your child is late returning home or meeting you somewhere, it is important that you notify the police (after checking with your network backup person). Be honest with the police about your child's behavior prior to this incident because it is often (unfortunately) assumed that a child who is missing is a runaway. However, be adamant if you know that the possibility of your child running away is essentially zero. You, not the police, are the best judge of your child's patterns and behavior and whether or not this incident is a deviation from the norm of your child. The older your child's age, the more likely you are to encounter resistance. Do not be bullied by bureaucratic statements that missing persons are not acted on for 24 or 48 or 72 hours. Most police departments are becoming more sensitive and responsive to this issue. However, if you meet with a stone wall, do not hesitate to go to the local media for assistance. Many local stations will have a child's picture on the news within hours, which can lead to valuable clues.

Organize friends, relatives, school and church organizations to help locate your child. Print up and distribute fliers with a recent picture of your child and a description of his or her physical characteristics, what he or she was wearing when last seen, where he or she was last seen, the phone number where you can be contacted, and the reward being offered, if any. (Another testimony to the seriousness of this concern is that Continental Insurance is now providing insurance to cover reward money for just such situations.) Also contact the "National Center for Missing and Exploited Children" so that organization can advise you and cooperate with you in your attempts to locate your child. Whatever you do, do not wait for a number of days before getting the public involved. The sooner people know about the problem, the sooner you can have information coming in which may significantly help you in finding your child.

Immunizations

Infectious diseases were historically the most significant wasters of human life but now have been controlled by many methods, including improved nutrition for the population, improved sanitation, improved housing, and, more recently, the development of immunizations and antibiotics. Immunizations are undoubtedly the most significant contribution of science to the control of disease. Smallpox has been eliminated from the face of the earth in the past decade by the worldwide campaign of the World Health Organization. Polio, which used to cripple 20,000 a year in the early 1950s, now affects only a handful of children a year in this country, thanks to immunization campaigns.

We are in a critical period in this nation's history with regard to immunizations. Many people have become complacent about immunization because of the rarity of the diseases that immunizations have aided in eliminating. In addition, the swine influenza controversy of the 1970s undermined confidence in immunizations and raised questions in the minds of many parents. There have always been problems in the use of immunizations. However, the risks of the vaccines are usually minimal when compared to the risks of the diseases they prevent. Because we feel immunizations are so vital to the health of children, we have chosen to elaborate on the issues rather fully in the following pages.

Avoiding an illness often presents many practical problems. Because the bacteria and viruses responsible for causing infectious illnesses may be spread by contact with people, with animals, or with airborne droplets, avoidance is often impossible. On the other hand, protection by the use of immunizations is a practical solution.

PRINCIPLES The principle of immunization is to develop your body's defense system to a point where it is capable of foiling any attack by a particular agent. To accomplish this, a small dose of the modified infectious agent is given. For some immunizations, the virus has been modified in a laboratory so that it causes a very mild infection. Other immunizations use a dead virus that is chemically the same but cannot infect. Others inject just a part of the viral organism or only a chemical that is part of the chemical makeup of the bacteria. The body's defense system builds up an immunity to this part, which in effect creates an immunity to the whole infecting agent. Still another method of immunizing involves substituting a related virus, which causes a much milder infection. This technique was used for smallpox, where the individual was inoculated with cowpox in order to prevent infection with smallpox.

The practice of immunization dates back several thousand years to attempts by a Chinese monk to prevent smallpox. Immunization methods were developed long before the nature of the infecting agents was known. Jenner, who is credited with the modern development of immunization in the late 18th century, borrowed the idea from English farmers who had been using cowpox to protect themselves from smallpox for centuries. Jenner merely performed a study validating what the farmers had been doing for years.

Immunizations have been developed to combat some of the most devastating illnesses affecting human life. Remember, however, that immunizations are only one of the ways of defending against illness. People in good health are less likely to be devastated by infectious disease than people in poor health. Many diseases, such as tuberculosis, have been steadily decreasing in frequency because of improved nutrition, sanitation, and medical care.

Modern virological techniques allow the manufacture of vaccines effective against many illnesses. But all immunizations carry with them the possibility of side effects due to the immunization itself. No immunization is 100% effective or 100% safe.

QUESTIONS

If you have questions about a particular immunization, your doctor or health department should be willing to explain in detail the current risks, benefits, and recommendations for that immunization. If you are not satisfied with the explanation, you may turn to any of the following groups:

- The Academy of Pediatrics, Committee on Infectious Diseases, Evanston, IL 60204
- Centers for Disease Control, Atlanta, GA 30333
- Council on Environmental Health, American Medical Association, Chicago, IL 60610

Questions you may have about a particular immunization include the following:

- How effective is this immunization?
- What are the possible side effects?
- Who is likely to experience a side effect?
- Will people in contact with the immunized person also become immunized?
- How long does the immunity last?
- Is the immunization safe for pregnant mothers?
- Is the immunization safe to give children if their mother is pregnant?
- Is the immunizing agent free of contaminating substances?

THE DPT, OR THREE-IN-ONE, SHOT

The DPT shot is a single injection that gives immunity against diphtheria, pertussis (whooping cough), and tetanus. The combination injection works as effectively as three separate shots and hurts less. We discuss the three diseases separately.

Diphtheria

Nature of the Illness   At this time, there are several hundred cases of diphtheria reported each year in the United States and probably far more than that are not reported. Diphtheria affects the throat, nose, and skin and is contagious. Complications of diphtheria include paralysis (in approximately 20% of patients) and heart damage (in approximately 50%). Diphtheria can be treated by a combination of penicillin and serum injections. However, despite treatment, approximately 10% of persons who acquire diphtheria will die from it. Only immunization can successfully prevent diphtheria.

Nature of the Immunization   Diphtheria immunization has been available since 1920. A portion of the toxin (the chemical product of the bacteria that causes damage) is altered with formalin to render it harmless. This modified toxin, known as a toxoid, stimulates the body's defense system to produce an anti-toxin. Immunity persists for many years after several inoculations. Although the person may come in contact with bacteria and even become infected with it, the toxin will be neutralized by the anti-toxin within the person's body and the infection will cause no harm.

Reactions to diphtheria toxoid are extremely uncommon. Generally, they are limited to a slight swelling at the injection site. This reaction can be decreased in older children and adults by giving lower concentrations of toxoid, known as "adult-type" diphtheria toxoid.

Diphtheria toxoid should be begun (in combination with pertussis and tetanus) at the age of two months. The primary schedule consists of three immunizations given several months apart in the first year of life. A booster shot should be given one year later. An additional booster shot should be given when the child enters school and every ten years thereafter.

Pertussis (Whooping Cough)

Nature of the Illness   Pertussis is a bacterial infection with a high mortality among infants. It causes extremely long and severe bouts of coughing. The prolonged coughing so robs the child of air that the child breathes in violently, causing a whooping sound. Pertussis is contagious. It can be treated with antibiotics with some success.

Nature of the Immunization   Pertussis vaccine has been administered since the late 1940s. The vaccine consists of killed pertussis bacteria. There have been recent changes in the vaccine that may increase its effectiveness.

There is some controversy about the pertussis vaccine because of the frequency of side effects. Many minor reactions are common; about half of all children will develop a fever. A similar number will experience pain or fretfulness. Nearly a third will become drowsy, while 1 in 5 children will temporarily show a decreased appetite and 1 in 12 will have local swelling at the injection site. Fortunately, more severe reactions are uncommon, but they do occur. About 1 in 300 children will develop a high fever (over 105°F), while 1 in 1600 will experience a brief seizure or short period of collapse. It is important to remember, however, that in a recent study, all the children who had these complications recovered *completely* from the reaction. Other potential neurological complications occur but the risk is extraordinarily low (less than 1 in 100,000).

We feel that pertussis is an effective though not an ideal vaccine and that the risks of the disease exceed the risks of the vaccine. Recent outbreaks of pertussis in England and Japan have been a poignant reminder that pertussis remains a contemporary and serious disease. We advise parents to go ahead with pertussis vaccinations, after discussion with their own doctor, and to follow future reports on the safety and effectiveness of this vaccine. Children who develop high temperatures (greater than 104°F), convulsions, extreme somnolence (sleepiness), or who collapse after one pertussis vaccination should not have any more.

Concerns about the safety of the pertussis vaccine led to the development of a vaccine called the "acellular" pertussis vaccine, that is not made from the whole pertussis bacteria cell. This vaccine appears to work well, and produces only half the local and nervous system reactions as the whole-cell vaccine. There is also about a 50% reduction in the number of children who develop fever. It is currently approved for use only for the fourth dose (at 15 months) and fifth dose (at four to six years).

Tetanus (Lockjaw)

Nature of the Illness   Tetanus is a dangerous illness. The tetanus bacteria and its spores are found everywhere; they are present in dust, soil, pastures, and human and animal waste. Symptoms consist of severe muscle spasm, often of the neck and jaw muscles, causing lockjaw. The symptoms are caused by toxin produced by the bacteria. The tetanus bacteria grows only in the absence of air, so wounds that are created by punctures or sharp objects, possibly introducing bacteria underneath the skin, are the wounds with the greatest likelihood of causing tetanus. After tetanus develops, it may be treated with antibiotics and tetanus immune globulin, but mortality may still be as high as 40%. Tetanus can and should be prevented by immunization.

Nature of the Immunization   Tetanus is one of our best immunizations. The tetanus immunization is a toxin that is produced by the tetanus bacterium and modified in the laboratory so that it has very little of its toxic potential. Of all the vaccines available, tetanus comes closest to 100% effectiveness after the initial series of shots. Local reactions are rare and usually cause only mild discomfort for a short time. Severe local reactions can occur if too many shots are received; this phenomenon was frequently seen in military recruits who received unneeded immunizations.

After the initial series of immunizations, tetanus shots are needed only every ten years unless a particularly dirty wound is suffered. (See page 271, **Tetanus Shots.**) Maintain your own records of your child's immunizations; do not rely solely on the doctor, health department, or clinic where the immunizations are given.

POLIO

Nature of the Illness   Polio is a contagious viral illness and a devastating disease that was familiar to almost everyone a few years ago. Besides the paralytic form, which crippled tens of thousands as recently as the 1950s and caused more than 1000 deaths annually, polio also causes meningitis and respiratory infections. There are no antibiotics that are effective against polio. The only way to prevent polio is through immunization.

Nature of the Immunization   In 1955, Dr. Jonas Salk introduced the inactivated polio virus vaccine (IPV). This injected vaccine caused a dramatic decline in the frequency of polio in this country. In 1961, the Sabin vaccine was introduced, in which live polio virus was prepared in such a way as to markedly weaken it. This live polio virus vaccine is known as the oral polio virus vaccine (OPV) because it can be taken by mouth. It was felt to be preferable because it activated the body's defense system in the upper respiratory tract and reduced the chances of an immunized person spreading a live virus.

There are other advantages to the oral polio virus vaccine. Longer-lasting immunity eliminates the need for repeated booster doses once the basic series is completed. And the record of the oral polio virus vaccine is outstanding. In the past decade, only 6 to 12 cases of polio were reported annually.

The only problem with oral polio virus vaccine is its rare side effects. One or two people out of 10 million will develop paralytic polio either because of receiving the oral vaccine or by coming in contact with a person who recently received it. Some of the people who developed paralytic polio after the oral vaccine had rare diseases that predisposed them to developing the infection. The risk of this is so minimal that most doctors do not mention it, but court cases against vaccine manufacturers and doctors have prompted recent developments. There is currently a compensation fund for those rare cases in which an individual suffers an adverse effect after any immunizations, including the OPV.

The American Academy of Pediatrics now regards either vaccine as acceptable and offers the injectable vaccine as an alternative, particularly for families with immunosuppressed individuals. Injections themselves carry risks, although extremely small, of serious muscle infections or nerve damage, so we are not yet convinced the safety record of the OPV will be bested. The injected polio virus vaccine is not known to cause paralytic polio. Breast feeding does not interfere with the oral polio vaccine.

Have your child vaccinated against polio. The risks of either vaccine are small compared to the risks of the disease.

MEASLES

Nature of the Illness   Measles is one of the "usual" childhood diseases but is potentially much more serious than the others. It is by no means a mild disease to be passed off as a ritual of childhood. Before the 1966 introduction of an improved measles vaccination, there were 400 deaths annually due to measles. Far more often than causing death, measles causes devastating complications such as brain infection, pneumonia, convulsions, and blindness. Despite the extraordinary success of measles immunization programs (a 99% reduction), thousands of cases of measles are still being reported annually, a disturbing figure two decades after the introduction of a vaccine. In the 1980s, 3000 to 4000 cases of measles occurred each year. A resurgence of measles occurred in 1990, with nearly 30,000 cases reported and 130 deaths. The disease spread because there are still many unvaccinated people and because the vaccine at 15 months is 95% effective, not 100%. Consequently it is recommended that children and adolescents receive a second dose, and most college entrants will be required to have documentation of two doses.

Even uncomplicated cases of measles can be serious (see **Measles**, page 199). Fortunately, most infants are protected against measles during the first few months of life because of antibodies that they received from their mother's blood, if the mother has had measles. This immunity wears off by the fourth month of life.

Nature of the Immunization   The first measles virus vaccine became available in 1963. A high rate of side effects led to the necessity of simultaneously administering measles immune globulin. A killed-measles vaccine was also available and commonly used until 1967. This vaccine left children susceptible to reactions when exposed to live virus.

Finally, in 1966, an effective live virus vaccine with a low rate of side effects was released. Reactions to measles virus vaccine have been minimal and may not appear for five to ten days. Symptoms generally consist of fever, mild irritability, or rash.

Measles vaccinations are no longer suspected of causing exacerbation of tuberculosis. Therefore, a tuberculosis skin test no longer needs to be done prior to a measles vaccination. The most widely used vaccine today is the Schwarz strain, which causes a rash in an estimated 5% of children and fever in an estimated 15%. There has been concern over the possibility of allergic reactions in children allergic to eggs or neomycin. Consult with your doctor about allergy testing before the immunization if these have occurred in your child.

The vaccine is most effective when given after the first year of life. Previous recommendations were to vaccinate at 12 months; approximately 90% of one-year-old children receiving vaccine will be adequately protected. Current recommendations are to immunize children initially at age 15 months, when the percentage protected is even higher.

During epidemics, younger children need to be vaccinated against measles. Protective antibody is developed within seven days of vaccination, and the incubation period of measles is ten days; children brought to the doctor immediately after exposure to a case of measles may be afforded protection by immunization. However, most children are given passive immunization in the form of measles immune globulin. Children who receive immunization before the age of one year should be re-immunized after the age of 15 months. People who were immunized before 1967 were immunized with the killed strain and may still be susceptible to measles. These individuals could be re-vaccinated with the live attenuated strain, although the adverse reaction rate is high.

RUBELLA (GERMAN MEASLES)

Nature of the Illness   Rubella, as well as rubella immunization, provides us with a set of very difficult decisions. Rubella is a mild illness, usually with a mild fever and a mild rash. Many people do not even know they have had the illness. The danger from rubella lies in the risk to a developing fetus during early pregnancy.

In 1964, during the last rubella epidemic in this country, more than 20,000 children were born with deformities caused by the rubella virus. The deformities included heart defects, blindness, and deafness. Many fetuses escape these devastating effects, but the huge number who have suffered greatly make efforts at eliminating this disease worthwhile.

Nature of the Immunization The purpose of a rubella immunization program is to prevent pregnant women from becoming infected. Before rubella immunization became possible in 1969, over 80% of women had already been infected and therefore were not at risk of producing an infant deformed by rubella. Women who have had rubella infections can become re-infected with rubella the second, third, or even fourth time. It is doubtful, however, that these re-infections pose any threat to the fetus. It is currently assumed, though not proven, that only the initial infection with rubella can damage the fetus.

A rubella immunization program was undertaken in this country in 1969 for several reasons. The most compelling reason was that rubella epidemics run in seven- to ten-year cycles and the next epidemic was expected soon. To avert the epidemic, widespread community campaigns were conducted. By immunizing at least two-thirds of the population, it was hoped that the entire population would be protected (so-called herd immunity). Unfortunately, even in communities in which 95% of the population has been immunized, introduction of the virus has been shown to result in infection of the remaining 5%.

It was decided to immunize children over the age of one year, hoping that the mothers would be protected by immunizing the children. This was a unique policy in the annals of immunization; for the first time, individuals were being exposed to vaccination and its complications in order to protect other individuals. This may or may not have been a reasonable policy. Supporters point to the fact that the expected epidemic did not occur. However, we are currently not certain how long the protective level of the immunizations will last. Some studies indicate that immunity may eventually be lost. If the protection does not remain sufficiently high to interfere with the production of deformities, then we may be producing a generation of women who will be susceptible at childbearing ages. Remember that in the past, more than 80% of the population was immune at childbearing age. Is it possible that the next epidemic will be even more devastating because a larger number of women will be susceptible? There has recently been a slight increase in rubella incidence in the United States, but fortunately only 11 children had deformities due to rubella in 1990.

Other countries have adopted a different policy and only vaccinate teenage girls who are shown by blood test to have no protection against rubella. Vaccination of males doesn't seem to make much sense because "herd" immunity does not work (although in an individual case, an infant boy may be prevented from infecting a susceptible pregnant woman). The policy of vaccinating teenage women diminishes the chance that the immunity will fall to such a low level that it will not be protective in the childbearing years.

But there are problems with adopting a policy of late immunization. Side effects from the immunization are much more common at this age. Particularly troublesome are the joint pains that occur in over 10% of women receiving the vaccination during adolescence or later. Some of these women have had arthritis for as long as 24 months following immunization. In addition, a few women (between 1 in 500 and 1 in 10,000) experience a peripheral neuropathy (a sensation of tingling created by an inflammation in the nerves of the hands).

An additional risk in administering the vaccine in older women is the possibility of pregnancy at the time of vaccination. Although none have occurred yet, rubella vaccine may possibly cause birth defects if given to pregnant women! Thus far, no congenital anomalies have been reported in women whose *children* were given rubella vaccine. It is probably safe to immunize the children of pregnant women, although it is not advised. A policy of immunizing women of childbearing age requires adequate contraception for two to four months following immunization.

Fortunately, a major rubella outbreak in the United States has not occurred since 1964. Small outbreaks have happened in some countries relying on a late-immunization policy. We currently recommend the American Academy of Pediatrics policy of giving rubella vaccine along with measles and mumps vaccine at 15 months. In addition we support a second immunization for adolescent and young adult women who have no evidence of rubella immunity. We also support an alternative strategy of offering rubella immunization to all women of childbearing age, without screening.

MUMPS

Nature of the Illness   Mumps is a contagious viral illness. It infects and causes swelling of the salivary glands. Swelling can be on one or both sides and is often painful. Meningo-encephalitis (inflammation of the brain and its coverings) occurs but is far less frequent and less serious than with measles. The greatest concern about mumps is the possibility of inflammation of the testes in males and of the ovaries in females. Inflammation of the ovaries is far less frequent than inflammation of the testes. Inflammation of the testes is usually one-sided; although it is commonly believed that it can result in sterility, this is very seldom the case.

Nature of the Immunization   The mumps vaccine is a live weakened virus and is one of the more recently developed immunizations. Exposure to the mumps virus usually gives two types of protection. One type of protection is carried by the blood serum, and the other type is carried by the white blood cells. Mumps immunization provides protection through the blood serum antibodies for at least 12 years, and possibly much longer.

Infants should be immunized at 15 months. Adolescent males and females who never had mumps or the immunization where the potential for testicular or ovarian inflammation exists should also be immunized.

HEMOPHILUS INFLUENZA TYPE B

Nature of the Illness   *Hemophilus* influenza type B (HIB) is a bacteria capable of causing serious illnesses such as meningitis, epiglottitis, and arthritis in young children. Most of these serious infections occur in the first four years of life. A considerable number of deaths and permanently damaged children have resulted from HIB. A major breakthrough in the control of this leading cause of meningitis in young children occurred in 1985 with the release of a vaccine.

Nature of the Immunization   The vaccine is made from a purified chemical that is part of the bacteria's capsule. The presence of this chemical tricks the immune system into mounting a response that is reasonably effective in preventing future infections with HIB. To enhance effectiveness, the newest version of the vaccine is linked (conjugated) to other proteins. These conjugate vaccines are effective for preventing disease in young infants.

Nature of the Illness   There are several types of hepatitis. All are infections of the liver and result in *jaundice* (yellowing of the "whites" of the eyes and of the skin), nausea, and weakness. Most children recover, but some develop chronic liver problems. Each year the hepatitis B virus causes infection in more than 200,000 people in the United States, and about 4000 to 5000 people die as a result of chronic problems from hepatitis B. This virus is acquired principally through contact with infected blood or from intimate contact of moist body surfaces, as during sexual intercourse. Many infants are exposed to hepatitis B as they are born. Infants and children may also get it from close contacts in families.

Nature of the Immunization   The currently available hepatitis B vaccine is genetically engineered by inserting a gene into everyday bakers' yeast. Since this is *not* a live virus, it is quite safe. In November 1991, the Advisory Committee on Immunization Practices of the Centers for Disease Control and Prevention recommended vaccination of all infants in the United States. The AAP supports this recommendation. Special policies apply to infants of mothers found to have certain markers for hepatitis B detected in their blood during routine prenatal screening. One dose will be given in the newborn nursery, a second dose at 1 to 2 months, and a final dose at least 4 months later, at 6 to 18 months. An alternate schedule permits immunization to start after leaving the nursery but before age 2 months. Vaccination is also recommended for adolescents in communities where drug use and sexually transmitted diseases are common or where there is known high-risk exposure in the family or environment (for example, working in a hospital).

Along with many other physicians, we are skeptical about the wisdom of this recommendation. Part of our concern is the cost. While the vaccine does work in preventing disease and appears safe, there are certainly other, more pressing health problems of children that would cost less to fix. For parents willing and able to pay, or having insurance policies that cover the cost, an additional concern is that the hepatitis B immunization is a separate injection, bringing to three the number of shots given during some office visits. Many parents object to their children becoming pin cushions, and we understand this feeling. Returning at another time is certainly an option, but can be both an inconvenience and an additional expense. We are also concerned that the pin cushion experience might make families reluctant to have other immunizations. Combining hepatitis B with other immunizations into a single shot would resolve this issue.

Finally, hepatitis B is principally a disease of adults (with the exception of some high-risk newborns whose mothers can transmit the virus and adolescents engaging in high-risk behaviors). Consequently, many physicians are currently not routinely giving this immunization to all infants and children. Adolescents who are sexually active or engaging in high-risk behaviors should be offered the immunization.

SMALLPOX

Smallpox, a disease that has killed millions of people over the centuries, *has been eliminated through immunization!* Smallpox vaccinations are no longer given.

CHICKEN POX

Nature of the Illness   Chicken pox, generally a mild illness of childhood, produces a rash and slight fever. But in rare instances, it can be complicated by more serious problems such as arthritis, meningitis, Reye's syndrome, or bacterial infection.

Nature of the Immunization   A vaccine, developed nearly 20 years ago, is licensed in Japan, where over 1.5 million doses have been given. The United States has conducted trials for more than 11 years and may soon license the vaccine.

INFLUENZA

Nature of the Illness   Influenza is a respiratory tract infection, caused by the influenza virus, that produces severe muscular symptoms as well. There are many illnesses that may mimic influenza, and the term "influenza" is applied liberally to many syndromes involving cough, cold, and muscle aches.

Nature of the Immunization   Influenza virus vaccines are live weakened viral vaccines. They are recommended only for children who have chronic illnesses, particularly respiratory and cardiac diseases.

One of the long-term benefits of the media coverage of the swine influenza controversy of 1976 was to sensitize the public to the complexities involved in making rational medical decisions. On the other hand, it is our hope that individuals will not be dissuaded from obtaining vaccines with a long history of demonstrated safety and effectiveness in preventing serious illness.

RABIES

Nature of the Illness   Rabies is a viral illness that is transmitted in the saliva of wild and domestic animals with rabies; it is fatal if not treated. Animals known to be significant carriers of rabies include raccoons, skunks, bats, and foxes. Rabies in dogs and cats does occur but has become rare. Rabies is rare in rodents such as squirrels, chipmunks, mice, and rats. Fortunately, there are only one or two cases of rabies in humans reported in this country each year. See **Animal Bites**, page 267.

Nature of the Immunization   Both a rabies vaccine made from human cells and rabies immune globulin are available. The complex methods of administering these vaccines will not be discussed here. Immediate cleansing of any animal wound is the first line of defense. Bites from wild animals or unknown domestic animals should be reported to a doctor immediately.

PNEUMOCOCCAL DISEASE

Nature of the Illness   The pneumococcus is a type of bacteria that is responsible for some common infections, such as ear infections, as well as some more unusual and severe diseases such as pneumonia and meningitis. There are many strains of the pneumococcus, making immunization a difficult problem. However, some children, including those with sickle-cell disease or immune deficiency, are particularly susceptible to serious pneumococcal infection and benefit from the vaccine.

Nature of the Immunization   The vaccine is prepared from chemical portions of the bacteria (antigens), which stimulate a protective (antibody) response. The current vaccine protects against 23 strains of the pneumococcus, is not recommended for children less than two, and is given to children without spleens or with sickle-cell disease, kidney (nephrotic) disease, or immune deficiency. It is not currently recommended to prevent recurrent ear infections.

**MENINGOCOC-
CAL DISEASE**

Nature of the Illness   The meningococcus is a bacteria that causes meningitis as well as a widespread blood infection, which often ends in shock and death. Fortunately, the disease is rare, but outbreaks occur.

Nature of the Immunization   The vaccine consists of purified parts of the bacteria's cell wall. It is recommended only for children with certain types of immune disorders and during epidemic outbreaks.

**IMMUNIZATION
SCHEDULE**

Those are the details; here is the bottom line. Table 3 lists the current immunization schedule that we recommend.

TABLE 3   *Immunization Schedule*

Age	*Immunization*
Newborn	Hepatitis B (later as directed by doctor)
2 months	DPT (diphtheria, pertussis, tetanus), OPV, (oral polio virus), and HIB (*hemophilus* influenza type B)
4 months	DPT, OPV, HIB
6 months	DPT, HIB
15 months	Measles, mumps, rubella
18 months	DPT and OPV
4–6 years	DTap (diphtheria, tetanus, acellular pertussis) and OPV
5–18 years	Measles, mumps, rubella
Every 10 years	T(d) (adult tetanus, diphtheria)

Dental Care

Children's teeth begin forming early during the fetal stage. The mother's diet provides the essential nutrients for tooth development. Calcium and phosphorus are the basic building blocks of teeth and are found abundantly in milk and dairy products. Vitamins C and D are also important, as are small amounts of the element fluoride. Drugs, especially tetracycline, should be avoided during pregnancy partly because of their capability for staining and weakening teeth.

TEETHING

Most children begin teething actions several months before the eruption of the first tooth. The first tooth usually appears on the lower jaw and is one of the front teeth, known as incisors. These incisors appear in most children by the age of 8 months, but may begin as early as 4 months or as late as 15 months. Usually, the two lower central incisors come first, followed by the four upper incisors. Many children will have these six teeth in place shortly after their first birthday. The next teeth, appearing between 18 and 24 months, are the first four molars and the remaining two bottom incisors. Before the child's second birthday, the pointed teeth, also known as the canine teeth, usually appear between the incisors and the molars. After the child is two, the remainder of the first set of teeth, the last four molar teeth, appear.

Many parents wonder about massaging children's gums before their teeth erupt. There is some evidence that the bacteria that cause tooth decay will stick to gums less often if the gums are gently massaged. Some preparations (e.g., Oragel) also help to prevent bacteria from sticking.

The eruption of teeth and the practice of teething has led to many folk tales. Some are true, but there are common misconceptions. For example, teething does *not* usually cause high fever. Children experience many mild viral infections within the first three years of life, and some of these viral infections will occur simultaneously with the eruption of teeth and during periods of increased teething activity. Teething hurts; however, it is not the cause of high fever, extreme irritability, or a marked change in your child's daily activities.

Teething is important and necessary and can be done on almost any hard rubber object. Certain pacifiers, such as the Nuk pacifier, can be used in teething infants and may have the advantage of ensuring proper tongue thrust for proper jaw development. Parents should be careful about allowing children to teethe on plastic objects, as splintering has resulted in injuries.

Children should *never* be allowed to teethe on a bottle full of milk. Milk is for the feeding of young infants and not for pacification or teething. Major problems may be caused by allowing an infant to fall asleep with a bottle of milk in his or her mouth. The milk remains in constant contact with the teeth, providing an ideal setup for tooth decay. And studies have proven that children who lie flat in bed with a bottle in their mouths drain some milk into their eustachian tubes and consequently may experience a higher incidence of ear infections.

TOOTH DECAY

Although dental problems are less dramatic than some other health problems, they nonetheless remain the most frequent health problem in children. Virtually every child in this country will need dental work at some point. More than 25 million adults in this country have no teeth at all. Dental cavities (caries) and, more important, gum disease can be prevented.

Genetics is one important factor in the development of cavities. There are people who have never had a cavity; they should in part thank their parents. More important than what we acquire from our parents, however, is what we do to ourselves: Food! If our teeth never came in contact with food, our teeth would never develop cavities. Given that all of us do eat, it is important to know the components of food that cause tooth damage: sugar and starch. Normal mouth bacteria require sugar and starch in order to survive. An excess number of these bacteria can lead to rapid decay. These bacteria produce an acid that is capable of eating through the hard enamel that covers our teeth.

Once the bacteria have eaten through the enamel, they then set up housekeeping inside the soft portion of the tooth. These bacteria are extremely resourceful and can live quite well in the absence of air. But they still need sugar and starch in order to survive. Their food supply now has to come to them. Only sugar and starch in extremely minute, dissolved form can enter a small hole in the surface of the enamel. This is why highly refined sugar is such a problem in fostering tooth decay. We now eat more than 15 times the amount of refined sugar as a century ago. A lost tooth of childhood, placed in a glass of sugary soda, will dissolve almost overnight! A certain amount of the acid that mouth bacteria produce is neutralized by the saliva; this is why our teeth do not rot quite as quickly from drinking soda as they do when allowed to sit and dissolve overnight.

A sound program of cleaning teeth, in order to remove trapped particles of food, is critical. This includes brushing, dental floss, and water jets. Finally, a diet that is high in roughage provides us with a natural toothbrush. It is the custom in many European countries to have salad at the end of the meal. This makes a good deal of sense because the salad is an extremely rough part of our diet, serves as a natural toothbrush, and clears away the sugars.

BRUSHING YOUR CHILD'S TEETH

Brushing should begin whenever your child seems receptive. Most children will enjoy brushing their teeth with their parents during the imitative years (from age one on). A two-year-old will not be very proficient yet at brushing teeth and will need help from mom or dad. However, by the age of three-and-one-half, children have acquired the fine skills necessary for doing a good job by themselves. Parents can help by encouraging brushing as a family activity and regarding tooth care as a pleasurable experience and not a chore.

Choice of Toothpaste

Several toothpastes have been recommended by the American Dental Association for their effectiveness in preventing cavities. The potential for preventing cavities lies in the fluoride content of these toothpastes. The fluoride content of our foods and water supplies is slowly increasing and the importance of using the right toothpaste is slowly diminishing. We still recommend a toothpaste that is ADA approved. Some other tips:

- So-called glamour toothpastes may contain caustic materials that may gradually wear out the tooth enamel.
- Your child should enjoy the taste of the toothpaste being used in order to re-inforce brushing as a pleasurable experience.

- Watch young children brush, and don't let them swallow or eat toothpaste. This can be a source of excessive fluoride.
- Regular non-fluoride brushing is better than sporadic fluoride brushing, but regular fluoride brushing is best.
- Soft toothbrushes are preferable.
- Although brushing vertically is currently not recommended, it is the thoroughness of the brushing that is most important.

DENTAL FLOSSING

Dental floss is an excellent tool for removing particles from between the teeth and encouraging healthy gums. Flossing is not recommended for children with their first teeth as it can be an unpleasant experience. However, once children have developed their permanent teeth, they should be encouraged to use dental floss. Again, the best way to encourage children to do things is to do them yourselves. Dental floss without the use of the wax covering is slightly preferable because particles of wax can become dislodged and remain between the teeth.

WATER JETS

Water jets are another effective method of removing particles from between the teeth and often are enjoyed by children who don't like dental floss. Dental floss does a better job of cleaning if used correctly, however. Some children are frightened by the noise of the water jet, however; a period of time to adjust to the water jet should be given.

FLUORIDE

Fluorine is a natural element that is extremely effective in preventing tooth decay in children. Some have promoted fluorine as a miracle drug, and others have attacked it because of its potential in extremely high doses of causing tooth damage. Fluoride compounds will decrease, but not eliminate, tooth decay.

Fluoride is found naturally in many water supplies and in many vegetables as well. In addition, a majority of cities in the United States have chosen to fluoridate their water supply. Because of these fluoridated water supplies, many packaged and canned foods in the supermarket also have measurable amounts of fluoride. Even if your community does not have a fluoridated water supply, you are undoubtedly receiving a fair amount of fluoride in the packaged foods that you buy.

In Areas Without Fluoridation

The current recommendations are for supplementing infant feedings with fluoride drops in areas where the water supply is not fluoridated. Children between the ages of three and ten should receive about 1 milligram (usually one tablet) each day if the water supply is not fluoridated. Fluoride is required through age ten or so until the permanent teeth are in. Too much fluoride can discolor the teeth, so do not exceed these dosages. The recommended fluoride dose is slowly being decreased because of the increasing amount of fluoride in other sources of food. Fluoride is a prescription item, and your doctor should be aware of changing developments in the use of fluoride.

THE FIRST TRIP TO THE DENTIST

The time to first visit the dentist depends on the nature of your child's teeth. Examine your child's teeth yourself periodically. It is not difficult to see small holes in the child's teeth. Usually, teeth will not begin developing cavities before the age of three. If at the age of three you think you see holes in your child's teeth, by all means make a dental appointment. Although baby teeth are lost and replaced with permanent teeth beginning at about age six, severe damage to baby teeth can cause both dental pain and potential loss of adult teeth. Losing baby teeth prematurely may result in permanent teeth coming in crooked. If your children's teeth appear all right, a dental visit may be postponed until age four or so when the child will be less frightened by the dental examination. We recommend yearly visits to the dentist thereafter.

PERMANENT TEETH

Permanent teeth begin forming in the first few months of life, but do not erupt until the fifth or sixth year. The permanent teeth come in approximately the same order as the baby teeth. In addition, the number of teeth will increase from 20 primary teeth to 32 permanent teeth. The eruption of permanent teeth generally does not end until an individual is in his or her 20s and the third molars (wisdom teeth) appear. There is no reason to have wisdom teeth removed routinely. If wisdom teeth become impacted or grow into other teeth, dental care may be necessary, but many people do have all 32 teeth.

Preventing Diseases and Injuries

The origins of the most serious medical problems of adults can usually be traced to habits and expectations developed in the early years. Tooth decay, accidents, and many infectious diseases can be effectively prevented, and the importance of doing so is obvious. Heart attack, stroke, high blood pressure, obesity, diabetes, cancer, emphysema, drug side effects, alcoholism, suicide, and homicide are the major national killers. AIDS is increasing rapidly among adolescents. There are effective ways of reducing the likelihood of developing every one of these problems; good health is a way of life. It is your responsibility to transmit the knowledge and attitudes that will serve your children over a lifetime.

AIDS

It is natural to be concerned about the risk of HIV (human immunodeficiency virus infection) and AIDS (acquired immune deficiency syndrome) to our children. However, there are far, far greater risks to the health of children. The risks of injury while riding in an automobile or on a bicycle are substantially greater than the risk of HIV infection. Similarly, the risk of acquiring serious infectious illness such as meningitis while attending day care or school is substantially greater than the risk of HIV infection. Fortunately, there are things that parents can do to minimize a child's risk of acquiring any of these problems. Using infant safety seats, seat belts, and bicycle helmets would save far more lives each year than have been claimed in the AIDS epidemic. Similarly, newer immunizations, such as the HIB, are reducing the risks of children's acquiring meningitis in day care.

There are also things that you can do and need not do to minimize your child's risk of HIV infection. First, you need not panic. By mid-1992, the number of AIDS cases in children in the United States was still under 4000. The vast majority of these children acquired AIDS in the course of their mother's pregnancy.

While some women acquired HIV infection through behaviors known to place them at risk, such as injecting drugs, many unknowingly have acquired HIV from partners. It is estimated that about half the women of childbearing age with HIV infection have no knowledge of any behaviors placing them at risk for HIV. Because of the availability of effective medical treatments it is now recommended that all pregnant women in high-risk areas (i.e., large urban centers) be screened for HIV infection. Women in other areas should be offered testing and decide on their own. We support a similar approach for newborns.

A number of children have acquired AIDS because of blood transfusions that they received that carried the HIV virus, which causes AIDS. Most of these children acquired their infection prior to April 1985, when blood screening for HIV began. It is currently estimated that the risk of blood containing HIV that has not been detected by today's testing procedures is 1 in 40,000, a real but nevertheless very small risk.

To date there has been no known transmission through casual contact (for example, being coughed on by a person infected with HIV). Because HIV is transmitted through blood, it is theoretically possible that toddlers biting hard enough to draw blood might acquire AIDS. For older children participating in sports, there is also a theoretical possibility of being exposed to infected blood in close body contact (possibly in wrestling) and transmission of the AIDS virus. Although these situations are theoretically possible, there have been no cases reported to date. Again, we wish to emphasize that the risk of children encountering serious infectious disease and injuries in day care and school is far, far greater.

It is with this information in mind that we therefore endorse the recommendations of numerous national policy-making groups that children with HIV infection be allowed to attend day care and schools. The risk to other children in these schools is thus far zero. Even should news of a transmission be smeared across tomorrow's headlines, it is important for parents to remember that this risk will be substantially less than a whole host of other things going around schools.

There are definitely steps that you can actively take to minimize your child's risk of HIV infection. Because HIV is a sexually transmitted disease (STD), we anticipate the continuing spread in adolescents engaging in sexual activity. This trend is truly alarming. Thirteen- to 24-year-olds with AIDS now outnumber children with AIDS by a ratio of more than three to one. AIDS is the 17th most common killer in this age period. Most individuals diagnosed with AIDS in the 20- to 24-year-old age group acquired HIV infection in their teenage years. Now, more than ever, it is imperative for parents to develop an openness about discussing sexual behavior with their children. As we have discussed in our section on sexuality, these discussions can begin in early childhood. Both your tone of voice and what you say about sexual development and sexual behavior are very important. Children have a natural curiosity about anatomy, physiology, and reproduction. Their questions should be answered matter-of-factly at an early age. When they approach the years of sexual activity, it will then be far easier to discuss with them both your values about sexual behavior and frank information about the variety of options for preventing STDs.

Virtually all our children will become sexually active at a certain stage of their adolescent or young adult life. Responsible behavior is what we advocate. Fostering good decision making about all aspects of life is essential to good decision making about sexual behavior. At a certain stage of life, saying "no" is responsible. At a later stage in life, safer sexual behavior (that is, protected intercourse with condoms and foam) will eventually save many lives. See "Adolescent Sexuality," page 539.

HEART ATTACK AND ARTERIO-SCLEROSIS

Once rare before middle life, heart attacks are now sometimes seen even in the 20s and 30s. A heart attack, sometimes called a *coronary thrombosis* or *myocardial infarction*, results when a clot forms in one of the small blood vessels that supply blood and oxygen to the heart muscle. The clot forms on a narrowing in the blood vessel called an **arteriosclerotic plaque**. Thus, a heart attack is a form of arteriosclerosis, or "hardening of the arteries." Arteriosclerosis happens to all of us, and the process begins in the teens. In some people, it happens faster than in others. You can control the rate at which serious arteriosclerosis develops.

The risk factors for heart attacks are well established. They include:

- Obesity
- Cigarette smoking
- High blood pressure
- Inactivity
- Diet
- Stress

Each of these factors relates to the others, each is under the control of the individual, and each is susceptible to the influence of parents.

Obesity and Diet

Both obesity and dietary fat intake are associated with the early onset of arteriosclerosis. The total body weight increases the load on the heart, and the dietary fats appear to cause the arteriosclerotic plaques to form more quickly. At any rate, for every pound greater than ten that an adult is over his or her ideal weight, life will be, on the average, one month shorter. Diabetes can be avoided in many individuals by maintaining a normal weight.

We describe on page 320 (**Overweight**) how fat children may become fat adults; the number of fat cells is determined very early in life, usually in the first year. More important, however, is the fact that the attitudes that result in adult obesity begin at the dinner table, where tastes for foods develop. Here is some specific advice:

- The cholesterol controversy continues, and we advise moderation with eggs, ice cream, and butter, and low-fat rather than whole milk for school children.
- A well-balanced meal is still the best policy. Children don't buy groceries; if they are eating too much junk food, you are responsible.
- Rules do not have to be rigid; arteriosclerosis develops over a lifetime, and good habits are not altered much by temporary indiscretions.

Stress also contributes to the high incidence of heart disease. Stress in our lives cannot be avoided, but it can be dealt with if it is recognized. Many current techniques (from massage to meditation) are being developed to help individuals cope with stress. Parents can help their children cope and thereby teach them that they need not confront all problems alone and head-on.

Exercise

Controlling heart attacks requires more than dietary control. Adequate exercise also protects us from serious heart attacks. Exercise must be regular and lifelong.

We are a nation of television watchers, videogame players, and desk sitters, and it is bad for us. Our most serious diseases, such as arteriosclerosis and hypertension, are sometimes called "diseases of civilization." Our stomachs ulcer at the stress of it all, and our bowels won't work by themselves. Everything from hemorrhoids to stomach ulcers to middle-aged spread can be related to our lack of serious physical exercise.

Good physical exercise is a workout for the heart and blood vessels, not just the muscles. To work the heart, the exercise must be steady, moderately strenuous, and must last at least 10 to 12 minutes. Bicycling, swimming, and jogging meet these requirements. Good exercise to fight the diseases of civilization must be regular, at least four or five times weekly, and must be a lifetime habit.

Healthy exercise will reduce the resting pulse rate, a sign of greater efficiency of the heart, and will reduce the blood pressure. And at any particular exercise level, the heart will not have to work as hard. Exercise, when regular and sustained, makes the heart more efficient. The best way to encourage your children to exercise is to exercise with them. It will benefit the whole family.

Other preventive measures are discussed below.

HIGH BLOOD PRESSURE AND STROKE

A stroke is like a heart attack, except that the clot forms in a blood vessel that goes to the brain, and some of the brain tissue dies. Hardening of the arteries, arteriosclerosis, is almost always the cause. Prevention of stroke is like prevention of heart attacks, and dietary fat intake should be moderate. High blood pressure increases the chances of a heart attack and increases the chances of a stroke even more.

Arteries in a person with high blood pressure are like overinflated tires on an automobile: they don't last as long. High blood pressure is unusual in children but is very common in adult life. And you can't tell if you have it because the high pressure doesn't usually cause any symptoms. How can you prevent it if you don't know that you have it? Easy, but you have to do three things.

- Control your weight. Fat people need a stronger pumping action of the heart, and the blood is supplied at a higher pressure. Weight control will improve or prevent high blood pressure in a good percentage of people.

- Engage in regular exercise to build "reserves" into your circulatory system. This can reduce your blood pressure as well as make you feel better.

■ Get a blood pressure check every year or so in adult life, starting at about age 21. But treatment of adult high blood pressure begins with development of the exercise habit and weight control in childhood.

SMOKING AND CANCER

Unfortunately, many cancers can't be prevented. However, you do have control over preventing some cancers. Lung cancer is the best example; this is the most common fatal cancer, and one of the hardest to treat; it results in about 20% of all cancer deaths. And almost all cases are caused by cigarette smoking. The cigarette tars and gases irritate the surface of the breathing tubes; after a period of time, a cancer may begin to grow in the irritated area. Of extremely heavy smokers, one out of seven die of this cancer. To prevent it, don't smoke cigarettes.

The examples and the attitude of the parents are critically important here; if you don't smoke, the chances are that your children won't either. If you do smoke, don't do it in front of your children. Children respect their parent's judgments and often imitate them. You improve their chances of not smoking by not smoking around them. More than one-half of lung cancer and emphysema victims began as teenage smokers. We also strongly advise seeing your physician about referral to a smoking-cessation program. This is *as* important for your child's health as for your own. Children of smokers have *much* higher rates of respiratory infections and asthma.

Nicotine also causes the blood vessels to constrict, and hardening of the arteries occurs at an earlier age in smokers. The tragic lung disease, emphysema, is almost always related to inhaled cigarette smoke. So there are three very important medical reasons why cigarette smoking is hazardous to your health: cancer, arteriosclerosis, and emphysema. Even moderate cigarette smokers lose two years of life expectancy. It is not only that cigarettes can kill you; it is the slow, lingering, painful way in which they do it.

Other Cancers

Your doctor will begin important cancer-prevention techniques in adolescents by showing young women how to check for breast lumps and young men how to screen for lumps in the testicles.

HOMICIDES

In the late teens and early 20s, only accidents cause more deaths than suicide and murder do. Murder is constantly dramatized as a purposeful crime, but it usually is not. Most frequently, it is spontaneous and results from temporary rage or miscalculation, often aggravated by alcohol or other drugs. Fully 75% of all murder victims are known to their attackers, and a high percentage of best friends and family members are involved.

The murderer often has not learned to assume responsibility for his or her acts, has grown up in a family where violent rage is common and sanctioned, and possesses a weapon capable of causing rapid death. Often there is a handgun in the home; many times there have been legal problems for the adolescent before the violent act. There is a pattern to senseless destructive behavior, and its origins are often early.

SUICIDE

Suicide is becoming a more common problem in our society. It is no longer unusual to find seven- and eight-year-olds attempting and committing suicide. Among adolescents, it is the third most common cause of death.

Most people think of committing suicide at some time. It is sometimes, but seldom, a sign of mental illness. Because suicide is against the law and violates religious and moral codes, it is often a problem that families keep hidden. A suicide gesture or attempt, no matter how minor, is extremely serious and is a signal for a family to obtain professional help.

There are many reasons why a person attempts suicide. Often a major change or stress will occur in the person's life. The person must cope with this situation. A variety of coping strategies may be explored, including suicide. It is in this stage when people can be helped the most by your support.

Adolescents who attempt suicide are often isolated. They are loners, have few friends, and are often withdrawn. Frequently, they will have only one close friend. Many have serious problems in school. A significant number are failing or experiencing discipline problems. Many live alone.

Following a seemingly minor altercation with a boyfriend, girlfriend, or parent, a suicide attempt will be made. At other times an attempt may follow an extremely angry exchange with a parent. The motivation behind such an act is complex. It may be an attention-seeking device, a manipulative act, an attempt at hurting a parent, or a reaction to a loss, anger, or guilt. It is always a cry for help. If an adolescent is crying for help, it does not mean that parents have failed; it means that help from a doctor, social worker, or other skilled counselor is needed at that moment.

Any child or adolescent who talks of suicide should be taken seriously. "You'll be sorry when I'm gone" should be listened to. Most suicide attempts are unsuccessful, but 10% of these individuals will later kill themselves. Children and adolescents involved in frequent serious accidents should also receive professional counseling.

SUBSTANCE ABUSE

Nicotine, caffeine, and alcohol are the traditional American drugs. But there are others. And everyone likes to think that the other person's drug is terrible, while his or her own habit is not. We see self-righteous zealots who are "drunk" their whole adult life on Valium or other tranquilizers. We see the alcoholic who finds his son's marijuana un-American. And the daughter who despises her parents' alcohol but uses "speed" to get up and "reds" (barbiturates) to get down. Medically, all drug habits are psychologically equivalent.

It is clear that children emulate their parents, so developing appropriate attitudes and behaviors about alcohol and other substance abuse must begin at home. Parents should begin open discussions about street drugs as soon as they leave their children alone on the street. Peer pressure can be enormous for school-age children and adolescents, and parents must promote appropriate decision making by discussion and example. Parents should also ensure that a school program is operating that reinforces this message.

Illegal Drugs

When parents worry about "drugs," they're usually concerned with illegal "street" drugs, such as cocaine, amphetamines, marijuana, and heroin. All of these substances are either banned or restricted to supervised medical use. Taking such drugs is dangerous for three reasons:

- They are by and large highly addictive. People who take these drugs feel compelled, physically or psychologically, to keep on taking them whatever the consequences: loss of money, loss of job, loss of friends, health problems, etc.

- It is possible to suffer severe, permanent harm from overdosing on these substances.

■ Because they are illegal, obtaining these drugs is associated with violent crime.

Parents should warn children about these drugs, and show them a good example by never using any. If a child does start taking illegal drugs, parents should help them into treatment. Breaking an addiction is never easy, but it can be done.

Remember that for children, alcohol and tobacco are illegal drugs. It is illegal to sell these substances to children under a certain age; usually 21 for alcohol and a younger age for cigarettes. Nevertheless, teenagers get their hands on these substances, and such surreptitious, risk-taking behavior too often leads users on to "harder" substances.

ALCOHOL

Alcohol is a useful sedative and has some social purposes. As a drug, it is even relatively safe when used as directed. If we were to prescribe it as a prescription drug, the label would read: "*Caution*: May cause drowsiness and altered judgment. Do not attempt productive work or driving after taking this drug. Do not exceed recommended dosage of one-and-one-half ounces daily. May be habit-forming. Long-term or excessive use has been associated with ulcer, gastro-intestinal hemorrhage, cirrhosis of the liver, inflammation of the pancreas, and atrophy (wasting) of the brain." Again, parents set the example, and again, we counsel moderation.

As the pattern of drug usage is shifting, more and more adolescents are turning to alcoholic beverages. Chronic alcoholism is now becoming common in adolescents. A chronic alcoholic needs help at any age.

Medication Abuse

There is another side to the national drug problem that is of major medical importance. One out of six hospitalizations is for a drug side effect, and one out of seven hospitalized patients will have a drug reaction *while in the hospital*. We think that the history of this century will record its drug dependence as one of its worst features and will remark on the curious concept that there is a pill for everything. You will note from the discussions in Part IV that, except for bacterial infections, there is a satisfactory pill for almost nothing. The average

American goes too often to the doctor and too often to the medicine chest. We spend fortunes on overpromoted patent medicines and on prescription drugs of low value. We get taken. And, unless we are careful, our children will get taken also.

Consider two boys, each of whom scrapes his arm. The parents of one clean it and apply a bandage. The parents of the other take him to the emergency room of the hospital, where an intern, trying to be helpful, thinks it might be infected and gives an antibiotic. After five days, both scrapes are healed. Consider the psychological repercussions: the family that used the emergency room has lost more than time and money. The antibiotic is credited with the cure! And the next time, the emergency room has become a necessity—to treat the infection that never was.

The other boy has learned the real lesson: that a scrape will heal if cleaned and left alone and that a certain measure of confidence in the natural healing properties of the body is warranted. Children learn drug dependence from their parents; you should never automatically encourage your child to use any drug, no matter how harmless.

Additional Reading

Thomas McGuire, D.D.S., *The Tooth Trip*. New York: Random House, 1972.

Donald M. Vickery, M.D., and James F. Fries, M.D., *Take Care of Yourself: The Complete Guide to Medical Self-Care*. Reading, Mass.: Addison-Wesley, 1993.

Donald M. Vickery, M.D., *LifePlan: Your Own Master Plan for Maintaining Health and Preventing Illness*. Reston, Va.: Vicktor, 1990.

CHAPTER 10

The Home Pharmacy

A s a society, we have relied far too long and far too often on drugs to solve our problems. Drugs have, of course, saved many lives. Medications for bacterial infections, epilepsy, cancer, heart failure, rheumatoid arthritis, diabetes, and other major illnesses have certainly changed the lives of those afflicted. However, there is almost unbelievable misuse of medications because of our social rituals, patients' demands, doctors' encouragement, and massive advertising.

The vast majority of all medical problems resolve themselves without the use of any medication. These illnesses include common colds, influenza, diarrhea, rashes, upset stomachs, headaches, and most of the other problems discussed in Part IV. However, we are bombarded with advertisements implying that these problems can only be cured, or can be cured much quicker, by the use of some medication. One out of every eight television advertisements is for an over-the-counter medication. These claims are simply not true. For nearly every advertised drug, there is a cheaper alternative that is just as good or better; in many instances, the preferred alternative is nothing at all. You will find our recommendations under the "Home Treatment" section for every problem in Part IV.

Finding a product on the shelf of a supermarket or pharmacy does not mean that it is either safe or effective. In 1966, the Food and Drug Administration (FDA) commissioned a study to evaluate 400 common over-the-counter (non-prescription) drugs. More than 75% of these drugs were found to be less than effective for some or most of their claims. Products that have been on the market since before 1938 have never been subject to requirements for demonstrating safety or efficacy! Therefore, most of the estimated 250,000 to 500,000 (!) products available on the shelves are less than effective, and many are unsafe. Well-intentioned use of these products results in many deaths annually.

The FDA is currently taking steps to review available over-the-counter products. But it will still be many years before consumers can purchase medications with the knowledge that they are both safe and effective. The definition of safety can be only a relative one. All medications can cause serious side reactions. Aspirin, a potent and useful drug, causes more serious reactions and more deaths than any other medication. Be careful with *every* drug.

In this chapter, we discuss several medicines to use in treating common symptoms. Many of these are covered more extensively under discussion of the individual problem in Part IV.

Essential Home Medical Supplies

We believe that only two types of medication are essential for the home pharmacy when children are in the home. First is the fever and pain reliever **acetaminophen**. Aspirin and ibuprofen also relieve fever and pain, and reduce inflammation as well. They present the risk of more severe side effects, however: Reye's syndrome and stomach inflammation. Even these medications are potentially fatal and should be kept out of the reach of children. (Very few childproof caps are childproof. Often the best way to open a childproof cap that you are having trouble with is to ask your child to open it for you.)

Second, every home should have **syrup of ipecac** on hand. Syrup of ipecac can induce vomiting in children who have swallowed dangerous medications. It can be used immediately when a child has swallowed pills, liquid medication, or plants. For other poisonous substances, it may be dangerous to induce vomiting. Write the name of your poison control center on your telephone and in Chapter C (**Poisoning**, page 258) now! Your poison control center can advise you if a household product that your child has taken can be safely removed by inducing vomiting.

CORRECTION

Please use this decision chart for Croup instead of that on page 381 of *Taking Care of Your Child*.

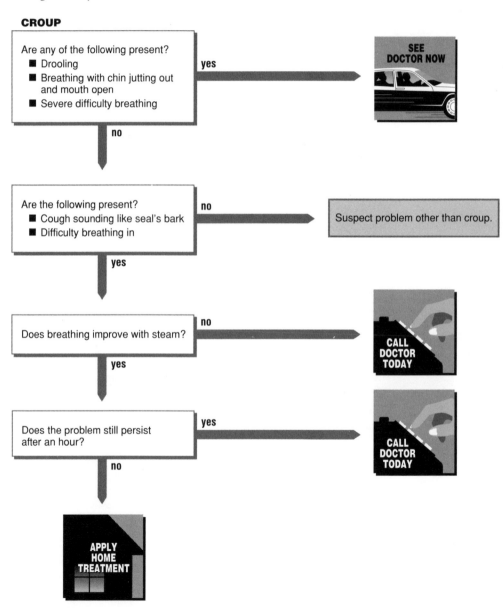

CROUP

Are any of the following present?
- Drooling
- Breathing with chin jutting out and mouth open
- Severe difficulty breathing

yes → **SEE DOCTOR NOW**

no ↓

Are the following present?
- Cough sounding like seal's bark
- Difficulty breathing in

no → Suspect problem other than croup.

yes ↓

Does breathing improve with steam?

no → **CALL DOCTOR TODAY**

yes ↓

Does the problem still persist after an hour?

yes → **CALL DOCTOR TODAY**

no ↓

APPLY HOME TREATMENT

In addition to these two medicines, have on hand:

- A thermometer
- Band-Aids
- A vaporizer
- Elastic bandages
- Sunscreen
- Hydrogen peroxide

Other common household drugs are described below to indicate their limitations. A drug on the shelf, whether intended for you or your child, is like keeping a loaded gun there. Dispose of leftover drugs now! By family discussion and by parental example, show that you understand what drugs do not do. Symptoms are not only nature's way of telling us that something is wrong, but also may be a part of the healing process. Often the cough, runny nose, or diarrhea is already helping the body get rid of the problem. Your child lives in a body brilliantly designed to restore health once it has been disturbed. Don't fool around with it.

For Allergy

ANTIHISTAMINE COMPOUNDS

These compounds are useful in children who have well-documented cases of allergic skin reactions to insect bites, allergic rhinitis, hay fever, or chronic hives. They are of dubious value for anything else. The discomfort of the hay fever should always be balanced against the risk of problems caused by the antihistamine compounds, which are usually dispensed in combination with a decongestant compound. The most common antihistamines are chlorpheniramine (Chlor-Trimeton), brompheniramine (Dimetane), triprolidine (Actidil), diphenhydramine (Benadryl), and carbinoxamine found with pseudophedrine (Rondec), and with hydroxyzine (Atarax, Vistaril). Newer preparations, such as terfenadine (Seldane), have fewer side effects.

Side Effects   Drowsiness is the most common side effect and can interfere with a child's schoolwork. Some parents feel that antihistamines are useful in helping children to get to sleep at bedtime, but children really never go into the sleep stage that is most restful while on antihistamines. In fact, such children have an insufficient amount of the proper type of sleep. Antihistamines occasionally can cause hyperactivity in children. Follow package instructions for dosage.

For Colds

With a cold or allergy, a runny nose is often the worst symptom. Although runny noses are a nuisance and not very aesthetic, they are seldom a serious problem and do help to carry the virus outside the body. One of the concerns with colds is that the swelling and secretions of a cold may block either the sinus outlets or the eustachian tube. Because sinuses are very poorly developed in very young children, there is not as much need to worry about sinusitis in children.

However, if the eustachian tube, which drains the normal secretions from the middle ear into the child's nasal cavity, remains plugged for a day or so, a middle-ear infection may begin. It is easier for this tube to swell and close in younger children because it is shorter, narrower, and at a more horizontal angle than in older children. Whether this tube can be kept open sufficiently during a cold to prevent an ear infection is uncertain, but it is the basis for therapy with either nose drops or decongestants. Use of the following agents is frequent but controversial.

NOSE DROPS   Nose drops usually contain decongestants such as phenylephrine (Neosynephrine), ephedrine, or oxymetazoline (Afrin). These drugs work by causing the muscle in the walls of the blood vessels to constrict, decreasing blood flow. After many applications, these small muscles become fatigued and fail to respond. Finally, they are so fatigued that they relax entirely, and the situation becomes worse than it was in the beginning. Generally, this fatigue process does not occur in the first three days, so most doctors recommend the use of nose drops only for a short temporary problem. The advantage of Afrin is that it is reputed to have a longer lasting effect than Neosynephrine, but it is more costly. Neosinephrine comes in ⅛%, ¼%, ½%, and 1% solutions. Package instructions should be followed for dosage.

Perhaps the cheapest and safest nose drops can be made by mixing one-half teaspoon of salt in a glass of water. Many doctors feel these to be as effective as medicated nose drops. But our favorite remedy for the runny nose is the handkerchief, used frequently and gently!

ORAL DECONGESTANTS

Pseudoephedrine (Sudafed, Actifed) and ephedrine (Rondec) are commonly used decongestants. Again, their use in stuffy noses is to cause constriction of the blood vessels in the nose and therefore to decrease the stuffiness. Unfortunately, doses of these medications high enough to cause constriction of the blood vessels in the nose are also high enough to cause constriction of other blood vessels in the body and can produce high blood pressure. Although these drugs are unlikely to produce high blood pressure at the recommended dosage levels, they are also unlikely to produce the necessary blood vessel constriction to give significant relief. At this time, there is no convincing evidence that the use of these decongestants prevents the complications of sinusitis or middle-ear infection. Because of their relative safety in appropriate doses and because of the theoretical possibility that they do assist once a person has developed sinusitis or an earache (otitis media), they are commonly used for this purpose.

Children who have recurrent bouts of ear infections may be placed on decongestants at the first sign of a cold with the hope that perhaps another ear infection may be avoided. We consider middle-ear infections serious enough to justify the use of these relatively safe medications in some such circumstances, although their effectiveness is unproved. We do not recommend the routine use of decongestants for all colds or stuffy noses.

COUGH SYRUPS

The cough is a natural reflex that helps clear the child's lung of mucous secretions that are accumulating because of an infection. The cough then is one of the body's defenses. There are two types of commonly available cough medications, but they are prepared in dozens of ways. Many familiar products now come in a range of formulas. For instance, Robitussin includes other ingredients as Robitussin-DM and Robitussin-AC. Pediatric Formula 44 is also available as Pediatric Formulas 44d, 44e, and 44m. Such variety can confuse parents. Read the labels carefully to be sure that the preparation contains the correct essential ingredient, and avoid drugs combining essential ingredients unless they have been specifically recommended by your doctor.

The first type of cough medication is commonly known as an **expectorant**. An example is guaifenisen. Its purpose is to help liquefy the secretions in the lungs and help the cough reflex to remove these secretions from the lungs. The principle of liquefaction of secretions is extremely important, especially in such illnesses as croup. However, there is little evidence to indicate that any cough syrup is very effective in producing this liquefaction. Vaporizers are far more useful for liquefying secretions.

The second type of cough medication is a **cough suppressant.** The cough reflex is a natural defense that usually should be encouraged. There are times when coughing seems to interfere with a child's getting better, such as when a cough keeps that child from getting to sleep. Unfortunately, a recent study has shown that the most common suppressants, dextromethorphan and codeine, are no more effective at reducing coughs and related symptoms than placebos. Dextromethorphan is sometimes sold in combination with other drugs, which have also not been shown to be effective against coughing. An antihistamine, diphenhydramine, has also been used occasionally for cough suppression, but it is associated with drowsiness. Until studies demonstrate some benefit to cough suppressant medicines, the only way to suppress a cough may be time.

Occasionally during the daytime, children may have coughing that is so prolonged and so severe that it begins to cause chest pain; this may be another indication for suppressing the cough.

For Constipation

There is hardly ever a time when you should treat constipation in children with drugs. Prune juice, every grandmother's favorite remedy, is still effective in relieving constipation. And constipation (see page 318) is very seldom a real problem. Prune juice acts by drawing a large amount of water into the intestines, thereby helping to soften hard stools. Food high in fiber or bran content is also extremely effective. Although laxatives (such as Maltsupex, Milk of Magnesia, Ex-lax, Colace, mineral oil, and Metamucil) are used commonly in children, they are only very rarely required. Mineral oil is perhaps the cheapest and most effective but is dangerous in infants and toddlers because of the potential problems it can cause if vomited and inhaled into the lungs.

There is virtually no indication for giving a child an enema. An enema for a child can be extremely frightening, and rare serious complications have occurred.

For Diarrhea

The proper management of diarrhea is discussed elsewhere (see page 507). Certain solutions are quite valuable for a child to drink and are available in supermarkets and drugstores (e.g., Pedialyte, Ricelyte, Lytren). Medication is of little use in the treatment of ordinary diarrhea. Compounds such as attapulgite (frequently found in preparations such as Kaopectate) or kaolin and pectin will help change a liquid stool into a more gelatinous stool. However, during periods of diarrhea, the total amount of water lost and the total number of stools produced will be about the same. We do not see the necessity of using these types of preparations merely to change the form of the stool. They do not decrease the diarrhea or the amount of water lost. Some parents, however, find that leakage out of diapers is less of a problem with a more formed bowel movement.

Recently bismuth subsalicylate (the active ingredient in Pepto-Bismol) has been shown to reduce diarrhea if given in oral solutions every four hours over three days. However, the children in this study who were not given medicine for diarrhea had the same weight loss. Again, giving this medicine seems like a lot of effort for minimal changes.

Paregoric-containing preparations (such as Parapectolin and Parelixir) are also not recommended for use in children. Paregoric is a narcotic that decreases the activity of the digestive tract. This increased activity of the digestive tract is a defense mechanism that usually should not be suppressed. In addition, narcotic overdose with paregoric can occur, and drowsiness and nausea can be caused. And, like most other narcotics, paregoric can produce constipation. Lomotil contains a narcotic-like compound that is also not recommended for children.

For Eye Irritations

Eye irritations in children seldom require the use of over-the-counter preparations (Visine, Murine). "Pink eye" and its treatment are discussed in **Eye Burning, Itching, and Discharge** (page 346).

For Pain

Children may experience pain following an injury such as a burn or trauma; during the course of an acute illness (sore throat) or chronic disease (arthritis); or even following a medical intervention such as immunization or surgery. For chronic conditions or after surgery you, your child, and your physician will need to establish a pain management plan. For most everyday conditions, pain can effectively be managed at home.

Talk is often the best medicine. Most minor injuries will not even require a trip to the medicine cabinet. In fact, a parent who reaches for a pain reliever every time a tear accompanies a fall is teaching a child to rely on bottled solutions. For most injuries, distraction works well. Discussing plans for later in the day or talking about imaginary situations are often effective and have no side effects. Even children with chronic pain from serious medical conditions benefit enormously from verbal techniques such as guided imagery and hypnosis. Children who are about to undergo painful procedures in a medical setting will benefit from adequate preparation, including age-appropriate discussions of what will happen and opportunities to ask questions. The accompanying reduction in anxiety and pain has been well documented.

If medicine is necessary to relieve a child's pain, we recommend a three-tiered pain management plan that begins with adequate and timely doses of mild (and inexpensive) pain relievers such as acetaminophen. If pain continues, the next step would be to move on to mild opioid (addicting) drugs such as codeine, moderate opioids such as meperidine (Demerol), and finally the extraordinarily potent drugs used in hospitals such as morphine. All but the mildest pain relievers require physician contact and prescriptions. We shall focus here on the principal choices: acetaminophen, aspirin, and ibuprofen.

ACETAMINOPHEN

This is a mainstay of treatment for pain as well as fever. It is very safe and works by blocking the pain sensations carried through a child's nerves. It is available in a number of familiar brand names (Tylenol, Liquiprin, Tenlap) as well as generically. Unfortunately, it is also available in many forms, particularly liquids, that are confusing and often lead to consistent under- and overdosing by parents. Always read the label. Each dose should give a child 10 mg per kilogram (a bit more than 2 pounds) of body weight. A 20-pound child should receive 100 mg per dose. A 65-pound child should receive 325 mg (one regular adult tablet). Acetaminophen is a very good pain reliever (analgesic) but lacks the ability of aspirin and other drugs to manage inflammation that often occurs with joint or bone pain.

ASPIRIN

Aspirin is one of the oldest, and most useful pain medications available. However, its potential side effects cause us instead to recommend acetaminophen when children or teenagers have fever.

Certain individuals are particularly sensitive to aspirin. They may experience gastro-intestinal disturbances, including bleeding, following the use of aspirin. Very rarely, patients may actually have wheezing induced by aspirin; this is an allergy.

There are side effects that all individuals will experience if aspirin is taken in too high a dosage. The most common complaint of aspirin overdosage in older children is ringing of the ears, although this is an unreliable sign in younger children. In younger children, aspirin overdosage initially causes very rapid breathing, followed after a period of time by slowed breathing, lethargy (laziness), and unresponsiveness. This is seen only when large amounts of aspirin have been taken. Giving an infant twice the recommended dosage for one dose only is an accident that should not cause alarm. However, if adult aspirin is mistakenly administered rather than baby aspirin, *four times* the dosage has been given, and your doctor or the poison control center should be contacted immediately.

Because of a recent association with a rare but serious problem known as Reye's syndrome, aspirin should not be used for children who may have chicken pox or influenza. Because it is not always possible to know which virus is responsible for an illness, we no longer recommend aspirin for fevers.

It is a common mistake to believe that there is liquid aspirin. Aspirin is not available in liquid preparations; it is available only in tablets or in Aspergum. Liquid acetaminophen has sometimes erroneously been called "liquid aspirin."

IBUPROFEN

Ibuprofen is an example of the relatively new and increasingly popular drugs known as a non-steroidal anti-inflammatory drugs. It possesses many of the features characteristic of steroids in reducing pain from inflammation in the bones or joints. However, some of the side effects seen with aspirin (such as stomach upset) also occur, though less frequently. Unfortunately, the cost is considerably higher. Nevertheless, it remains a very important second-line drug should acetaminophen not be effective and aspirin not considered acceptable. It is available over the counter for children over 12 years. The liquid form, for children under 12, requires a prescription.

CODEINE

This weak narcotic requires a prescription and instructions from your physician. It is effective in pain relief and often combined with acetaminophen for increased relief.

For Fever

ACETAMINOPHEN

Acetaminophen (Tylenol, Tempra, Liquiprin, Valadol) is a very effective medication for reducing fever in children. It is the primary drug used for fever in children because it is safer than aspirin and because it is easier to administer as a liquid. It is also less expensive than ibuprofen. However, in extremely high doses, even acetaminophen can be fatal by causing massive liver damage.

Dosage   The dosages of acetaminophen, aspirin, and ibuprofen are discussed on pages 231–233. *Caution:* Acetaminophen is available in many concentrations. A teaspoon of one preparation (drops) can contain four times as much drug as a teaspoon of another preparation (elixir). Read the label carefully!

IBUPROFEN   Ibuprofen is considered as effective as acetaminophen in treating fevers in children older than six months. The fever reduction is actually somewhat better, and a single dose lasts longer than a single dose of acetaminophen for children older than two years with temperatures over 102.5°F. Nevertheless, it is still considered a second line of defense after acetaminophen because it has more side effects, is more expensive, and requires a prescription for children younger than 12 years. It shares a positive feature with acetaminophen in that they both reduce fever without reducing the production of interleukin-1, an important body chemical that fights infections.

In some situations doctors may advise combining acetaminophen with ibuprofen. This has not yet been tested, although tests of acetaminophen/aspirin combinations proved efficacious.

ASPIRIN   Because aspirin is associated with an increased risk of a very rare condition known as Reye's syndrome (particularly in children exposed to chicken pox or flu), it is no longer used to reduce fever during infections. Fortunately, there are excellent alternatives. However, in situations where fever is from a known source and there is virtually no likelihood of chicken pox or influenza, aspirin may have a role.

For Poisoning

For any item other than medication, a call to the poison control center or your doctor should be made *before* attempting to induce vomiting (see **Poisoning**, page 258); with medicines, call right after inducing the vomiting.

SYRUP OF IPECAC

To induce vomiting, use syrup of ipecac. Any time your child swallows a large number of pills or liquid medication, ipecac can be given. The dose is one tablespoon for young children, two tablespoons for older children. Give the child a glass of water or milk immediately after giving a dose of syrup of ipecac to induce vomiting. If there has been no vomiting in 20 minutes, repeat with another dose. If after 40 minutes there is no vomiting, it will be necessary for the child to be seen by a doctor in order to have his or her stomach emptied. Ipecac itself can produce problems if it is not vomited up.

There are other methods of inducing vomiting in children such as mixing mustard with warm water and forcing them to drink it, or by touching the back of the child's throat; these are less aesthetic but sometimes as good.

Shake Lotions

These lotions, such as calamine, have a cooling effect and are useful in diminishing the itching of a variety of rashes—from poison ivy to chicken pox. They should not be applied to raw or weeping areas of skin.

Sterilizing Agents and Antiseptics

The best way to clean a dirty wound is to scrub it with soap and water.

HYDROGEN PEROXIDE

Hydrogen peroxide, a sterilizing agent that foams and cleanses as you work it into the wound, is also a good cleansing agent. It should not be used in strengths greater than 3%, so watch out for bottles sold at higher strengths for bleaching hair.

IODINE
Although iodine, an antiseptic, is a good agent and kills germs, it is irritating to the skin and sometimes winds up in little children's mouths; it is a dangerous poison. In addition, some people are allergic to iodine. Betadine is a non-stinging iodine preparation but reasonably expensive. Soap, water, and hydrogen peroxide will take you a long way in dealing with the common scrapes and sores that all children get. Mercurochrome has a pretty color but is not effective. In only rare instances are antibiotic ointments more effective than soap and water.

BANDAGES
Band-Aids are for children what medals are for adults. They are worn proudly as symbols of surviving major confrontations with the ground. As such, they may be awarded when the child feels they are necessary. Here is another example where you can encourage your child to make a decision about the application of a treatment. For wounds in areas likely to have heavy exposure to dirt, Band-Aids (and occasionally gauze dressings) may be necessary in order to prevent further contamination and the possibility of infection.

Sunscreen

Products with a variety of ingredients effectively block out some of the sun's rays and can reduce the risk of sunburning and skin cancer. See **Sunburn**, page 444. Get a product with an SPF (sun-protection factor) of at least 15, which blocks out 94% of ultraviolet light. SPF 30 blocks out 97%.

Vaporizers

A vaporizer is one of the best investments you can make. It efficiently provides the fog necessary for the relief of croup and is soothing for many other coughs. A cold-steam vaporizer is preferable because there is no possibility of a child being burned by hot steam. It is not necessary to add any medication to the vaporizer. These preparations make the room smell nice but do not add to the therapy offered by the fog alone.

Vitamins

We mention vitamin pills only to emphasize that vitamin supplementation is *not* required for most children. An ordinary diet, balanced with foods from each of the major groups, contains far more vitamins than the growing body requires. Vitamin-D supplementation is recommended for infants during the period of breast feeding, but that is all. Minerals are also abundantly present in common foods. (See **Weakness and Fatigue,** page 329, for a list of iron-containing foods.) Fluoride needs to be provided if the water supply is deficient, as outlined in Chapter 9. The American child taking vitamin and mineral supplements secretes the most expensive urine in the world because that is where excess materials end up.

Giving Medicine

The medical encounter frequently results in instructions to give the child some medication.

Patients often leave the offices of doctors and pharmacists with considerable confusion and many unanswered questions. One study of patients leaving their doctors' offices revealed that over 50% made at least one error when describing what their doctors expected. This is not surprising, for recordings of the medical visits revealed that the doctors did not even discuss 20% of the medicines they prescribed. For 30% of these medications, the doctors gave no information about the name or purpose of the drug. Of all the patients in this study, 90% were not told by their doctors how long to take the medicines, and less than 5% of the prescription bottles contained this information. Clearly, then, there is a need for both better communication between doctors and patients and information to make young patients and their parents better informed.

There are a number of factors to keep in mind whenever you give your child medicine; here are some of the more important ones.

- Do you understand the instructions? Check before leaving the office. If the instructions on the medication bottle differ from what the doctor or pharmacist said, call your doctor immediately. If you are confused, call the doctor or the pharmacist.

- Be sure of the strength of the medication. Some common medications appear in many different concentrations, and the wrong strength may be dangerous.

- Be sure your child is not allergic to the medication. Even the most careful doctor occasionally forgets that a child may be allergic to penicillin and may prescribe it. Do not give your child anything that you know he or she is allergic to.

- Be as precise as possible in your measurements. Teaspoons vary greatly in size. When most doctors prescribe a teaspoon, they mean to prescribe 5 cubic centimeters (cc) of medication. Kitchen measuring spoons are more accurate. Many pharmacies sell small plastic measuring devices or give them away.

- Never give a child medication intended for another person.

- Never give a child medication if the expiration date has passed.

See Table 4 for a list of questions you should always ask yourself before giving any medicine.

Some of your most interesting moments with your children will be spent trying to give them medications. An average child seems able to spit an average medication a distance of 15 feet. While this may be good practice for the annual North Carolina Watermelon Pit Spitting Contest, medication on the walls has seldom been known to do the child any good. Getting medication into your children will be a great test of your ingenuity. Remember, you are older, wiser, more clever, and ultimately bigger. But here are some hints so that you don't have to use force.

- Never tell a child that medication is candy. As soon as your back is turned, children will sometimes try to get as many of these candies into their mouths as possible.

- Do not tell a child a medication tastes good when you know it doesn't. This will help get the first dose into the child, but you will have an impossible time with the second.

- For younger infants, you can mix some medications in with applesauce or ice cream.

TABLE 4   *Parent Medication Checklist*

BEFORE GIVING your child ANY medication make sure that the following questions are answered and understood.

- What is the medicine's name?
- What does it do?
- How much do I give?
- How often must I give it?
- How long do I need to continue giving the medicine?
- Are there special preparation instructions
  (e.g., do I need to shake it vigorously?)
- Are there special times to take the medicine?
- Must I refrigerate the medicine?
- Are there common side effects I can expect?
- Are there rare adverse risks that I should be aware of?
- If my child has a particular allergy,
  might he or she also be allergic to this drug?
- How much does this medicine cost?
- Is a generic form of comparable quality available?
- Does my child really need this medicine?
  Do its benefits outweigh its risks and costs?

- Medications usually do not give a pleasant flavor to milk and we discourage this practice; most children are familiar with how their milk tastes and are suspicious of funny-tasting milk.
- Toddlers often prefer tablets ground up in ice cream to seemingly tasty syrup preparations. Cranberry juice is another good place to hide medication.
- For infants younger than six months, use a syringe or calibrated eyedropper.

Ultimately, every parent will participate in a knock-down, drag-out fight with a toddler or pre-school-aged child over taking medicine. The child in this situation, with his or her ability to spit, vomit, and clench teeth, will win in a showdown every time. In fact, the more intense the struggle, the more the child will relish the fight.

In these situations, it pays to draw back for a few minutes, let the struggle defuse itself, and try again. This method may provide the toddler with enough sense of control that he or she will acquiesce to taking the medicine the next time around. The truly recalcitrant child is a difficult problem that should be discussed with your doctor, who may be able to suggest an easier way to deliver the medicine (for example, by syringe) or other approaches to giving that medication.

Finally, older children should be required to take medication as a matter of course. They should not need to be threatened or bribed any more than they need to be bribed or threatened when it is their bedtime. Children over the age of three-and-one-half can begin to be treated as adults with regard to taking medications. Development of proper respect for medication is important at this age. So, start talking to them about the importance of medication to help them through their illness; do not talk of medication as either magic or rewards. Ultimately, emphasize the importance of taking medication as directed, not more or less. Enlist older children's help in remembering to take medicine.

Tincture of Time

Used prudently, time is the most important medicine. It is the only known cure for the common cold, as well as most of the other problems of everyday life. With time, things get better. In the remainder of this book, we try to tell you how to use time and how long it should take.

Why do we, as a society, use drugs rather than time, even though time usually works and the drugs usually don't? Sure, we are impatient, confused by the complexities of science, and hustled by the advertisers. But let's not ignore the biggest reason. We use drugs, and sometimes the doctor, to prove that we care for our child. The statements "I'll run down and get something from the drugstore" or "You're going to the doctor first thing in the morning" are part of our everyday life. We must have something to do, if we care, for a sick child; in our society, we have equated caring with the giving of drugs.

Consider the consequences of such actions. The child receives a pill rather than a parent. He or she learns that a symptom requires a drug. Colds, scrapes, headaches, and constipation are associated with the need to imbibe a pill or some odious fluid. Later, the parent is disturbed when the child wants pills, shots, or fluids to cure boredom, unhappiness, or agitation, or just to interact socially with friends. And, although time will take the symptom away, the drug will take the credit. The child fails to learn that the body is strong and thinks instead that health is frail and only precariously maintained by an intake of chemicals.

If you care for your child, your instruction in health maintenance must express these truths. With a sick child, you can care by spending time instead of money. Non-drug treatments, such as encouraging fluids, running the vaporizer, and cleaning and soaking the wound will give you plenty to do. Most medical problems are learning experiences. If you and your child react and interact appropriately, the lessons can be very positive and can lead to emotional growth and physical confidence. The choice is yours, and the consequences are immense. It is drug dependence or personal independence. We hope that the guidelines of the remainder of this book will help your family toward the goal of personal self-reliance and independent living.

The Child and the Common Complaint

CHAPTER A

Interpreting Childhood Complaints

How to Use Part IV

In this part of *Taking Care of Your Child* you will find information and decision charts that show you how to deal with more than 100 common medical problems of children. The general information describes possible causes of the problems, methods for treating them at home, and what to expect at the doctor's office if you need to take your child there. The decision charts summarize this information, helping you decide whether to use home treatment or to consult a physician.

Follow these steps to use these sections.

1. Is emergency action necessary?

Usually the answer is obvious. The most common emergency signs are listed in the top box on page 246. More advice on these emergency problems is found starting on page 250. Other signs of illness that might indicate a need to take your child to the doctor are discussed on pages 251–253.

It is a good idea to read Chapter B, "Emergencies," *now* so that you're prepared for an emergency if one occurs. Fortunately, the great majority of children's complaints don't require emergency treatment.

2. Find the section that covers your medical problem.

Determine your child's chief complaint or symptom—for example, a cough, an earache, dizziness. Use the list of problems in the bottom box on page 246 to find the appropriate chapter. The first page of each chapter lists the problems it covers, organized by type of complaint or by area of the body: neck pain, shoulder pain, arm pain, and so on.

The illustration on page 247 will help you to find the right section for problems that are localized in one part of the child's body. You can also look up a symptom in the Table of Contents as well as in the Index. They may help you skip directly to the section you need.

3. Find the section for your child's worst problem first.

Your child may have more than one problem, such as abdominal pain, nausea, and diarrhea. In such cases, look up the most serious complaint first, then the next most serious, and so on.

You may notice some duplication of questions in the decision charts, especially when the symptoms are closely related. If you use more than one chart, use the most "conservative" outcome: if one chart recommends home treatment and the other advises a call to the doctor, then call the doctor.

4. Read all of the general information in the section.

The general material gives you important information about interpreting the decision chart. If you ignore it, you may inadvertently select the wrong course of action.

5. Go through the decision chart.

Start at the top. Answer every question, following the arrows indicated by your answers. *Don't* skip around: that may result in errors. Each question assumes that you've answered the previous question.

6. Follow the treatment indicated.

Sometimes there will be an instruction to go to another section, or a description or diagram of the action to take. More often you will find one of the instruction boxes shown at right and on the next page.

Don't assume that an instruction to use home treatment guarantees that the problem is trivial and may be ignored. Home therapy must be used conscientiously if it is to work. Also, as with all treatments, the home therapy may not be effective in a particular case so don't hesitate to visit a doctor's office if the problem doesn't improve or for other reasons.

Similarly, if the chart indicates that you should consult a doctor, it does not necessarily mean that the illness is serious or dangerous. Often a physical examination is necessary to diagnose the cause of the problem, or you will benefit from certain facilities at the doctor's office.

The charts usually recommend one of the following actions:

- **Apply Home Treatment**
Follow the instructions for home treatment closely, and keep it up. These steps are what most doctors recommend as a first approach to these problems.

If over-the-counter medications are suggested, look them up in the Index, and read about dosage and side effects in Chapter 10 before you use them.

There are times when home treatment is not effective despite conscientious application. Think the problem through again, using the decision chart. The length of time you should wait before calling your doctor can be found in the general information in most sections. If you become seriously worried about your child's condition, call the doctor.

- **See Doctor Now**
Go to your doctor or health-care facility right away. In the general information we try to give you the medical terminology related to each problem so that you can "translate" the terms your doctor may use during your conversation.

- **See Doctor Today**
Call your pediatrician's office and say that you are bringing your child in. Describe your child's problem over the phone as clearly as you can.

- **Make Appointment with Doctor**
Schedule a visit to your doctor's office any time during the next few days.

- **Call Doctor Today**
Often a phone call will enable you and your child's doctor or nurse to avoid unnecessary visits, using medical care more wisely. Remember that most doctors do not charge for telephone advice but regard it as part of their service to regular patients. Don't abuse this service in an attempt to avoid paying for necessary medical care.

If every call results in a recommendation for a visit, your pediatrician is probably sending you a message: Come and don't call. This is unfortunate, and you may want to look for a doctor willing to put the telephone to good use.

With these guidelines, you will be able to use the following sections to locate quickly the information you need while not burdening yourself with information that you don't require. Examine some of the charts now; you will swiftly learn how to find the answers to health problems.

On the following pages we list several indications that your child may be ill. Noticing most of these signs depends on your knowledge of how your child usually behaves. Above all, trust your own judgment. Remember that you know your child best. If your child appears quite sick, be sure to get the necessary help; common sense is your best guide in such matters.

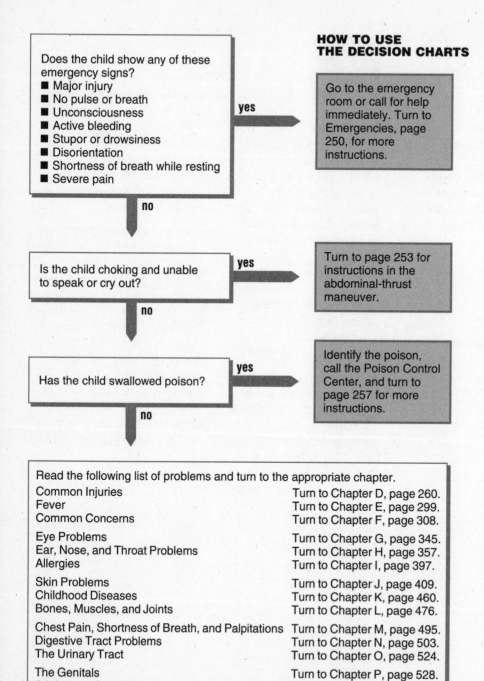

Does the child show any of these emergency signs?
- Major injury
- No pulse or breath
- Unconsciousness
- Active bleeding
- Stupor or drowsiness
- Disorientation
- Shortness of breath while resting
- Severe pain

yes

no

Is the child choking and unable to speak or cry out?

yes

no

Has the child swallowed poison?

yes

no

HOW TO USE THE DECISION CHARTS

Go to the emergency room or call for help immediately. Turn to Emergencies, page 250, for more instructions.

Turn to page 253 for instructions in the abdominal-thrust maneuver.

Identify the poison, call the Poison Control Center, and turn to page 257 for more instructions.

Read the following list of problems and turn to the appropriate chapter.

Common Injuries	Turn to Chapter D, page 260.
Fever	Turn to Chapter E, page 299.
Common Concerns	Turn to Chapter F, page 308.
Eye Problems	Turn to Chapter G, page 345.
Ear, Nose, and Throat Problems	Turn to Chapter H, page 357.
Allergies	Turn to Chapter I, page 397.
Skin Problems	Turn to Chapter J, page 409.
Childhood Diseases	Turn to Chapter K, page 460.
Bones, Muscles, and Joints	Turn to Chapter L, page 476.
Chest Pain, Shortness of Breath, and Palpitations	Turn to Chapter M, page 495.
Digestive Tract Problems	Turn to Chapter N, page 503.
The Urinary Tract	Turn to Chapter O, page 524.
The Genitals	Turn to Chapter P, page 528.
Adolescent Sexuality	Turn to Chapter Q, page 539.

Key to Localized Problems
See Table of Contents for a Complete Listing of Problems

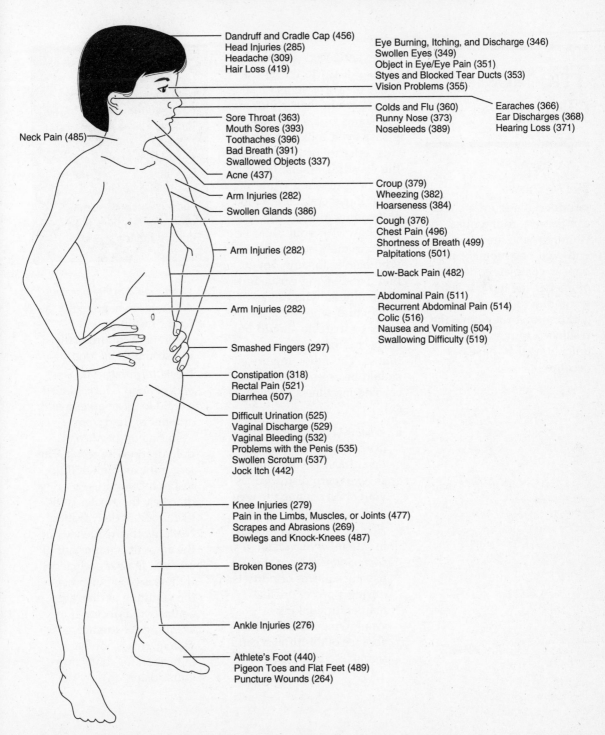

Dandruff and Cradle Cap (456)
Head Injuries (285)
Headache (309)
Hair Loss (419)

Eye Burning, Itching, and Discharge (346)
Swollen Eyes (349)
Object in Eye/Eye Pain (351)
Styes and Blocked Tear Ducts (353)
Vision Problems (355)

Colds and Flu (360)
Runny Nose (373)
Nosebleeds (389)

Earaches (366)
Ear Discharges (368)
Hearing Loss (371)

Sore Throat (363)
Mouth Sores (393)
Toothaches (396)
Bad Breath (391)
Swallowed Objects (337)

Neck Pain (485)

Acne (437)

Arm Injuries (282)

Swollen Glands (386)

Croup (379)
Wheezing (382)
Hoarseness (384)

Cough (376)
Chest Pain (496)
Shortness of Breath (499)
Palpitations (501)

Arm Injuries (282)

Low-Back Pain (482)

Abdominal Pain (511)
Recurrent Abdominal Pain (514)
Colic (516)
Nausea and Vomiting (504)
Swallowing Difficulty (519)

Arm Injuries (282)

Smashed Fingers (297)

Constipation (318)
Rectal Pain (521)
Diarrhea (507)

Difficult Urination (525)
Vaginal Discharge (529)
Vaginal Bleeding (532)
Problems with the Penis (535)
Swollen Scrotum (537)
Jock Itch (442)

Knee Injuries (279)
Pain in the Limbs, Muscles, or Joints (477)
Scrapes and Abrasions (269)
Bowlegs and Knock-Knees (487)

Broken Bones (273)

Ankle Injuries (276)

Athlete's Foot (440)
Pigeon Toes and Flat Feet (489)
Puncture Wounds (264)

The Sick Child

Sudden illness in a child can be very frightening. Children who are playing and well one moment may appear completely devoid of energy the next. It is a testimony to the strength of children that they have the resiliency to recover as quickly as they have become ill.

All experienced parents quickly learn to recognize what the early signs of illness are in their children. For some, it is a dazed or glassy-eyed look; for others, it is lethargy or bags under the eyes; for others, it is a pale or "pre-vomit white" color. In general, observation and common sense will tell you how sick your child is. An extremely active child who begins to slow down may be showing early signs of an illness, whereas a quiet child who becomes fussy or irritable should be suspected of having an illness. The following areas should be assessed whenever considering illness in your child.

- *What is your child's activity like compared to usual at this time of day? Is your child's sleep pattern disturbed? Is your child playing the way he or she usually plays?*

- *How does your child respond to pleasant or unpleasant stimulation?* If he or she usually squirms or protests during a procedure like temperature-taking or swallowing medicine, the absence of this protest can signal a serious illness.

As a general rule, all ill children under the age of six weeks should be brought immediately to the doctor; illness in the first three to four months generally warrants a phone call. Illness in this age group is potentially more serious because it may progress far more rapidly than in older children.

- *Is your child eating normally?* All children have some food finickiness, but severely ill children will refuse almost all food.

- *If vomiting or diarrhea is present, what is its nature?* If a child loses an excessive amount of fluid from vomiting or diarrhea, dehydration can result. The larger the amount of fluid lost in the vomitus or diarrhea, the greater is the likelihood of dehydration. Not only the frequency but the amount is important to consider in your evaluation. If there is blood in either the vomiting or diarrhea, contact your doctor. If the vomiting is extremely violent, this is another indication for contacting your doctor.

- *Has your child urinated?* Infrequent urination or dark yellow urine are signs that your child is becoming dehydrated.
- *What is your child's skin turgor like?* Gather the skin on your child's stomach together using your five fingers. When you release it, it should immediately spring back. Dehydrated skin does not have the elasticity of normal skin. If there is a question in your mind, compare the sick child's skin with another child's or your own. The skin of a dangerously dehydrated child is like the skin of a very old person.
- *What is your child's skin color?* Children often become flushed or may even look pale, but a bluish color should prompt an immediate doctor consultation.
- *Are your child's eyes and mouth moist?* A dry mouth or eyes that appear sunken are signs of dehydration that require immediate attention by your doctor.

- *What is your child's temperature?* Fever is discussed extensively in Chapter E, page 299. A high fever can make your child feel quite uncomfortable and increase fluid requirements. A fever not associated with physical exertion is a sign of illness in a child.
- *What is your child's heart rate?* Children have a higher heart rate than adults, and the heart rate increases further with fever. It may decrease after severe head injury. In general, pulse rates over 130 or under 60 when a child is resting warrant an immediate doctor visit.
- *How fast is your child breathing?* The rate at which your child breathes decreases as the child becomes older. Breathing rates are far higher after activity; when evaluating your child's breathing rate, the child should be resting. While many newborns have breathing rates of 50 to 60, by the age of one year resting rates are usually between 25 and 35. A resting rate over 40 is of concern except in children under one year old. By age six, resting respiratory rates should be below 30

and by age 10, below 25. Fever is a common cause of an elevated breathing rate, so assess your child's breathing rate at rest after you have attempted to reduce the fever.

As you become more experienced with illnesses in your children, these observations, and many of your own that are far more subtle, will become intuitive. You will soon learn that you are the best judge of illness in your child. Doctors can only help in diagnosing the specific causes of the illness. The purpose of Part IV of this book is to assist you in managing many of the more commonly recognized illnesses on your own.

CHAPTER B

Emergencies

Emergencies require prompt action, not panic. What action you should take depends on the nature of the problem and the facilities available.

If there are massive injuries or if your child is unconscious, you must get help immediately. Go to the emergency room if it is close by. Have someone call ahead if you can.

If you can't get to the emergency room quickly, you can often obtain help over the phone by calling the emergency room or the paramedic rescue squad. Calling for help is especially important if you think that your child has swallowed poison. Poison control centers and emergency rooms can often tell you over the phone how to counteract the poison, thus beginning treatment as early as possible. See Chapter C, "Poisons," on page 257.

Work out a procedure for medical emergencies. Develop and test it before an actual emergency arises. If you plan your actions ahead of time, you will decrease the likelihood of panic and increase the probability of receiving the proper care quickly.

The most important thing is to be prepared. Record the phone numbers of the nearest emergency room, poison control center, and ambulance or paramedic rescue squad in the front of this book and on page 257. Know the best way to reach the emergency room by car. Develop these procedures *before* an actual emergency arises.

WHEN TO CALL AN AMBULANCE

Usually, the slowest way to reach a medical facility is by ambulance. It must go both ways and is not twice as fast as a private car. If your child can readily move or be moved and a private car is available, use the car and have someone call ahead.

The ambulance brings with it a trained crew, who know how to lift a patient to minimize the chance of further injury. Oxygen is usually available, splints and bandages are carried, and, in some instances, lifesaving resuscitation may be used en route to the hospital. Thus, the care afforded by the ambulance attendants may most benefit a child who:

- Is gravely ill
- Has a back, neck, or head injury
- Is severely short of breath

Ambulances are too expensive to use as taxis. The type of accident or illness, the facilities available, and the distance involved are all important factors in deciding whether an ambulance should be used.

EMERGENCY SIGNS

The decision charts in the rest of this book assume that no emergency signs are present. Emergency signs "overrule" the charts and dictate that medical help should be sought immediately. Be familiar with the following emergency signs.

Major Injury

Common sense tells us that the child with an obviously broken leg or a large chest wound deserves immediate attention. Emergency facilities exist to take care of major injuries. They should be used, and promptly.

No Pulse or Breath

Again, a child whose heart or lungs are not working needs help right away. Call 911 for help. If you know CPR, start it after you call for help.

Unconsciousness

Obviously, any child in a state of coma or semi-consciousness should be brought immediately to the nearest medical facility. Coma is most often due to a medication or other toxic product taken by mouth, a seizure, drowning, severe head trauma, or a severe allergic reaction. Any medication or other suspected material that might have been taken should be brought to the medical facility with you. Children breathing with difficulty should have their mouths cleared. Artificial respiration can be given at the rate of ten breaths per minute either through the mouth or the nose.

Active Bleeding

Most cuts will stop bleeding if pressure is applied to the wound. Unless the bleeding is obviously minor, a wound that continues to bleed despite the application of pressure requires attention in order to prevent unnecessary loss of blood. The average adult can tolerate the loss of several cups of blood with little ill effect, but children can tolerate only smaller amounts, proportional to their body size. Remember that active and vigorous bleeding can almost always be controlled by the application of pressure directly to the wound. This is the most important part of first aid for such wounds.

Stupor or Drowsiness

A decreased level of mental activity, short of unconsciousness, is termed "stupor." A practical way of telling whether the severity of stupor or drowsiness warrants urgent treatment is to note the child's ability to answer questions. If he or she is not sufficiently awake to answer questions concerning what has happened, then urgent action is necessary. Children are difficult to judge, but the child who cannot be aroused needs immediate attention.

Disorientation

In medicine, disorientation is described in terms of time, place, and person. This simply means that the child cannot tell the date, the location, or who he or she is. The child who does not know his or her own identity is in a more difficult state than one who cannot give the correct date.

Disorientation may be part of a variety of illnesses and is especially common when a high fever is present. The child who becomes disoriented and confused deserves immediate medical attention.

Shortness of Breath

Shortness of breath is described more extensively on page 499. As a general rule, a child deserves immediate attention if there is shortness of breath while resting. However, in young adults the most frequent cause of shortness of breath at rest is the hyperventilation syndrome, which is not a serious concern (see page 326). Nevertheless, if it cannot be confidently determined that shortness of breath is due to the hyperventilation syndrome, then the only reasonable course of action is to seek immediate aid.

Severe Pain

Surprisingly enough, severe pain is rarely the symptom that determines that a problem is serious and urgent. Most often it is associated with other symptoms that indicate the nature of the condition; the most obvious example is pain associated with a major injury, such as a broken leg, which itself clearly requires urgent care.

The severity of pain is subjective and depends on the particular child; often the magnitude of the pain has been altered by emotional and psychological factors. Nevertheless, severe pain demands urgent medical attention, if for no other reason than to relieve the pain.

Much of the art and science of medicine is directed at the relief of pain, and the use of emergency procedures to secure this relief is justified even if the cause of the pain eventually proves to be inconsequential. However, the person who frequently complains of severe pain from minor causes is in much the same situation as the boy who cried "wolf"; calls for help will inevitably be taken less and less seriously by the doctor. This situation is a dangerous one, for there may be more difficulty in obtaining help when it is most needed.

Poisoning

Poisoning is described on page 258. Seldom does the delay of a few moments make any difference in the eventual outcome. However, making a hasty wrong decision can be dangerous.

Many poisons do their damage while being swallowed (acids, strong alkalis, drain and oven cleaners), and vomiting should *not* be induced. Other poisons (turpentine, gasoline, furniture polish) cause damage from their vapors, and, again, vomiting should *not* be induced. Medication can be safely vomited. *Always bring the poison with you to the doctor or emergency room.*

Seizures (Convulsions)

Seizures are discussed on page 334. During the seizure, it is most important to protect the child from injury. Except for children known to have recurrent seizures, a prompt medical visit is required.

CHOKING

If an object has become lodged in your child's windpipe, choking may ensue. A child who can still speak or cry out is still breathing; in this case, it is best to go to the emergency room to have the object extracted. Do not try to dislodge the object yourself if the child can breathe, because you might inadvertently cause complete obstruction. Violent coughing will often dislodge the object naturally.

If the child is choking on an object and not breathing, see the next page for advice.

M ost communities now offer courses in infant cardiopulmonary resuscitation (CPR). Costs are usually minimal, and this may be a most valuable time investment for your child and your peace of mind.

Choking

Choking is an emergency situation, but emergency medical services—doctors, EMTs, ambulances, emergency rooms, hospitals —play virtually no role in its treatment. In almost every case, the child's fate will be decided by the time they respond. Someone must step forward and relieve the choking.

The best approach to choking is preventive. Don't feed hard or large pieces of food to small children. Be especially careful with hot dogs, grapes, peanuts, and hard candy. Unfortunately, infants seem to put everything in their mouths.

Reading the advice in this book is not the best preparation for a real choking emergency. Learn how to deal with choking by taking one of the courses that teach CPR.

HOME TREATMENT

The most effective way to relieve choking in adults, adolescents, and older children is with the abdominal-thrust, or Heimlich, maneuver. Pushing on the lungs from below rapidly raises the air pressure inside the lungs and behind the foreign object causing the choking. This results in the forceful expulsion of the food from the throat back into the mouth. Done properly, an abdominal-thrust maneuver does not pose great risk of doing harm. Still, it's not the kind of thing that you want to do to a child who will not benefit from it. If the child in difficulty can speak, forget about the abdominal-thrust maneuver.

FOR ADOLESCENTS AND OLDER CHILDREN

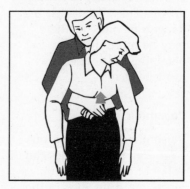

1. Stand behind the child and place your arms around him or her. Make a fist and place it against the child's abdomen, thumb side in, between the navel and the breastbone.

2. Hold the fist with your other hand, and push upward and inward, four times quickly.

If the victim is a pregnant or obese adolescent, place your arms around his or her chest and your hands over the middle of the breastbone. Give four quick chest thrusts.

If the child is lying down, roll the child over onto his or her back. Place your hands on the abdomen and push in the same direction on the body that you would if the victim were standing (inward, and toward the upper body).

3. If the child does not start to breathe, open the mouth by moving the jaw and tongue, and look for the swallowed object. *If you can see the object,* sweep it out with your little finger. If you try to remove an object you can't see, you may only push it in more tightly.

4. If the victim does not begin to breathe after the object has been removed from the air passage, use mouth-to-mouth resuscitation.

5. Call for help, and repeat these steps until the object is dislodged and the victim is breathing normally.

FOR SMALL CHILDREN

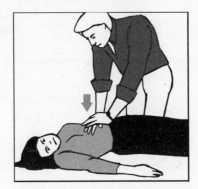

1. Kneel next to the child who should be lying on his or her back.

2. Position the heel of one hand on the child's abdomen between the navel and the breastbone. Deliver six to ten thrusts inward and toward the upper body.

3. If this doesn't work, open the mouth by moving the jaw and tongue and look for the swallowed object. *If you can see the object,* sweep it out of the throat using your little finger.

4. If the child does not begin to breathe after the object has been removed, use mouth-to-mouth resuscitation.

5. Call for assistance, and repeat these steps until the object is dislodged and the child is breathing normally or until help arrives.

FOR INFANTS

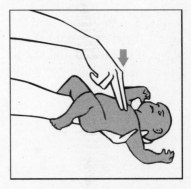

1. Hold the infant along your forearm, face down, so that the head is lower than the feet.

2. Deliver four rapid blows to the back, between the shoulder blades, with the heel of your hand.

3. If this doesn't work, turn the baby over and, using two fingers, give four quick thrusts to the chest.

4. If you're still not successful, look for the swallowed object in the throat the same way you would for an older child. *If you can see it*, try to sweep it out gently with your finger.

5. If the infant doesn't begin to breathe after the object has been removed, use mouth-to-nose-and-mouth resuscitation.

6. Call for assistance, and repeat these steps until the object is dislodged and the patient is breathing normally.

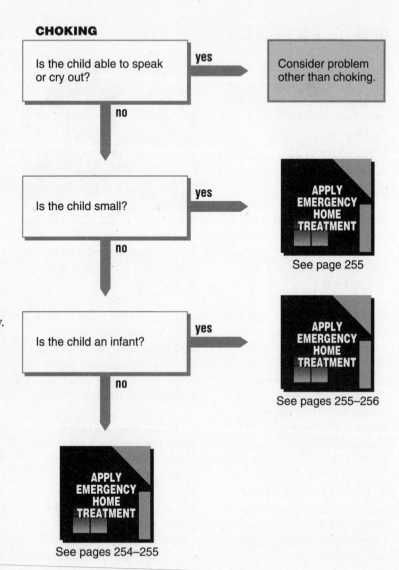

CHOKING

Is the child able to speak or cry out? — yes → Consider problem other than choking.

no

Is the child small? — yes → **APPLY EMERGENCY HOME TREATMENT**
See page 255

no

Is the child an infant? — yes → **APPLY EMERGENCY HOME TREATMENT**
See pages 255–256

no

APPLY EMERGENCY HOME TREATMENT
See pages 254–255

CHAPTER C

Poisons

Poisoning *258*
What to do on the way to the doctor.

Write down these phone numbers now:

POISON CONTROL CENTER

_____

EMERGENCY ROOM

_____

Write these numbers down in the front of this book as well, and keep them by your phone.

Many communities have established poison control centers to identify poisons and give advice. These are often located in emergency rooms. Find out if such a center exists in your community and, if so, record the telephone number here and in the front of this book. Quick first aid and fast professional advice are your best chance to avoid a tragedy.

Poisoning

Although poisons may be inhaled or absorbed through the skin, for the most part they are swallowed. The term *ingestion* refers to oral poisoning.

Most poisoning can be prevented. Children almost always swallow poison accidentally.

Don't allow children to reach potentially harmful substances such as the following:

- Medications
- Insecticides
- Caustic cleansers
- Organic solvents
- Fuels
- Furniture polishes
- Antifreezes
- Drain cleaners

The last item is the most damaging; drain cleaners such as Drano are strong alkali solutions that can destroy any tissue they touch.

Identifying the Problem

Treatment must be prompt to be effective, but identifying the poison is as important as speed. *Don't panic.* Try to identify the swallowed substance without taking up too much time. If you cannot or if the victim is unconscious, go to the emergency room right away.

If you can identify the poison, call the doctor or poison control center immediately and get advice on what to do. Always bring the container with you to the hospital. Life-support measures come first in the case of an unconscious victim, but the ingested substance must be identified before proper therapy can begin.

Suicide attempts cause many significant medication overdoses in teenagers. Any suicide attempt is an indication that help is needed. Such help is not optional, even if the patient has "recovered" and is in no immediate danger. Most successful suicides are preceded by unsuccessful attempts.

HOME TREATMENT

All cases of poisoning require professional help. Someone should call immediately. If the child is conscious and alert and the ingredients swallowed are known, there are two types of treatment: those in which vomiting should be induced and those in which it should not.

Do *not* induce vomiting if the child has swallowed any of the following:

- *Acids*—battery acid, sulfuric acid, hydrochloric acid, bleach, hair straightener, etc.
- *Alkalis*—Drano, drain cleaners, oven cleaners, etc.
- *Petroleum products*—gasoline, furniture polish, kerosene, lighter fluid, etc.

These substances can destroy the esophagus or damage the lungs as they are vomited. Neutralize them with milk while contacting the physician. If you don't have milk, give the child water or milk of magnesia.

Vomiting is a safe way to remove medications, plants, and suspicious materials from the stomach. It is more

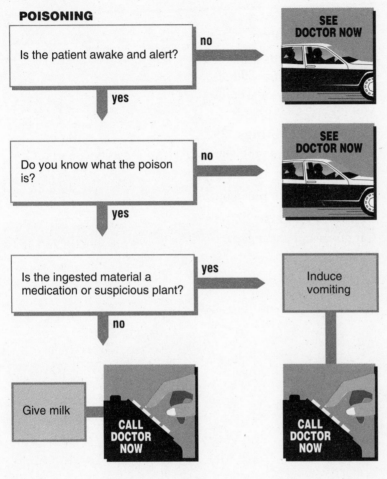

POISONING

Is the patient awake and alert? — no → **SEE DOCTOR NOW**

↓ yes

Do you know what the poison is? — no → **SEE DOCTOR NOW**

↓ yes

Is the ingested material a medication or suspicious plant? — yes → **Induce vomiting**

↓ no

Give milk ← **CALL DOCTOR NOW**

Induce vomiting → **CALL DOCTOR NOW**

Vomiting usually follows within 20 minutes. Mustard mixed with warm water also works. If there is no vomiting within 25 minutes, repeat the dose. Collect what comes up so that it can be examined by the doctor.

Before, during, and after first aid for poisoning, contact a doctor.

If an accidental poisoning has occurred, make sure that it doesn't happen again. Refer to Chapter 9 for information on "childproofing" your house.

WHAT TO EXPECT AT THE DOCTOR'S OFFICE

Significant poisoning is best managed at the emergency room. Treatment of the conscious child depends on the particular poison and whether vomiting has been achieved successfully. If indicated, the stomach will be evacuated by vomiting or by the use of a stomach pump. Children who are unconscious or have swallowed a strong acid or alkali will require admission to the hospital. With those who are not admitted to the hospital, observation at home is important.

effective and safer than using a stomach pump and does not require the doctor's help. Vomiting can sometimes be achieved immediately by touching the back of the throat with a finger. Don't be squeamish! This is usually the fastest way, and time is important.

Another way to induce vomiting is to give two to four teaspoons of **syrup** (not extract) **of ipecac**, followed by as much liquid as the child can drink.

Cuts (Lacerations)

Most cuts affect only the skin and the fatty tissue beneath it. Usually, they heal without permanent damage. However, injury to internal structures such as muscles, tendons, blood vessels, ligaments, or nerves presents the possibility of permanent damage. Your doctor can decrease this likelihood.

Deeper Damage

You may find it difficult to determine whether major blood vessels, nerves, or arteries have been damaged. These signs all call for examination by a doctor:

- Numbness
- Bleeding that cannot be controlled with pressure
- Tingling
- Weakness in the affected limb

Signs of infections—such as pus oozing from the wound, fever, extensive redness and swelling—will not appear for at least 24 hours. Bacteria need time to grow and multiply. If these signs do appear, a doctor must be consulted.

Stitches

Stitching (suturing) a laceration is a ritual in our society. The only purpose in suturing a wound is to pull the edges together to hasten healing and minimize scarring. If the wound can be held closed without the use of stitches, they are not recommended because they themselves injure tissue to some extent.

Stitching must take place within eight hours of the injury, because germs begin to grow in the wound and can be trapped under the skin to fester. Thus the chart says "See doctor now." Decide immediately whether to see a doctor or treat at home. Also refer to **Tetanus Shots** (page 271).

Difficult Cuts

A cut on the face, chest, abdomen, or back is potentially more serious than one on the legs or arms (extremities). Luckily, most lacerations do occur on the extremities. Cuts on the trunk or face should be examined by a doctor unless the injury is very small or extremely shallow. If you see fat protruding from the wound, see the doctor.

In a young child who drools, facial wounds are often too wet to treat with bandages, so the doctor's help is usually needed. Because of potential disfigurement, all but minor facial wounds should be treated professionally. Often, stitching is required in young children, who are apt to pull off bandages, or in areas that are subject to a great deal of motion, such as the fingers or joints. Cuts in the palm are also prone to infection, so do not attempt home treatment unless the cut is shallow.

See **For Pain** (page 230).

HOME TREATMENT

Cleanse the wound. Soap and water will do, but be vigorous. Hydrogen peroxide (3%) can also be used. Make sure that no dirt, glass, or other foreign material remains in the wound. This is very important. Antiseptics such as Mercurochrome and Merthiolate are unlikely to help, and some are painful. Iodine will kill germs but is not really needed and is also painful. (Betadine is a modified iodine preparation that is painless but costly.)

The edges of a clean, minor cut can usually be held together by "butterfly" bandages or "steri-strips" (preferred)—strips of sterile paper tape. Apply either of these bandages so that the edges of the wound join without "rolling under."

See the doctor if the edges of the wound cannot be kept together, if signs of infection appear (pus, fever, or extensive redness and swelling), or if the cut is not healing well within two weeks.

Your doctor will tell you when the stitches are to be removed. Unless there is some other reason to return to the doctor, you can perform this simple procedure.

1. Clean the skin and the stitches. Sometimes a scab must be removed by soaking.

2. Gently lift the stitch away from the skin by grasping a loose end of the knot.

3. Cut the stitch at the far end as close to the skin as possible and pull it out. A pair of small, sharp scissors or a fingernail clipper works well. It is important to get close to the skin so that a minimum amount of the stitch that was outside the skin is pulled through. This reduces the chance of contamination and infection.

CUTS

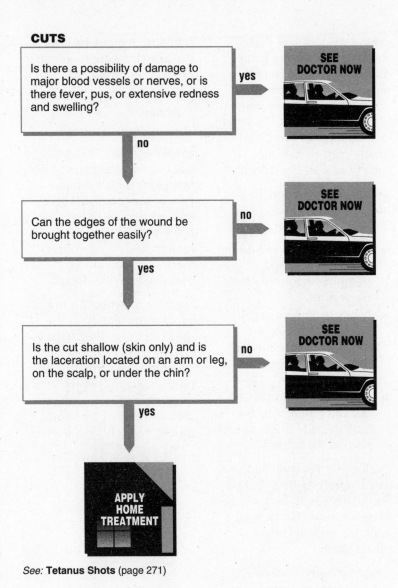

Is there a possibility of damage to major blood vessels or nerves, or is there fever, pus, or extensive redness and swelling?

yes → SEE DOCTOR NOW

no

Can the edges of the wound be brought together easily?

no → SEE DOCTOR NOW

yes

Is the cut shallow (skin only) and is the laceration located on an arm or leg, on the scalp, or under the chin?

no → SEE DOCTOR NOW

yes

APPLY HOME TREATMENT

See: **Tetanus Shots** (page 271)

WHAT TO EXPECT AT THE DOCTOR'S OFFICE

The wound will be thoroughly cleansed and explored to be sure that no foreign particles are left in the wound and that blood vessels, nerves, and tendons are undamaged. The doctor may use an anesthetic to deaden the area. Be aware of any allergy to lidocaine (Xylocaine) or other local anesthetics; report any possible allergy to the doctor. The doctor will determine need, if any, for a tetanus shot and decide whether antibiotics are needed (usually not).

Lacerations that may require a surgical specialist include those with injury to tendons or major vessels, especially when this damage has occurred in the hand. Facial cuts may also require a surgical specialist if a good cosmetic result appears difficult to obtain.

Puncture Wounds

Puncture wounds are those caused by nails, pins, tacks, and other sharp objects. The most important question is whether a tetanus shot is needed. Consult **Tetanus Shots** (page 271) to determine this. Occasionally, puncture wounds do occur in which further medical attention is required.

Signs to Call the Doctor

Most minor puncture wounds are located in the extremities, particularly in the feet. If the puncture wound is located on the head, abdomen, or chest, a hidden internal injury may have occurred. Unless a wound in these areas is obviously minor, see a doctor.

Many doctors feel that puncture wounds of the hand, if not very minor, should be treated with antibiotics. Once started, infections deep in the hand are difficult to treat, and many lead to loss of function. Call the doctor.

Injury to a nerve or to a major blood vessel is rare but can be serious:

- Injury to an **artery** may be indicated by blood pumping vigorously from the wound.
- Injury to a **nerve** usually causes numbness or tingling in the wounded limb beyond the site of the wound.
- Injury to a **tendon** causes difficulty in moving the limb (usually finger or toe) beyond the wound.

Major injuries such as these occur rarely with a narrow implement such as a needle; they are more likely with a nail, ice pick, or larger instrument.

Unfortunately, because drug abuse remains common in our society, children encounter discarded needles in playgrounds and on buses and beaches. A puncture wound from a discarded needle requires prompt medical care.

Infection

To avoid infection, be absolutely sure that nothing has been left in the wound. Sometimes, for example, part of a needle will break off and remain in the foot. If there is any question of a foreign body's remaining, the wound should be examined by the doctor.

Signs of infection do not occur immediately at the time of injury; they usually take at least 24 hours to develop. The formation of pus, a fever, or severe swelling and redness are indications that the wound should be seen by a doctor.

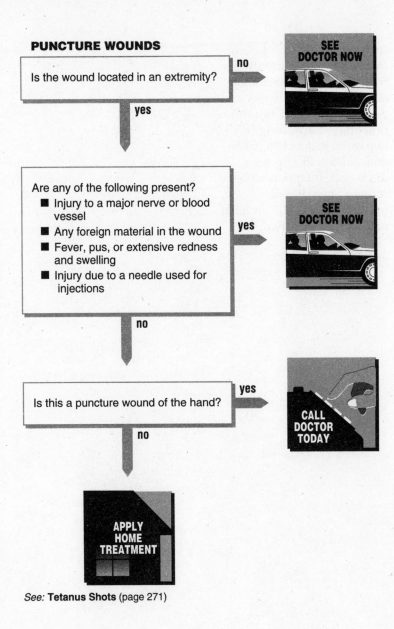

PUNCTURE WOUNDS

Is the wound located in an extremity? — **no** → **SEE DOCTOR NOW**

↓ **yes**

Are any of the following present?
- Injury to a major nerve or blood vessel
- Any foreign material in the wound
- Fever, pus, or extensive redness and swelling
- Injury due to a needle used for injections

— **yes** → **SEE DOCTOR NOW**

↓ **no**

Is this a puncture wound of the hand? — **yes** → **CALL DOCTOR TODAY**

↓ **no**

APPLY HOME TREATMENT

See: **Tetanus Shots** (page 271)

HOME TREATMENT

Cleanse the wound with soap and warm water or hydrogen peroxide (3%) to prevent infection. Let it bleed as much as possible to carry foreign material to the outside, because you cannot scrub the inside of a puncture wound. Do not apply pressure to stop the bleeding unless there is a large amount of blood loss and a "pumping," squirting bleeding.

Soak the wound in warm water several times a day for four to five days. The object of the soaking is to keep the skin puncture open as long as possible, so that any germs or foreign debris can drain from the open wound. If the wound is allowed to close, an infection may form beneath the skin but not become apparent for several days. Consult **Tetanus Shots** (page 271).

See **For Pain** (page 230).

See the doctor if there are signs of infection or if the wound has not healed within two weeks.

WHAT TO EXPECT AT THE DOCTOR'S OFFICE

The doctor will answer the questions on the decision chart from the patient's history and examination. The wound will be surgically explored if necessary. More frequently, it will be observed for a reaction to a foreign body over the next few days. If a metallic foreign body is suspected, X-rays may be taken. Be prepared to tell the doctor the date of the last tetanus shot.

Most doctors will recommend home treatment. Antibiotics will only rarely be suggested. In puncture wounds caused by buckshot, the shot may be left in the skin. Occasionally, glass or wood may be left in for a period of time to give the body time to push it to the surface.

In the unlikely event of an injury with a needle used for injections, your doctor may consider measures to protect your child from the potential of acquiring hepatitis. Risks of hepatitis and AIDS will also be discussed. Infants who have received hepatitis-B immunization will not be at risk for this infection.

Animal Bites

The question of rabies is uppermost following an animal bite. Although 3000 to 4000 animals are found each year with rabies, only one or two humans annually contract the disease in the United States.

The main carriers of rabies are wild animals, especially skunks, foxes, bats, raccoons, and possums. Rabies is also carried, though rarely, by cattle, dogs, and cats, and it is extremely rare in squirrels, chipmunks, rats, and mice.

Rabid animals act strangely, attack without provocation, and may foam at the mouth. Be concerned if the attacking animal has any of these characteristics.

Any bite by an animal other than a pet dog or cat requires consultation with the doctor as to whether the use of anti-rabies vaccine will be required. If the bite is by a dog or a cat, if the animal is being reliably observed for sickness by its owner, and if its immunizations are up to date, then consultation with the doctor is not required.

If the bite has left a wound that might require stitching or other treatment, consult **Cuts** (page 261) or **Puncture Wounds** (page 264). You should also check **Tetanus Shots** (page 271). Facial wounds should be checked by a doctor because of potential cosmetic disfigurement.

HOME TREATMENT

An animal whose immunizations are up to date is, of course, unlikely to have rabies. However, arrange for the animal to be observed for the next 15 days to make sure that it does not develop rabies. Most often, the owners of the animal can be relied on to observe it. If the owners cannot be trusted, then the animal must be kept for observation by the local public agency charged with that responsibility. Many localities require that animal bites be reported to the health department. If the animal should develop rabies during this time, a serious situation exists, and treatment by a doctor must be started immediately.

For the wound itself, use soap and water. Treat bites as cuts (see page 261) or puncture wounds (page 264), depending on their appearance. The best approach to animal bites is to avoid getting them. We have suggested some ways in which parents can teach their children to get along with dogs and avoid bites on pages 267–268.

See **For Pain** (page 230).

WHAT TO EXPECT AT THE DOCTOR'S OFFICE

The doctor must balance the usually remote possibility of exposure to rabies against the hazards of rabies vaccine or anti-rabies serum. An unprovoked attack by a wild animal or a bite from an animal that appears to have rabies may require both the rabies vaccine and the anti-rabies serum. The extent and locality of the wounds also play a part in this decision; severe wounds of the head are the most dangerous.

ANIMAL BITES

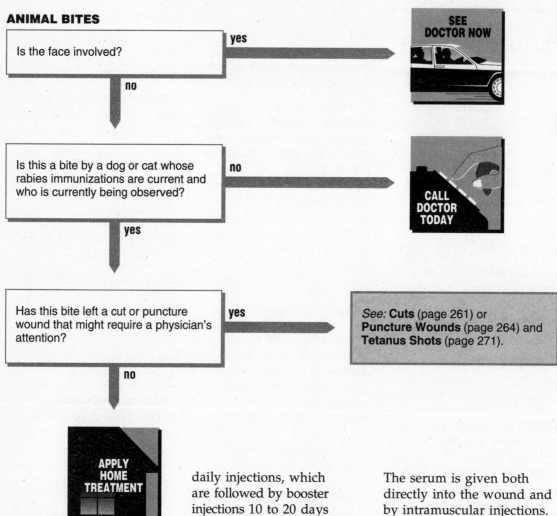

Is the face involved?

yes → SEE DOCTOR NOW

no ↓

Is this a bite by a dog or cat whose rabies immunizations are current and who is currently being observed?

no → CALL DOCTOR TODAY

yes ↓

Has this bite left a cut or puncture wound that might require a physician's attention?

yes → *See:* **Cuts** (page 261) or **Puncture Wounds** (page 264) and **Tetanus Shots** (page 271).

no ↓

APPLY HOME TREATMENT

A bite caused by an animal that has then escaped will often require the rabies vaccine. This is one of the most difficult decisions in medicine. Rabies vaccine is administered in 14 to 21 daily injections, which are followed by booster injections 10 to 20 days after the initial series. The vaccine will often cause local skin reactions as well as fever, chills, aches, and pain. Severe reactions to the vaccine are rare. The anti-rabies serum, unfortunately, has a high risk of serious reactions.

The serum is given both directly into the wound and by intramuscular injections.

Many doctors give a tetanus shot if the child is not "up to date," because tetanus bacteria can (rarely) be introduced by an animal bite. Be sure you know when your child last received a tetanus shot.

Scrapes and Abrasions

Scrapes and abrasions are shallow. Several layers of the skin may be torn or even totally scraped off, but the wound does not go far beneath the skin. Abrasions are usually caused by falls onto the hands, elbows, or knees, but skateboard and bicycle riders frequently find ways to get abrasions on just about any part of their bodies. Because abrasions expose millions of nerve endings, all of which send pain impulses to the brain, they are usually much more painful than cuts.

HOME TREATMENT

Remove all dirt and foreign matter. Washing the wound with soap and warm water is the most important step in treatment. Hydrogen peroxide (3%) can also be used to cleanse the wound. Most scrapes will "scab" rather quickly; this is nature's way of "dressing" the wound. Using Mercurochrome, iodine, and other antiseptics does little good and is usually painful.

Adhesive bandages may be necessary for a wound that continues to ooze blood, but they should be discontinued as soon as possible to allow air and sun to reach the wound. Remove or replace bandages if they get wet.

The main benefit of some antibacterial ointments (Neosporin, Bacitracin, etc.) is in keeping bandages from sticking to the wound. They are optional.

Loose skin flaps, if they are not dirty, can be left to help form a natural dressing. If the skin flap is dirty, cut it off carefully with nail scissors. (If it hurts, stop! You're cutting the wrong tissue.)

Watch the wound for signs of infection—pus, a fever, or severe swelling or redness—but don't be worried by redness around the edges; this indicates normal healing. Infection will not be obvious in the first 24 hours; serious infection without fever is rare.

Pain can be treated for the first few minutes with an ice pack in a plastic bag or towel applied over the wounds as needed. The worst pain subsides fairly quickly, and pain medication can then be used if necessary (see page 230).

See the doctor if the scrape is not healed within two weeks.

WHAT TO EXPECT AT THE DOCTOR'S OFFICE

The doctor will make sure that the wound is free of dirt and foreign matter. Soap and water and hydrogen peroxide (3%) will often be used. Sometimes a local anesthetic is required to reduce the pain of the cleansing process.

An antibacterial ointment is sometimes applied after cleansing the wound. Mupirocin is especially effective. Betadine is a painless iodine preparation that is also occasionally used.

Tetanus shots are not required for simple scrapes, but if your child is overdue, it is a good chance to get caught up.

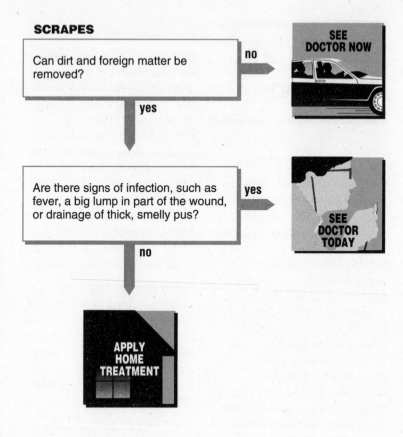

SCRAPES

Can dirt and foreign matter be removed? — **no** → **SEE DOCTOR NOW**

yes ↓

Are there signs of infection, such as fever, a big lump in part of the wound, or drainage of thick, smelly pus? — **yes** → **SEE DOCTOR TODAY**

no ↓

APPLY HOME TREATMENT

Tetanus Shots

Parents often bring their children to the doctor's office or emergency room simply to get a tetanus shot. Often the wound is minor and needs only some soap and water. If the shot is not needed, you don't need a doctor. The chart on page 272 illustrates the essential recommendations. It can save you and your children several visits to the doctor.

The question of whether a wound is minor may be troublesome. Wounds caused by sharp, clean objects, such as knives or razor blades, have less chance of becoming infected than those in which dirt or foreign bodies have penetrated and lodged beneath the skin. Abrasions and minor burns will not result in tetanus. The tetanus germ cannot grow in the presence of air; the skin must be cut or punctured for the germ to reach an airless location.

IMMUNIZATION

If your child has never had a basic series of three tetanus shots, then you should see the doctor. Sometimes a different kind of tetanus shot is required if you have not been adequately immunized. This shot is called **tetanus immune globulin** and is used when immunization is not complete and there is a significant risk of tetanus. This shot is more expensive, more painful, and more likely to an allergic reaction than is the tetanus booster. So keep a record of your family's immunizations in the back of this book and know the dates.

During the first tetanus shots (usually a series of three injections given in early childhood), immunity to tetanus develops over a three-week period. This immunity then slowly declines over many months. After each booster, immunity develops more rapidly and lasts longer. If your child has had an initial series of five tetanus injections, immunity will usually last at least ten years after every booster injection. Nevertheless, if a wound has left contaminated material beneath the skin and not exposed to the air and if your child has not had a tetanus shot within the past five years, a booster shot is advised to keep the level of immunity as high as possible.

TETANUS SHOTS

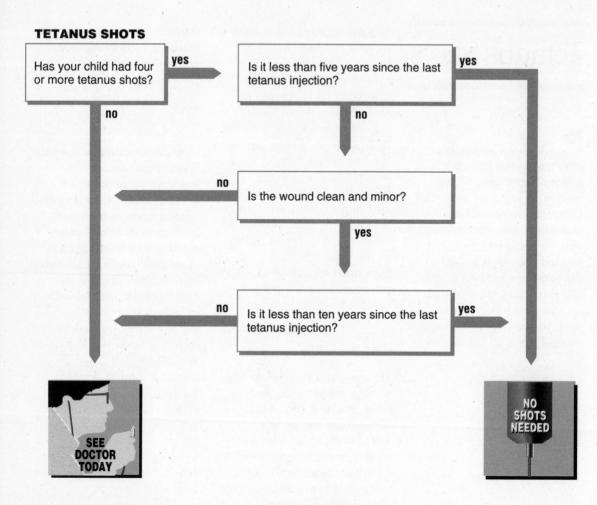

Tetanus immunization remains very important, because the tetanus germ is quite common and the disease (lockjaw) is so severe. Be absolutely sure that each of your children has had the basic series of three injections and appropriate boosters. Because the immunity lasts so long, adults usually get away with a long period between boosters, but with children it should be "by the book." (See pages 193–207.)

Broken Bones

Neither parent nor doctor can always see whether or not a bone is broken. So you often need an X-ray if there is a reasonable suspicion of a fracture. The chart on page 275 is a guide to "reasonable suspicion."

In the majority of fractures, the bone fragments are already aligned for good healing. Thus, prompt manipulation of the fragments is not necessary. If the injured part is protected and resting, a delay of several days before casting does no harm. Remember that the cast does not have healing properties; it just keeps the fragments from getting joggled too much during the healing period. Possible fractures are discussed further at **Ankle Injuries** (page 276), **Knee Injuries** (page 279), and **Arm Injuries** (page 282).

Serious Fractures

A fracture can injure nearby nerves and arteries. If the limb is cold, blue, or numb, see the doctor now! Fractures of the pelvis or thigh are particularly serious. Fortunately, these fractures are relatively rare except when great force is involved, as in automobile accidents. In these situations the need for immediate help is obvious. For head injuries, see page 285.

Paleness, sweating, dizziness, and thirst can indicate shock, and immediate attention is needed.

A crooked limb is an obvious reason to check for fracture. Pain that prevents use of the injured limb suggests the need of an X-ray. Soft-tissue injuries (skin, muscle, fat, tendons) usually allow some use of the limb.

Although large bruises under the skin may be caused by soft-tissue injuries alone, marked bruising in a limb that may have a fracture means that you should see the doctor.

Common sense tells us that when great force is involved, such as in an auto accident, the possibility of a broken bone is increased. The child who has fallen 20 feet out of a tree is much more likely to have a broken limb than is the child who has stumbled and fallen.

Children's bones are younger and hence more flexible and resilient than those of adults. Instead of outright breaks, young bones often bend or splinter like young tree limbs; these breaks are called **greenstick fractures**. Young bones are also still growing. The growth plates of bones are near the ends. As a result, an injury to a bone near the end must be treated more cautiously because damage may stop limb growth.

HOME TREATMENT

Apply ice packs. The immediate application of cold will help to decrease swelling and inflammation. If a broken bone is suspected, the involved limb should be protected and rested for at least 48 hours. To rest a bone effectively, the joint above and below the bone should be immobilized. For example, you suspect a fracture of the lower arm, the splint should prevent the wrist and elbow from moving. Magazines, cardboard, and rolled newspaper will all serve. Do not wrap tightly or circulation will be cut off.

Any injury that is still painful after 48 hours should be examined by a doctor. Minutes and hours are *not* crucial unless there is misalignment or injury to arteries or nerves. A limb that is adequately protected and rested is likely to have a good outcome even if casting is delayed. See **For Pain** (page 230).

WHAT TO EXPECT AT THE DOCTOR'S OFFICE

Usually, an X-ray will be required. In a small number of cases, it is possible to be relatively sure that an X-ray is not needed from the history and physical examination. A crooked limb must be "set." Plaster casts are rapidly being replaced with plastic splints. Prescription pain medications may be used. Sometimes general anesthesia is required to realign a bone. Pinning the fragments together surgically is required for certain fractures, such as elbow fractures.

BROKEN BONES

Are any of the following conditions present?
- The limb is cold, blue, or numb.
- The fracture is in the pelvis or thigh.
- The child is sweaty, pale, dizzy, or thirsty.

yes → **SEE DOCTOR NOW**

no

Is the limb crooked?

yes → **SEE DOCTOR NOW**

no

Is the limb not usable or unable to bear weight?

yes → **SEE DOCTOR TODAY**

no

Is there a great deal of bleeding and bruising in the area, was the injury the result of a severe blow, or is the possible fracture near a joint?

yes → **SEE DOCTOR TODAY**

no

APPLY HOME TREATMENT

Ankle Injuries

Ligaments are tissues that connect the bones of a joint to provide stability during the joint's action. When the ankle is twisted severely, either the ligament or the bone must give way. If the ligaments give, they may be stretched (strained), partially torn (sprained), or completely torn (torn ligaments). If the ligaments do not give, then one of the bones around the ankle must break (fracture).

Strains, sprains, and even some minor fractures of the ankle will heal well with home treatment. Even some torn ligaments may do well without a great deal of medical care. Immediate attention is necessary when the injury has been severe enough to cause fracture to the bones or a torn ligament. This is indicated by a deformed joint with abnormal motion. Fractures are more likely in a fall from a considerable height or in automobile or in bicycle accidents. They are *not* likely to happen when the ankle is twisted while walking or running. (See page 273.)

Swelling

The typical ankle sprain swells either around the bony bump at the outside of the ankle or about two inches in front of and below it. The usual sprain does not need prolonged rest, casting, or X-rays. Home treatment should be started promptly. Detection of damage to the ligaments is difficult immediately after the injury because of the amount of swelling that may be present. Because it is easier to do an adequate examination of the foot after the swelling has gone down and because no damage is done by resting a mild fracture or torn ligament, there is no need to rush to the doctor.

Sprains and torn ligaments usually swell quickly, because there is bleeding into the tissue around the ankle. The skin will turn blue-black in the area as the blood is broken down by the body. The amount of swelling will not differentiate between sprains, tears, and fractures.

Pain

Pain tells you what to do with ankle injuries. If what the child is doing hurts, don't do it. If pain prevents standing on the ankle for more than 24 hours, see the doctor. If little progress is being made so that pain makes weight bearing difficult at 72 hours, see the doctor.

ANKLE INJURIES

Is the ankle deformed or bending in an abnormal fashion?

yes → SEE DOCTOR NOW

no

Has pain prevented the ankle from bearing any weight for more than 24 hours?

yes → SEE DOCTOR TODAY

no

Has pain made weight bearing difficult for more than 72 hours?

yes → SEE DOCTOR TODAY

no

APPLY HOME TREATMENT

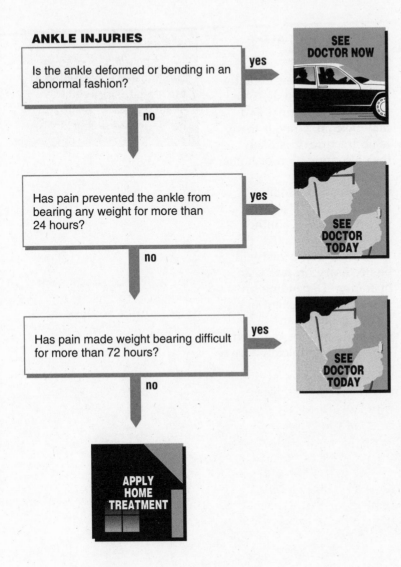

HOME TREATMENT

Home treatment is adequate for all ankle injuries except for some fractures and complete ligament tears. Even if a fracture is present, if the ankle is rested and protected, no harm will be done by waiting and watching.

Wrapping with an elastic bandage may prevent some swelling and damage. Elevate the ankle and keep it elevated. Do *not* let the child return to play as soon as the pain becomes bearable. Apply ice in a towel to the injured area and leave it there for at least 30 minutes. If there is any evidence of swelling after the first 30 minutes, then ice should be applied for 30 minutes on and 15 minutes off through the next few hours. If pain subsides completely in the elevated position, then weight bearing may be attempted cautiously. If pain is present when bearing weight, then weight bearing should be avoided for the first 24 hours.

If at the end of the first 24 hours the pain prevents weight bearing, then see the doctor. If pain is present but not severe, use crutches until walking can be accomplished with little discomfort, usually two or three days. During this time, an elastic bandage may be used, but this will not prevent re-injury if full activity is resumed. Do not stretch the bandage so that it is very tight and interferes with blood circulation. The ankle should feel relatively normal by about ten days. Full healing will not take place for from four to six weeks. See **For Pain** (page 230).

WHAT TO EXPECT AT THE DOCTOR'S OFFICE

The doctor will examine the motions of the ankle to see if they are abnormal and may take an X-ray. If there is no fracture, it is likely that a continuation of home treatment will be recommended. Home treatment may also be recommended if a minor chip fracture is noted. For other fractures, a cast will be necessary or, rarely, an operation. Depending on the nature and extent of a ligament injury, an operation may be required to repair a completely torn ligament.

Adjust **crutches** so that the shoulder support is two finger-breadths short of the armpit. The weight should be taken on the hand, not the armpit.

Knee Injuries

The ligaments of the knee may be stretched (strained), partially torn (sprained), or completely torn (torn ligament). Unlike the ankle, torn ligaments in the knee need to be repaired surgically as soon as possible after the injury occurs. If surgery is delayed, the operation is more difficult and less likely to be successful. For this reason, the approach to knee injuries is more cautious than for ankle injuries. If there is any possibility of a torn ligament, go to the doctor.

Fractures in the area of the knee are less common than around the ankle, and all need to be cared for by a doctor.

Knee injuries usually occur during sports, when the knee is more likely to experience twisting and side contact. (Deep knee bends stretch knee ligaments and may contribute to knee injuries; they should be avoided.) Serious knee injuries occur when the leg is planted on the ground and a blow is received to the knee from the side. If the foot cannot give, the knee will. There is no way to totally avoid this possibility in athletics. The use of shorter spikes and cleats help, but elastic knee supports and wraps give virtually no protection.

Abnormal Motion

When ligaments are completely torn, the lower leg can be wiggled from side to side when the leg is straight. Compare the injured knee to the opposite knee to get some idea of what amount of side-to-side motion is normal. Your examination will not be as skilled as that of the doctor, but if you think that the motion may be abnormally loose, see the doctor.

If the cartilage within the knee has been torn, then the normal motion of the knee may be blocked, preventing it from being straightened out. Although a torn cartilage does not need immediate surgery, it deserves prompt medical attention.

Pain and Swelling

The amount of pain and swelling does not indicate the severity of the injury. The ability to bear weight, to move the knee through the normal range of motion, and to keep the knee stable when wiggled is more important.

Typically, strains and sprains hurt immediately and continue to hurt for hours and even days after the injury. Swelling in strains and sprains tends to come on rather slowly over a period of hours, but they may reach rather large proportions. When a ligament is completely torn, there is intense pain immediately, which subsides until the knee may hurt little or not at all for a while. Usually, there is significant bleeding into the tissues around the joint when a ligament is torn so that swelling tends to come on quickly and be impressive in its quantity.

The best policy when there is a potential injury to the ligament is to have the child avoid any major activity until it is clear that this strain or sprain is minor. Home treatment is intended only for minor strains and sprains.

HOME TREATMENT

RIP is your memory key—

- Rest
- Ice
- Protection

Get the child off the knee and elevate it. Apply ice in a towel for at least 30 minutes to minimize swelling. If there is more than slight swelling or pain, despite the fact that the knee was put immediately to rest and ice was applied, see the doctor. If this is not the case, then ice should be on the knee for 30 minutes and then off for 15 minutes for the next several hours. Wrapping with an elastic bandage may prevent some swelling and damage. Limited weight bearing may be attempted during this time with a close watch for increased swelling and pain.

Heat may be applied after 24 hours. By 24 hours, the knee should look and feel relatively normal; after 72 hours, this should clearly be the case. Remember, however, that a strain or sprain is not completely healed for four to six weeks and that it requires protection during this healing period. Elastic bandages will not give adequate support but will ease symptoms a bit and remind the child to be careful with the knee. See **For Pain** (page 230).

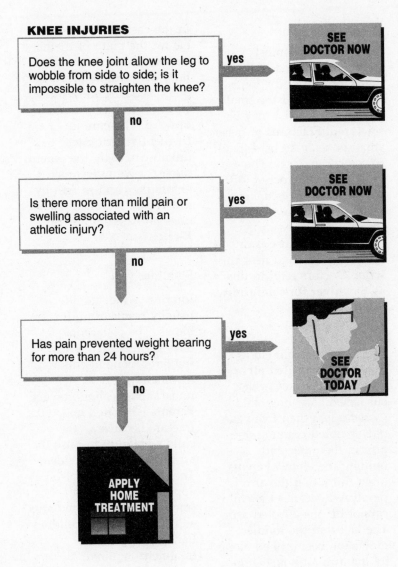

KNEE INJURIES

Does the knee joint allow the leg to wobble from side to side; is it impossible to straighten the knee?

yes → SEE DOCTOR NOW

no

Is there more than mild pain or swelling associated with an athletic injury?

yes → SEE DOCTOR NOW

no

Has pain prevented weight bearing for more than 24 hours?

yes → SEE DOCTOR TODAY

no

APPLY HOME TREATMENT

WHAT TO EXPECT AT THE DOCTOR'S OFFICE

The knee will be examined for range of motion, and the lateral stability will be tested by stressing the knee from side to side. A massively swollen knee may have blood removed from the joint with a needle. Torn ligaments need surgical repair. X-rays may be taken but are not always helpful. For injuries that appear minor, home treatment will be advised. Pain medications are sometimes, but not often, required.

Arm Injuries

The ligaments of the wrist, elbow, and shoulder joints may be stretched (strained) or partially torn (sprained), but complete tears are rare in children. This is because the weakest points of long bones in children are the soft cartilage growth plates at the bone ends. Trauma will often result in injury to these growth plates; injuries to bone ends must be treated cautiously. Fractures may occur at the wrist, are less frequent around the elbow, and are uncommon around the shoulder. Injuries to wrist and elbow occur most often during a fall, when the weight of the body is caught on the outstretched arm.

Wrists

The wrist is the most frequently injured of these joints. Strains and sprains are common and the small bones in the wrist may be fractured. Fractures of these small bones may be difficult to see on an X-ray. The most frequent fracture of the wrist involves the ends of the long bones of the forearm and is easily recognized because it causes an unnatural bend near the wrist. Doctors refer to this as the **silver fork deformity**.

Elbow

The most frequent elbow injury is the **pulled elbow**, which is often not even suspected. A young child (usually less than five years old) is noted cradling one arm in the other and holding the elbow. Parents often think that the arm is paralyzed because the child cannot lift the affected arm. The fact that the toddler may have been pulled along by the arm an hour before is often not remembered, although this is probably what caused the injury.

In the case of a pulled elbow, the palm of the hand is facing down toward the floor or inward toward the belly. The cure for a pulled elbow is to turn the palm upward. The cure is frequently performed unknowingly by the parent, nurse, or X-ray technician before the child is seen by the doctor. A pulled elbow does not show up on an X-ray.

Shoulder

Injuries to the shoulder usually result from direct blows. The collarbone (clavicle) is a frequently fractured bone in children; fortunately, it has remarkable healing powers. Parents will often notice the fracture because of the child's inability to raise the arm on the affected side. The shoulders may also appear uneven. This fracture occurs in newborns as well as in older children. Bandaging is all the treatment required.

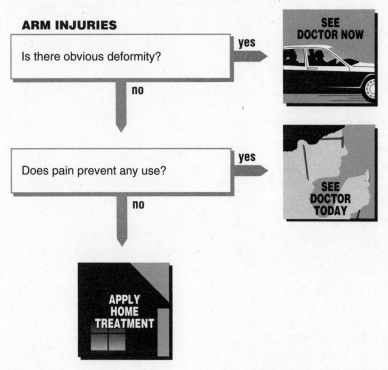

ARM INJURIES

Is there obvious deformity? — **yes** → SEE DOCTOR NOW

no ↓

Does pain prevent any use? — **yes** → SEE DOCTOR TODAY

no ↓

APPLY HOME TREATMENT

HOME TREATMENT

Pulled Elbow

Bend the elbow so that the forearm and upper arm form a right angle. With one hand, hold the elbow so it cannot move. With the other hand, grasp the child's hand and wrist and turn the palm upward, while gently pulling away from the body at the elbow. Some doctors prefer a quick, forceful twist, but we think a gentle turn does as well. Initially, this turning will cause discomfort. A click may then be felt or even heard. Repeating this maneuver, followed by bringing your child's palm up to the shoulder (bending the elbow), may enhance your chances for success. Stop the procedure if your child complains of severe pain.

If treatment occurs soon after the injury, then immediate relief of pain is usual. If treatment is delayed, then some soreness usually remains for a short period. Great force is not required for this treatment. If success does not come easily, see your doctor.

The shoulder separation often seen in high school athletes is perhaps the most common injury of the shoulder. It is a stretching or tearing of the ligament that attaches the collarbone to one of the bones that forms the shoulder joint. It causes a slight deformity and extreme tenderness at the end of the collarbone. Sprains and strains of other ligaments occur, but complete tearing is rare, as are fractures. Dislocations of the shoulder are rare outside of high school athletics, but are best treated early when they do occur.

In summary, severe fractures and dislocations are best treated early. These usually cause deformity, severe pain, and limitation of movement. Other fractures will not be harmed if the injured limb is rested and protected. Complete tears of ligaments are rare; strains and sprains will heal with home treatment.

Sprains and Strains

RIP is the memory key—

- Rest
- Ice
- Protection

Rest the arm and apply ice for at least 30 minutes. If the pain is gone and there is no swelling at the end of this time, the ice may be discontinued. A sling for shoulder and elbow injuries and a partial splint for wrist injuries will give protection and rest to the injury while allowing the child to move around.

If swelling appears, ice wrapped in a towel applied for 30 minutes on and 15 minutes off may be continued through the first 8 hours. Heat may be applied *after* 24 hours. The injured joint should be usable with little pain within 24 hours and should be almost normal by 72 hours. If not, see the doctor.

Complete healing takes from four to six weeks; activities with a likelihood of re-injury should be avoided by the child if possible during this time.

See **For Pain** (page 230).

WHAT TO EXPECT AT THE DOCTOR'S OFFICE

An examination and sometimes X-rays will be performed. A broken bone may require a cast. The pulled elbow will be fixed if you didn't fix it already. A sling may be devised. Strong pain medication is sometimes given, but aspirin or acetaminophen are often effective and are less hazardous.

Head Injuries

Every child will experience a bang on the head sometime in life, and many children seem to bump their heads every few days. Many of these injuries will be minor, such as those from walking into a table or falling from the couch.

A serious head injury is more likely to occur with more severe trauma, such as falling from a roof, being hit by a baseball, or automobile accidents. Injuries from such accidents are usually preventable.

- Bicyclists, skateboarders, roller skaters, and baseball players should always wear helmets.
- Children riding in a car should always wear seat belts or, if younger, be protected by a car seat.
- Parents should educate their children about safety.

All head injuries are potentially serious, but few lead to problems. The major concern in a head injury in which the skull is not obviously damaged is the occurrence of bleeding inside the skull. The accumulation of blood inside of the skull will eventually compress the brain and cause damage. Fortunately, nature has carefully cushioned the valuable contents of the skull. In infants, the fontanel, or soft spot, serves as a safety valve to help diminish the severity of head injuries.

Signs of Serious Injury

Careful observation is the most valuable tool for diagnosing serious head injury. This can be done as well at home as at the hospital; there is some risk either way, and it is your choice.

Observation of your child begins with the accident. If your child was knocked unconscious or cannot remember the events immediately before or after the accident, it is evidence of a concussion, and the child should be brought to the doctor.

The initial observation period is crucial. Bleeding into the head can be very rapid within the first 24 hours, and may continue for 72 hours or more. Some very slow bleeding may occur; this is called a **subdural hematoma** and may produce chronic headache, persistent vomiting, or personality changes months after the injury.

How does your child act? Increased lethargy (laziness), alternating alert and drowsy periods during the day, and unresponsiveness are all signs of possible bleeding within the skull. The child may be lethargic, seem to recover, and then become lethargic again.

Vomiting usually occurs at lease once after any significant head injury. If *repeated* vomiting occurs, see the doctor. The seriously affected child also cannot be easily aroused.

How does your child look? Children who appear persistently pale, sweaty, or weak should be brought to the doctor. Look also for unequal pupil size, which can be caused by pressure on the brain created by blood within the skull. Some children have pupils that are unequal all the time; this is normal for them. If the pupils become unequal after an injury, however, it is a serious sign.

A slow pulse (less than 60) or an irregular pulse is a sign of internal bleeding.

In a typical minor head injury, a bump may immediately develop. The child remains conscious and cries immediately. For a few minutes, the child is inconsolable and may vomit once or twice over the first few hours. Some sleepiness from the excitement may be noted; the child may nap but is easily aroused. Neither pupil is enlarged and the vomiting ceases shortly. The child does not appear pale, and the pulse is strong and regular. Within eight hours, the child is back to normal except for the tender and often prominent "goose egg."

HOME TREATMENT

For minor injuries, ice applied to a bruised area may minimize swelling. Children often develop "goose eggs" anyway. The size of the bump does not indicate the severity of the injury.

In serious accidents, injury to the chest, abdomen, or extremities must not be overlooked.

You should reassess your child's condition frequently. If there are any suspicious signs in your child, consult the doctor by phone at once. Because most accidents occur in the evening hours, children will be asleep several hours after most accidents; you can look in on them periodically to check their pulse, pupils, and arousability if you are concerned. With minor head bumps, nighttime checking is usually not necessary.

WHAT TO EXPECT AT THE DOCTOR'S OFFICE

The doctor will take an extensive history on the nature of the accident, assess the child's general appearance, and take repeated blood pressures and pulse rates. In addition, the head, eyes, ears, nose, throat, neck, and nervous system will be examined. The doctor will also check for other possible sites of injury, such as the chest, abdomen, and arms and legs.

The diagnosis of bleeding within the skull cannot be made with great accuracy. Skull X-rays are seldom helpful except in detecting whether a fragment of bone from the skull has been pushed into the brain, but this situation is rare. With severe injuries, neck X-rays may occasionally be required.

Where internal bleeding is of concern, special radiological tests (computerized tomography) may be ordered or the child may be hospitalized for observation. During this observation period, the child's pulse, pupils, and blood pressure will be checked periodically. Use of medications, which may obscure the situation, will be avoided.

HEAD INJURIES

Have any of the following occurred?
- Severe injury
- Child was knocked unconscious
- Child cannot remember injury
- Seizure

yes → SEE DOCTOR NOW

no ↓

Are any of the following present?
- Visual problems
- Bleeding from other than scalp
- Black eyes or blackness behind ears
- Change in child's behavior (sleepiness, irritability, laziness)
- Fluid draining from nose
- Persistent or slow vomiting
- Irregular breathing or heartbeat

yes → SEE DOCTOR NOW

no ↓

Is there a cut?

yes → See: **Cuts** (page 261).
See: **Tetanus Shots** (page 271).

no ↓

APPLY HOME TREATMENT

Burns

How bad is a burn? Burns are classified according to depth.

First-degree burns are superficial and cause the skin to turn red. A sunburn is usually a first-degree burn. First-degree burns may cause a lot of pain but are not a major medical problem. Even when they are extensive, they seldom give rise to lasting problems and seldom need a doctor's attention.

Second-degree burns are deeper and result in splitting of the skin layers or blistering. Scalding with hot water or a very severe sunburn with blisters are common instances of second-degree burns. Second-degree burns are also painful, and extensive second-degree burns may cause significant fluid loss. Scarring, however, is usually minimal, and infection usually is not a problem.

Second-degree burns can be treated at home if they are not extensive. Any second-degree burn that involves an area larger than the child's hand should be seen by a doctor. In addition, a second-degree burn that involves the face or hands should be seen by a physician; these might result in cosmetic problems or loss of function.

Third-degree burns destroy all layers of the skin and extend into the deeper tissues. They are *painless*, because nerve endings have been destroyed. Charring of the burned tissue is usually present. Third-degree burns result in scarring and present frequent problems with infection and fluid loss. The more extensive the burn, the more difficult these problems. All third-degree burns should be seen by a doctor, because not only do they possibly lead to scarring and infection but skin grafts also are often needed.

Prevention

Most burns are avoidable. Review potential fire hazards in your home with the help of Chapter 9. Teach your older child what to do if burned. Be sure your child knows *not* to run if clothing catches on fire as this will fan the flames. A child should stop, drop, and roll on the ground to put out flames.

HOME TREATMENT

Apply cold water or ice immediately. This reduces the amount of skin damage caused by the burn and also eases pain. The cold should be applied for at least five minutes and continued until pain is relieved or for one hour, whichever comes first. Be careful not to apply cold so long that the burned area turns numb because frostbite can occur! It may be re-applied if pain returns.

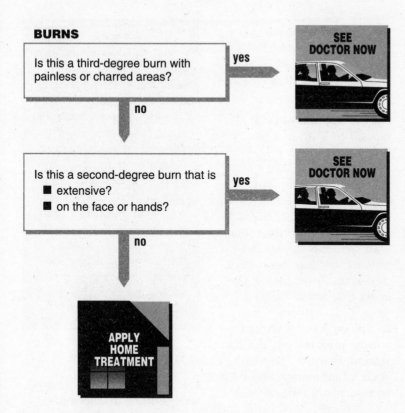

BURNS

Is this a third-degree burn with painless or charred areas?

yes → SEE DOCTOR NOW

no ↓

Is this a second-degree burn that is
- extensive?
- on the face or hands?

yes → SEE DOCTOR NOW

no ↓

APPLY HOME TREATMENT

WHAT TO EXPECT AT THE DOCTOR'S OFFICE

The doctor will establish the extent and degree of the burn and will determine the need for antibiotics and further treatment. An anti-bacterial ointment and dressing, with frequent changes and checks for infections, will often be recommended. Extensive burns may require hospitalization, and third-degree burns may eventually require skin grafts.

Acetaminophen can be used to reduce pain. Blisters should not be broken. If they burst by themselves, as they often do, the overlying skin should be allowed to remain as a wet dressing. The use of local anesthetic creams or sprays is not recommended, because they may slow healing. Also, some patients develop an irritation or allergy to these drugs. Do not use butter, cream, or Vaseline; they may slow healing and increase the possibility of infection. Certain antibiotic creams (such as Neosporin and Bacitracin) probably neither help nor hurt minor burns and are expensive.

Any burn that continues to be painful for more than 48 hours should be seen by a doctor.

Infected Wounds and Blood Poisoning

Blood poisoning, to a doctor, means bacterial infection in the bloodstream and is termed **septicemia**. Fever is a more reliable guide than red streaks to this rare occurrence. A local wound should only cause a very minor temperature elevation unless infected. If there is a fever, see the doctor.

An infected wound usually festers beneath the surface of the skin, resulting in pain and swelling. Bacterial infection requires at least a day, and usually two or three days, to develop. Therefore, a late increase in pain or swelling is a legitimate cause for concern. If the festering wound bursts open, pus will drain out. This is good, and the wound will usually heal well. Still, this demonstrates

that an infection was present, and the doctor should evaluate the situation unless it is clearly minor.

An explanation of normal wound healing will be helpful.

1. The body pours out serum into a wound area. Serum is yellowish and clear and later turns into a scab. *Serum is frequently mistaken for pus. Pus is thick, cheesy, smelly, and never seen in the first day or so.*

2. The edges of a wound will be pink or red, and the wound area may be warm. Such inflammation around a wound is normal.

3. The lymphatic system is actively involved in debris clearance, and pain along lymph channels or in the lymph nodes themselves may be present *without* infection.

There is a folk saying that red streaks running up the arm or leg from a wound are blood poisoning and that the patient will die when the streaks reach the heart. In fact, such streaks are only an inflammation of the lymph channels carrying away the debris from the wound. They will stop when they reach local lymph nodes in the armpit or groin and do not, by themselves, indicate blood poisoning. However, they usually are worth checking with your doctor.

HOME TREATMENT

Keep a wound clean. If it is unsightly or in a location where it gets dirty easily, bandage it, changing bandages daily; if not, leave it open to the air. Soak and clean it gently with warm water for short periods— three or four times daily to remove debris and to keep the scab soft. Children like to pick at scabs and often will fall on a scab. In these instances, bandages are useful and appreciated by most children.

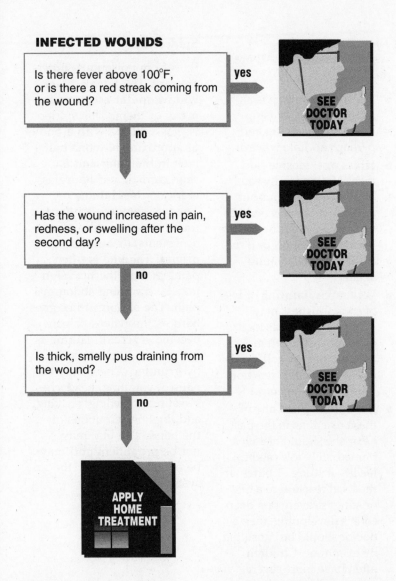

INFECTED WOUNDS

Is there fever above 100°F, or is there a red streak coming from the wound? — **yes** → SEE DOCTOR TODAY

no ↓

Has the wound increased in pain, redness, or swelling after the second day? — **yes** → SEE DOCTOR TODAY

no ↓

Is thick, smelly pus draining from the wound? — **yes** → SEE DOCTOR TODAY

no ↓

APPLY HOME TREATMENT

WHAT TO EXPECT AT THE DOCTOR'S OFFICE

An examination of the wound and regional lymph nodes will be done, and the child's temperature will be taken. Sometimes cultures of the blood or of the wound are performed, and sometimes antibiotics are prescribed, either for the skin (Mupirocin) or via oral medicine.

If there is a suspicion of bacterial infection, then cultures may be taken before the antibiotics are given. If a wound is festering, it may be drained either with a needle or a scalpel. This procedure is not very painful and actually relieves discomfort. For severe wound infections, hospitalization may be needed.

The simplest wound of the face requires three to five days for healing. The healing period is five to seven days for the chest and arms and seven to nine days for the legs. Larger wounds, or those that have gaped open and must heal across a space, take correspondingly longer to heal. Children heal more rapidly than adults do.

Insect Bites or Stings

Most insect bites are trivial, but some insect bites or stings may cause reactions either locally or in the basic body systems. **Local reactions** may be uncomfortable but do not pose a serious hazard. In contrast, **systemic reactions** occasionally may be serious and may require emergency treatment.

There are three types of systemic reactions. All are rare.

- The most common is an **asthma** attack, causing difficulty breathing and perhaps audible wheezing.
- **Hives** or extensive skin rashes following insect bites are less serious but indicate that a reaction has occurred and a more severe reaction might occur if the child is bitten or stung again.
- Very rarely, **fainting** or loss of consciousness may occur. If a child has lost consciousness, you must assume that the collapse is due to an allergic reaction. This is an emergency.
- If a child has had any of these reactions in the past, he or she should be taken immediately to a medical facility if stung or bitten. If the local reaction to a bite or sting is severe or a deep sore is developing, then a doctor should be consulted by telephone. Children often have more severe reactions than adults.

Spider Bites

Bites from poisonous spiders are rare. The female **black widow spider** accounts for many of them. This spider is glossy black, with a body of approximately one-half inch in diameter and a characteristic red hourglass mark on the abdomen. The black widow spider is found in wood piles, sheds, basements, or outdoor privies. The bite is often painless, and the first sign may be cramping abdominal pain. The abdomen becomes hard as the waves of pain become severe. Breathing is difficult and accompanied by grunting. There may be nausea, vomiting, headaches, sweating, twitching, shaking, and tingling sensations of the hand. The bite may not be prominent and may be overshadowed by the systemic reaction.

INSECT BITES

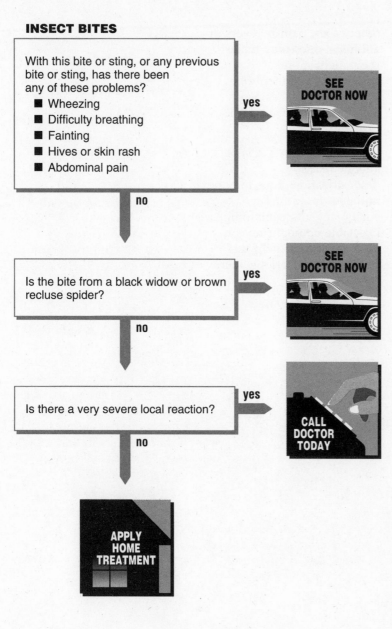

With this bite or sting, or any previous bite or sting, has there been any of these problems?
- Wheezing
- Difficulty breathing
- Fainting
- Hives or skin rash
- Abdominal pain

yes → **SEE DOCTOR NOW**

no

Is the bite from a black widow or brown recluse spider?

yes → **SEE DOCTOR NOW**

no

Is there a very severe local reaction?

yes → **CALL DOCTOR TODAY**

no → **APPLY HOME TREATMENT**

Brown recluse spiders cause painful bites and serious local reaction but are not nearly as dangerous as black widows. They are slightly smaller than black widows and have a white violin pattern on their backs.

Tick Bites

Tick bites are common. The tick lives in tall grass or low shrubs and hops on and off of passing mammals, such as deer and dogs. In some localities, ticks may carry Rocky Mountain spotted fever or Lyme disease, but most tick bites are not complicated by subsequent illness. Ticks will commonly be found in the scalp. Refer to **Ticks** (page 450).

HOME TREATMENT

Apply something cold promptly. Ice or cold packs can be used. Delay in application of cold results in a more severe local reaction.

Acetaminophen or other pain relievers can be used. Antihistamines, such as Benadryl, can be helpful by reducing the allergic response somewhat. If the reaction is severe, the doctor may be consulted by telephone.

WHAT TO EXPECT AT THE DOCTOR'S OFFICE

The doctor will ask what sort of insect or spider has inflicted the wound and will search for signs of systemic reaction. If a systemic reaction is present, adrenalin by injection is usually necessary. Rarely, measures to support breathing or blood pressure will be needed; these measures require the facilities of an emergency room or hospital.

If the problem is a local reaction, the doctor will examine the wound for signs of death of tissue or infection. Occasionally, surgical drainage of the wound will be needed. In other cases, pain relievers or antihistamines may make the patient more comfortable. Epinephrine injections are occasionally used for very severe local reactions.

If there has been a severe allergic reaction, desensitization shots may be initiated. In addition, emergency kits containing injectable epinephrine can be purchased to help the person with a serious allergy.

Fishhooks

The problem, of course, is the barb. While fish seem to get loose easily enough, children may stay hooked. If you and the child can keep calm, you can remove the fishhook, unless it is in the eye. No attempt should be made to remove hooks that have actually penetrated the eyeball; this is a job for the doctor. Hooks in the face should also be taken more seriously because of cosmetic issues.

You will need the child's confidence and cooperation in order to avoid a visit to the doctor. The advantage of the doctor's office is a local anesthetic and extra hands to help hold the child. Remember that the injection of the anesthetic will hurt some, so this is not a choice between pain and no pain. And you can save a lot of time by doing it at home.

HOME TREATMENT

Occasionally, the hook will have come all the way around so that it lies just beneath the surface of the skin. If this is the case, often the best technique is simply to push the hook on through the skin, cut it off just behind the barb with wirecutters, and remove it by pulling it back through the way it entered. This may be painful; some children may not tolerate it.

On other occasions, the hook will be embedded only slightly and can be removed by simply grasping the shank of the hook (pliers help), pushing slightly forward and away from the barb, and then pulling it out.

Sometimes, neither of these maneuvers will do the trick. In these cases, the method illustrated on page 296 usually removes the hook quickly.

1. Put a loop of fish line through the bend of the fishhook so that at the appropriate time a quick jerk can be applied and the hook can be pulled out directly in line with the shaft of the hook.

2. (Drawing a) Holding onto the shaft, push the hook slightly in and away from the barb so as to disengage the barb.

3. (Drawings b and c) Holding this pressure constant to keep the barb disengaged, give a quick jerk on the fish line and the hook pops out.

If you are successful, then be sure that the child's tetanus shots are up to date (see **Tetanus Shots**, page 271). Treat the wound as in the home treatment for **Puncture Wounds** (page 264). If all else fails, a visit to the doctor should solve the problem.

Note that a pair of needlenose pliers with a wire-cutting blade should be part of your fishing equipment.

FISHHOOKS

Is the hook in the eye?	**yes** →

no ↓

SEE DOCTOR NOW

Was the injury to the hand or face?	**yes** →

no ↓

CALL DOCTOR TODAY

APPLY HOME TREATMENT

See: **Puncture Wounds** (page 264)
See: **Tetanus Shots** (page 271)

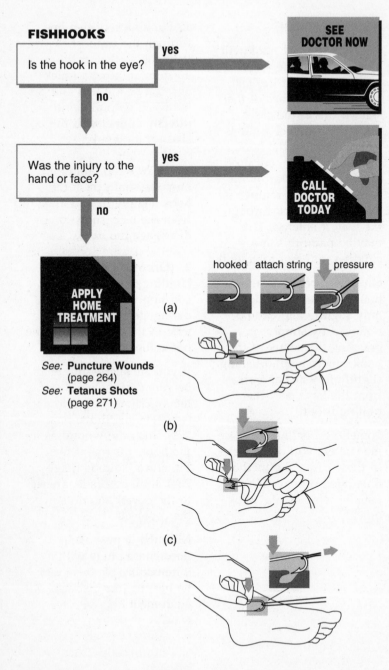

hooked   attach string   pressure

(a)

(b)

(c)

A splinter under the skin can often be pulled out with tweezers. If some material remains, you can usually dislodge it by picking away at the overlying skin with a clean needle. Sterilize the needle first by dipping it in rubbing alcohol or holding it in a match flame. Another option is to soak the area of skin twice a day in a cup of very warm, but not hot, water mixed with one teaspoon of baking soda; the splinter will probably come out by itself in a day or two. Don't let a splinter wound become infected.

WHAT TO EXPECT AT THE DOCTOR'S OFFICE

The doctor will use one of the methods above to remove the hook. The area around the hook may be infiltrated with a local anesthetic before the hook is removed.

If the hook is in the eye, it is likely that the help of an ophthalmologist (eye specialist) will be needed, and it may be necessary to remove the hook in the operating room.

Drawing adapted from George Hill, *Outpatient Surgery*. Philidelphia: W.B. Saunders, 1973.

Smashed Fingers

Children always seem to be smashing their fingers in car doors or desk drawers or with hammers or baseballs. If the injury involves only the end segment of the finger (the *terminal phalanx*) and does not involve a significant cut (see **Cuts**, page 261), then these injuries seldom need the help of a doctor. Blood under the fingernail (*subungual hematoma*) is painful but treatable.

Joint Fractures

Fractures of the bone in this end segment are not treated unless they involve the joint. Many doctors feel that it is unwise to splint the finger even if there is a fracture of the joint. While the splint will decrease pain, it may also increase the stiffness of the joint after healing. However, if the fracture is not splinted, then pain may persist longer, and your child may end up with a stiff joint anyway. You should discuss these advantages and disadvantages of splinting with your doctor.

Dislocated Nails

Fingernails are often dislocated in these injuries. Except in extraordinarily unusual accidents—injuring and destroying the nails' growth plate—fingernails always grow back. Some nails that are attached precariously may need to be clipped off to avoid catching painfully on other objects. Nail re-growth will take from four to six weeks. Dangling fingernails can be replaced over a clean nailbed, and a bandage applied. The new nail will grow underneath and lift the old nail.

HOME TREATMENT

If the injury does not involve other parts of the finger and if the child can move the finger easily, then apply home treatment of an ice pack for the swelling and use acetaminophen for the pain.

Blood Under a Nail

If a large amount of blood under the fingernail is causing pain, this problem can be relieved. However, the following procedure is potentially dangerous, for it can introduce infection to the finger. Care must be taken to use sterile technique. We recommend it not be used on younger children and in general be reserved for situations when pain is severe and medical help not accessible.

1. Bend open an ordinary paper clip, holding the paper clip with a pair of pliers.

2. Heat one end using a candle or a cigarette lighter.

3. When the tip is very hot, touch it to the nail, and it will melt its way right through the fingernail, leaving a small, clean hole. Steady the hand holding the pliers with the opposite hand so that the paper clip goes only through the nail and not into the flesh below.

SMASHED FINGERS

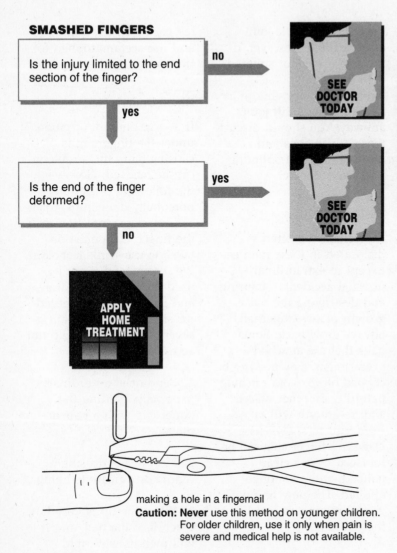

Is the injury limited to the end section of the finger?

no → SEE DOCTOR TODAY

yes ↓

Is the end of the finger deformed?

yes → SEE DOCTOR TODAY

no ↓

APPLY HOME TREATMENT

making a hole in a fingernail
Caution: Never use this method on younger children. For older children, use it only when pain is severe and medical help is not available.

Ingrown nails can be treated at home. Cut the nail straight across so that its corner can grow outside the skin. Let the nail grow free by firmly pushing the skin back from the corner with a cotton swab twice a day. Keep the area clean. For **hangnails**, keep them clean. Encourage your child not to chew on them.

WHAT TO EXPECT AT THE DOCTOR'S OFFICE

The finger will be examined; an X-ray is likely if it appears that more than the end segment is involved. If there is a fracture involving the last joint on the finger, you should expect a discussion of the advantages and disadvantages of splinting the finger. Often, splinting of one finger is accomplished by bandaging it together with the adjacent finger. If the finger is splinted, then periods of exercise should be included to preserve mobility. Severe injuries of fingers may occasionally require surgery in order to preserve function.

The blood trapped beneath the nail can now escape through the small hole, and the pain will be relieved as the pressure is released. If the hole closes and the blood re-accumulates, then the procedure can be repeated using the same hole once again.

CHAPTER E

Fever

Fever *300*
What goes up will come
down.

Fever

Many people, including doctors, speak of fever and illness as if they were one and the same. Surprisingly, an elevated temperature is not necessarily a sign of illness. Normal body temperature varies from individual to individual. If we measured a large number of healthy children's body temperatures while they were all resting, we would find a difference between the lowest and highest child of about 1.5°F. This is another reminder that children are individuals and that there is nothing either magical or accurate about the figure 98.6°F (37°C).

Normal body temperature varies greatly during the day. Temperature is generally lowest in the morning upon awakening. Many things will elevate body temperature including food, excess clothing, room temperatures, excitement, and anxiety. Vigorous exercise can raise body temperature to as much as 103°F. Severe exercise, without water or salt, can result in a condition known as **heat stroke**, with temperatures above 106°F. Other mechanisms also influence body temperature. Hormones, for example, account for a monthly variation of body temperature in ovulating women.

In general, children have higher body temperatures than adults do and seem to have greater daily variation because of their greater amounts of excitement and activity. Fevers get easier to control as your child gets older. The child's temperature regulatory center matures. In addition, the child loses a layer of brown fat, located between the backbones, that is responsible for a great deal of heat insulation in the infant.

Parents frequently ask us what temperature should be considered dangerous or at what temperature a child should be brought to the doctor if no other symptoms are present. Consult a doctor immediately for the following:

- A temperature of more than 100°F in a child less than four weeks.
- Fever of more than 101°F in a child less than three months.
- Fever of more than 103°F in a child less than two years.
- Fever of more than 105°F if the home treatment measures described below fail to reduce the temperature at least partly. Any temperature of 106°F should be evaluated by a doctor promptly.
- Fever persisting for more than five days.

CAUSES OF FEVER

The most common causes for persistent fevers in children are viral and bacterial infections such as colds, sore throats, earaches, diarrhea, roseola, occasionally pneumonia and urinary tract infections, and rarely appendicitis and meningitis.

A viral infection can result in a normal temperature or a temperature of 105°F. While the height of the temperature is *not* a reliable indicator of the seriousness of the underlying infection, higher temperatures do bring a slightly higher risk of serious problems, especially in children under two years.

TAKING YOUR CHILD'S TEMPERATURE

Either Fahrenheit or Celsius thermometers are acceptable. Rectal temperatures are usually more accurate and are about 0.5°F higher than oral temperatures. Oral temperature can be affected by hot or cold foods, routine breathing, and smoking. Generally, oral thermometers can be recognized by the longer bulb at the business end of the thermometer. The length of the bulb is to provide for a greater surface area and a faster, more accurate reading. Rectal thermometers have a shorter, rounder bulb to facilitate entry into the rectum.

Rectal thermometers can be used to take oral temperatures, but require a longer period in the mouth to achieve the same degree of accuracy as the oral thermometer. While oral thermometers can be used to take rectal temperatures, their shape is not ideal for younger children, and we do not recommend their use in children.

Lubricants can facilitate placement of rectal thermometers. You need not bury the thermometer. Only an inch or so need be inside the child's rectum. The mercury will rise within seconds on a rectal thermometer because the rectum closely contacts the thermometer. Remove the thermometer when the mercury is no longer rising after a minute or two is up. Children should be placed on their stomachs when rectal temperatures are being taken. You should place a hand on their bottom to prevent them from moving.

Underarm thermometers are reliable for young infants. Several plastic skin strips have been marketed, but they are not consistently reliable. New electronic thermometers can take a measurement through the ear. They are fast, generally (though not always) reliable, and expensive.

FEBRILE SEIZURES (FEVER FITS)

The danger of an extremely high temperature is the possibility that the fever will cause a seizure (convulsion). All of us are capable of "seizing" if our body temperatures become too high. Febrile seizures are relatively common in normal, healthy children; about 3 to 5% will experience a febrile seizure. However, although common, they should not be minimized and must be treated with respect.

Febrile seizures occur most often in children between the ages of six months and four years. Illnesses that cause rapid elevations to high temperatures, such as roseola, have been frequently associated with febrile seizure. Rarely, a seizure is the first sign of a serious underlying problem such as meningitis.

During a seizure, the brain, which is normally transmitting electrical impulses at a fairly regular rhythm, begins misfiring because of the overheating and causes involuntary muscular responses termed a seizure, convulsion, fit, or "falling out spell." The first sign may be a stiffening of the entire body. Children may have rhythmic beating of a single hand or foot or any combination of the hands and feet. The eyes may roll back and the head may jerk. Urine and feces may pass involuntarily.

Most seizures last only from a few seconds to a few minutes. There is very little evidence that such a short seizure is of any long-term consequence. On the other hand, prolonged seizures of more than 30 minutes are often a sign of a more serious underlying problem.

Less than half of all children who have a short febrile seizure will ever experience a second, and less than half who experience a second will ever have a third.

Although a "seizing" child is a terrifying sight to a parent, the dangers to the child during a seizure are small. The following commonsense rules should be followed during a seizure.

- Protect your child's head from hitting anything hard. Place the child on a bed.

- Considerable damage can be done by forcing objects into the child's mouth to prevent biting of the tongue. Surprisingly, cut tongues are both uncommon and quick to heal.

- Make sure the child's breathing passage is open. Forcing a stick in your child's mouth does not ensure an open airway. To facilitate breathing, (1) clear the nose and mouth of vomitus or other material and (2) pull the head backward slightly to "hyper-extend" the neck. Artificial respiration is almost never necessary. These techniques are best learned in demonstrations.

In a true emergency, hyper-extend the neck and breathe ten times each minute through the child's nose while keeping the mouth covered (or through the child's mouth while keeping the nose pinched with your finger). Only blow air in; the child will blow the air out naturally.

- Begin fever reduction (discussed below) and seek medical attention immediately. Do *not* give medicine by mouth to a seizing or unconscious child. Fortunately, once the seizure has stopped the child is usually temporarily resistant to a second seizure. However, because there are exceptions to this rule, medical attention is critical.

After the seizure has subsided, the child may be very groggy and have no recollection of what has occurred. Others may show signs of extreme weakness and even paralysis of an arm or leg. This paralysis is almost always temporary but must be carefully evaluated.

A good doctor will do a careful evaluation to determine the cause of a febrile seizure. For the first febrile seizure, this may include a spinal tap (lumbar puncture) and fluid analysis if the child appears ill, to make certain that the seizure was not caused by meningitis. Following the termination of the fever, the doctor will stress the importance of fever control for the next few days and will often place the child on anti-convulsant medications. For further discussion, see **Seizures**, page 334.

THE MEANING OF THE CHILL

A chill is another symptom of a fever. The feeling of being hot or cold is maintained by a complex system of nerve receptors in our skin and in a part of our brain known as the hypothalamus. This system is sensitive to the difference between the body temperature and the temperature outside. Cold can be sensed in two different ways, either by lowering the environmental temperature or by raising the body temperature.

The body responds in a similar manner to a fever as it would if the outside temperature dropped. All of the normal systems that increase heat production, such as shivering, become active.

Eating has already been mentioned as a means of increasing heat production, and hunger may be experienced. The body tries to conserve heat by causing constriction of the blood vessels near the skin. Children will sometimes curl up in a ball to conserve heat. Goose bumps are intended to raise the hairs on our body to form a layer of insulation. Don't bundle up your child in blankets if he or she shivers or becomes chilled; this will only cause the fever to go higher. Use home treatment as described below.

HOME TREATMENT

Remember that fever is the body's way of naturally responding to a variety of conditions including infection. A fever may signify the response of a body's immune system to an infection and thus be the visible manifestation of a

beneficial effect. Nonetheless, fevers do make children uncomfortable. Controlling a fever that is sufficiently high to interfere with a child's eating, drinking, sleeping, or other important activities will make the child feel better. In short, if your child seems to be suffering from the fever, treat it. If the fever is mild and the child shows no effects, it may be unnecessary to treat.

There are two ways in which to reduce a fever: sponging and medication. If fever remains above 103°F after an hour or so of home treatment, call the doctor.

Sponging

Evaporation has a cooling effect on the skin and hence on the body temperature. Evaporation can be enhanced by sponging the skin with water. Although alcohol evaporates more rapidly, it is somewhat uncomfortable for the child, and the vapors can be dangerous. Do *not* use alcohol! Generally, sponging with tepid water (water that is comfortable to the touch) will be sufficient.

Heat is also lost by conduction if a child is sponged or sitting in a tub. Conduction is the process in which heat is lost to a cooler environment (the bathwater or air) from the warmer environment of the body. A comfortable tub of water (70°F) is sufficiently lower than the body temperature to encourage conduction. Although cold water will work somewhat faster, the discomfort of the procedure makes this less desirable. The child will tolerate cold bathing and sponging for a much shorter period.

Some doctors believe sponging unnecessary and a potential for ultimately raising the body's temperature. Others disagree. For most fevers, sponging will not be required. For very high fevers, your doctor may suggest sponging.

Medication

Medication should not be given by mouth to a seizing or unconscious child. A child who has just had a seizure can be given an aspirin suppository. Most aspirin suppositories come in 5-grain sizes, and approximately 1¼ to 1½ grains per year of age can be given—somewhat higher than the recommended dose for oral aspirin. The suppository can be cut lengthwise using a warm knife, to give the proper dose.

Temperature can be controlled in the conscious, alert child with acetaminophen (Tylenol, Tempra, Liquiprin, Phenaphen, Datril, Tenlap). Aspirin and ibuprofen are equally effective in reducing fever when given in appropriate dosages. The safety of these products differs, so we recommend acetaminophen.

Dangers of Fever Medications

Children and teenagers who take aspirin when they have chicken pox or the flu stand a higher chance of later developing **Reye's syndrome**, a rare but serious problem of the brain and liver. Because it is hard to recognize chicken pox and flu in their early stages, we recommend that parents always give children and teenagers acetaminophen instead of aspirin. It does not carry the risk of Reye's syndrome and is slightly safer on the stomach than ibuprofen.

No medication should be given by mouth to a child who is seizing or unconscious. A child who has just had a febrile seizure can be given an aspirin suppository instead; to give the proper dose, the suppository can be cut lengthwise with a warm knife. The dosage can be slightly higher than the dose of aspirin taken by mouth.

In addition to the types of acetaminophen and ibuprofen sold over the counter, there are higher doses of these drugs available by doctor's prescription—potentially more than twice the strength of the non-prescription formulas. If you have both types of one drug in the house, do not mix up the two. Medication prescribed for one person should never be given to another, especially a child.

All drugs kept at home should be in childproof bottles. Because there are no totally childproof bottles, drugs should also be kept out of small children's reach. Aspirin in excessive doses has been responsible for more childhood deaths than any other medication. Acetaminophen and ibuprofen carry reputations for being safer than aspirin, but they, too, can cause damage in overdoses. There are no "safe" drugs.

Acetaminophen

Acetaminophen has sometimes been called "liquid aspirin," but it is a completely different medication. The advantage of acetaminophen is that it can be given in either liquid or tablet form. Acetaminophen is as effective as aspirin in fever reduction. It is not as effective as aspirin for other purposes, such as reducing inflammation. Fewer people are allergic to acetaminophen, and it does not cause as many gastrointestinal disturbances. However, if the patient has never had nausea, abdominal pain, or other gastrointestinal problems with aspirin, this is probably not an important consideration. Overdoses can cause liver damage and consequently can cause death.

Acetaminophen is available in drops, suspension, tablets, or capsules. A tablet or capsule contains 325 mg; more than a preschooler needs. See Table E for recommended dosages. The concentration of the drops is much higher than the suspension and therefore this form must be administered cautiously. An unsuspecting person used to a different preparation of acetaminophen can create a problem by using the wrong dosage on your child.

TABLE E	*Acetaminophen Dosages for Fever Relief*[*]
Weight of Child	*Dosage*
Up to 11 pounds (infants)	40 mg
12–17 pounds	80 mg
18–23 pounds	120 mg
24–35 pounds	160 mg
36–47 pounds	240 mg
School-age children	325 mg (one regular tablet or capsule)
Adolescents	500 mg (one "extra-strength" tablet or capsule)

[*]*The recommended dosage is 10 milligrams (mg) for every 2.2 pounds (1 kilogram) of the child's weight every four hours, up to five times in one day. The amounts given here are* **estimated** *dosages.*

Ibuprofen

Ibuprofen is the most recently released over-the-counter drug for fever control. A prescription is still needed to give ibuprofen to children under 12 years. It is available in tablet form under different brand names (Advil, PediaProfen, Nuprin, etc.) and appears in some other medications (Midol, etc.).

Allergic reactions to ibuprofen are rare, as are gastrointestinal side effects. If you have problems with stomach irritation, you can take ibuprofen after a meal without badly affecting its absorption. Ibuprofen is safer for children than aspirin, but it is not available in as many forms as acetaminophen. Liquid ibuprofen's concentration differs from the tablet form. Read the labels carefully!

Aspirin

Aspirin is universally familiar and effective. However, we do not recommend giving it to children and teenagers with fever because of the risk of Reye's syndrome. A few individuals are allergic to aspirin and may experience severe skin rashes or gastrointestinal bleeding. All people will suffer if they take too much aspirin. Early signs of excess aspirin include rapid breathing and ringing in the ears. An excessive dose of aspirin can be fatal.

"Baby aspirin" contains 1¼ grains (81 mg) per tablet, one-quarter the amount of aspirin in an adult tablet. By the time a child is five, an adult aspirin (5 grains) can be given. Toxic effects may begin to develop at less than twice the recommended dosage, so you must handle this medication carefully.

WHAT TO EXPECT AT THE DOCTOR'S OFFICE

This will depend on how long your child has had a fever and how sick your child appears. An examination to determine whether an infection is present will evaluate skin, eyes, ears, nose, throat, neck, chest, and belly. If no other symptoms are present and the exam does not reveal an infection, watchful waiting may be advised. In girls and uncircumcised boys less than one year old, urine is often required for analysis. If the fever has been prolonged, the child appears ill, or the infant is very young, tests of the blood and urine may be done. A chest X-ray or spinal tap may be needed. Specific infections will be treated appropriately; fever will be treated as discussed in Home Treatment.

STARVE A FEVER? This folk remedy probably came from individuals clever enough to notice the relationship between food and temperature elevation. However, there are many reasons why children should be fed during a fever. The increased heat increases caloric requirements because calories are being consumed rapidly at the higher body temperature. More important, there is an increased demand for fluid.

Liquids should never be withheld from a feverish child. If a child will not eat because of the discomfort caused by fever, it is still essential to continue to encourage that he or she drink fluids.

FEVER

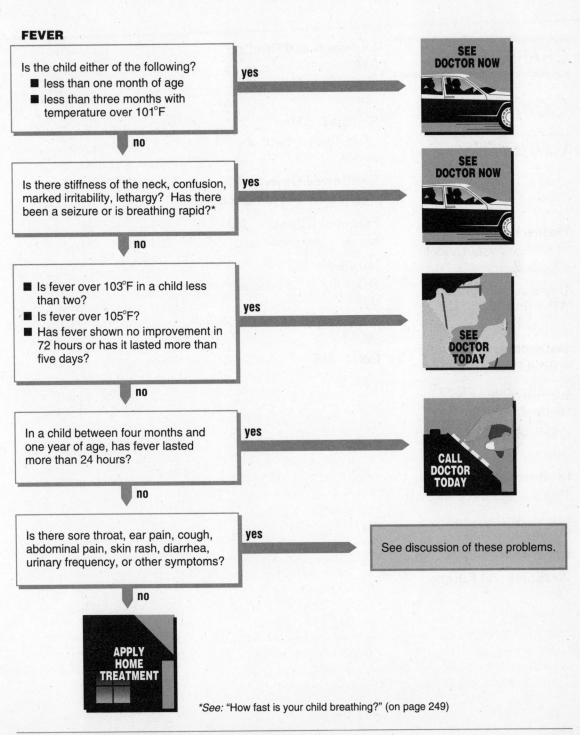

Is the child either of the following?
- less than one month of age
- less than three months with temperature over 101°F

yes → SEE DOCTOR NOW

no ↓

Is there stiffness of the neck, confusion, marked irritability, lethargy? Has there been a seizure or is breathing rapid?*

yes → SEE DOCTOR NOW

no ↓

- Is fever over 103°F in a child less than two?
- Is fever over 105°F?
- Has fever shown no improvement in 72 hours or has it lasted more than five days?

yes → SEE DOCTOR TODAY

no ↓

In a child between four months and one year of age, has fever lasted more than 24 hours?

yes → CALL DOCTOR TODAY

no ↓

Is there sore throat, ear pain, cough, abdominal pain, skin rash, diarrhea, urinary frequency, or other symptoms?

yes → See discussion of these problems.

no ↓

APPLY HOME TREATMENT

See: "How fast is your child breathing?" (on page 249)

CHAPTER F

Common Concerns

Headache

More than 40% of all children have had a headache by the age of 7, and 4% of 7-year-olds are troubled by frequent headaches. By the age of 15, 75% of children have had a headache, and 20% will have experienced frequent headaches.

In younger children, the cause of head pain is frequently stress; in older children *most* headaches are due to stress. Stress and tension can cause headaches even in 5-year-olds. Muscle spasms in the neck and scalp cause these pains, aggravated possibly by a widening of blood vessels inside the brain. Tension headaches can occur in any part of the head, produce a dull or swollen feeling, and usually come on slowly. Headaches are often the first symptom of stressful problems at school, at home, or with friends. A child who is functioning poorly in any of these areas needs help.

Many medications, including decongestants and antihistamines, can cause headaches. Headache is also common just before menstruation. Eyestrain is often blamed but is seldom a cause of headache; usually it is tension.

Migraines

Migraine headaches may start in childhood. Many toddlers who vomit frequently have migraine; as they become older, they can tell you that their head aches. Children with classic migraine usually have at least two of the following:

- Headaches on one side of the head only
- Nausea
- Visual disturbances before an attack
- Other family members with migraines

These headaches often begin suddenly and are throbbing in character. There are also other forms of migraine.

Other Causes

An isolated headache is often due to an earache, sore throat, toothache, or eye infection. Young children have poorly developed sinuses, so sinusitis is uncommon. When a child has a headache combined with stiff neck, irritability, a soft spot (fontanel) that is bulging, or vomiting with a fever, **meningitis** should be suspected.

Occasionally, headaches will be the only sign of a **seizure disorder**. These headaches usually begin suddenly and are followed by a period of drowsiness or sleep.

Many parents are concerned about **brain tumors**. Headaches associated with brain tumors are often persistent and progressive, present in the morning, and accompanied by other problems such as difficulty in walking, personality changes, vomiting, visual problems, limb weakness, and speech difficulties. Fortunately, brain tumors are very rarely the cause of a headache. Only 1 child in 40,000 is found to have a tumor of the brain or nervous system.

HOME TREATMENT

Headaches due to ear infections, toothaches, strep throats, acute sinusitis, or other serious infections require medical help.

For headache associated with colds, flu, or stress, acetaminophen may be effective. Headache may frequently be relieved by massage or heat to the back of the neck. Children with hay fever often have headaches during the pollen season and antihistamines and decongestants may help; but remember that these drugs *cause* headaches in some children!

Persistent headaches that do not respond to such measures should be called to the attention of a doctor. Headaches associated with weakness of the arms or legs or with slurring of speech, as well as those that are rapidly increasing in frequency and severity, also require a visit to the doctor.

Remember, tension is the usual cause of head pain. Whenever possible, identify the causes of the stress and work with your children in relieving them.

WHAT TO EXPECT AT THE DOCTOR'S OFFICE

The doctor will check the temperature, blood pressure, head, eyes, ears, nose, throat, and neck and also test nerve function. Laboratory testing will depend on what is found in the history and physical examination; usually none are needed. With an acute headache, a source of infection may be found. Even with most recurring headaches, a history and physical examination are probably all that is required. Occasionally, brain wave tests are performed if a seizure disorder is suspected. If migraine is interfering with your child's functioning, medication may be prescribed.

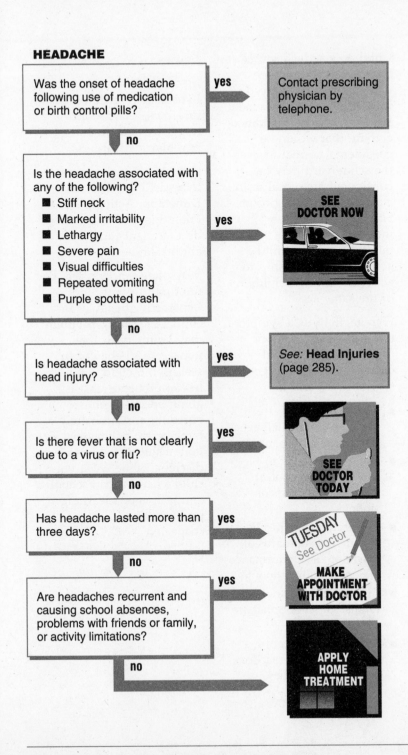

HEADACHE

Was the onset of headache following use of medication or birth control pills? — yes → Contact prescribing physician by telephone.

no ↓

Is the headache associated with any of the following?
- Stiff neck
- Marked irritability
- Lethargy
- Severe pain
- Visual difficulties
- Repeated vomiting
- Purple spotted rash

yes → SEE DOCTOR NOW

no ↓

Is headache associated with head injury? — yes → *See:* **Head Injuries** (page 285).

no ↓

Is there fever that is not clearly due to a virus or flu? — yes → SEE DOCTOR TODAY

no ↓

Has headache lasted more than three days? — yes → TUESDAY See Doctor / MAKE APPOINTMENT WITH DOCTOR

no ↓

Are headaches recurrent and causing school absences, problems with friends or family, or activity limitations? — yes →

no ↓ APPLY HOME TREATMENT

Hyperactivity or Short Attention Span

Controversy and confusion are the rule when it comes to the problem of hyperactivity; the word means "more than normal activity." The problem here is in knowing what is "normal." The normal activity level of a 3-year-old is certainly much greater than that of most 40-year-olds. Is this hyperactivity?

Children's activity levels change as they become older, as they are placed in new situations, or when they are excited. In addition, activity that would be considered normal in the schoolyard may be considered abnormal in the classroom or dining room. Hyperactivity merely indicates someone's perception of the child's activity in relationship to activities of other children of the same age.

Hyperactivity is a symptom, not a disease. It may be found in most normal children (especially ages two to four), and in:

- Older children of above-average intelligence with inquisitive behavior
- Children reacting to problems in school, at home, with friends or siblings
- Children unable to adjust to different standards of behavior and performance at home and in school
- Children who do not speak English in schools that are not bilingual
- Children with hearing difficulties, drug reactions, or visual difficulties

Causes

Medical causes of hyperactivity include seizure disorders, retardation, hyperthyroidism, and psychiatric disorders.

Common **drugs** may cause hyperactivity; these include Dimetapp, Actifed, Sudafed, Triaminic, and many other decongestants and antihistamines.

Very often teachers will be the first to point out hyperactive behavior in a child. A **learning problem** may be the basis for hyperactive behavior in school. If a child is not able to comprehend what is going on in the classroom, boredom and ultimately hyperactive behavior may result. If there is a discrepancy between your child's behavior in school and at home, a learning problem should be suspected.

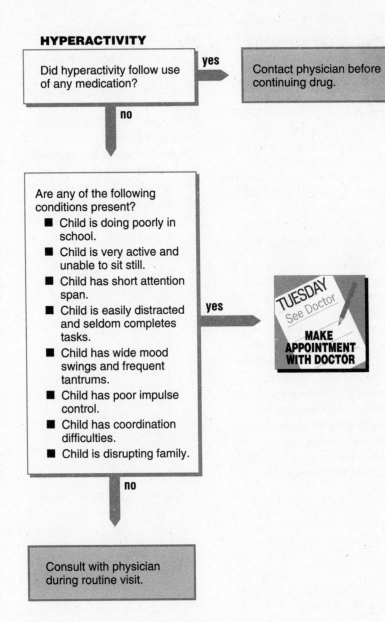

HYPERACTIVITY

Did hyperactivity follow use of any medication?

yes → Contact physician before continuing drug.

no ↓

Are any of the following conditions present?
- Child is doing poorly in school.
- Child is very active and unable to sit still.
- Child has short attention span.
- Child is easily distracted and seldom completes tasks.
- Child has wide mood swings and frequent tantrums.
- Child has poor impulse control.
- Child has coordination difficulties.
- Child is disrupting family.

yes → TUESDAY See Doctor MAKE APPOINTMENT WITH DOCTOR

no ↓

Consult with physician during routine visit.

Hyperactivity Syndrome

Finally there is a small group of children to whom the word *hyperactive* has been applied as the name of a syndrome. In medical terminology, a syndrome is a group of symptoms that occur together; often there is no known cause.

The hyperactivity syndrome, called attention deficit disorder or attention deficit hyperactivity disorder (ADHD), consists of greatly increased activity, easy distractability, wide mood changes from moment to moment, poor impulse control, short attention span, explosive moods, and often learning problems. Some of these children may be clumsy, with difficulties controlling fine movements.

The underlying cause for this syndrome is not known. It is even argued whether this is really a disorder; it may represent slower than usual maturation of attention.

HOME TREATMENT

Observation of other children is a good guide to your own child's activity. You will notice that there are many extremely active children of all ages. Some seem to be on constant "seek-and-destroy" missions.

If problems develop in school, a conference should be held with the teacher to determine the precise nature of the problem. Medical assistance in dealing with this type of problem can be helpful.

If your child's hyperactivity began following the use of a medication, discontinue using this medication immediately and contact the prescribing doctor. (See further discussion, pages 312–313.)

WHAT TO EXPECT AT THE DOCTOR'S OFFICE

Discussion can be initiated during a child health supervision visit. If a problem is suspected, a lengthy evaluation will be scheduled. Most doctors will wish to review a copy of the child's school records (you can save time by having these at the time of the visit) and may wish to talk with the teacher.

The history and physical examination will emphasize the nervous system. Your observations with respect to the child's behavior are important. Tests of muscle coordination, reading, spelling, and so on may be conducted. Hearing and vision will be tested. If a seizure disorder is suspected, brain wave tests may be performed.

The decision to begin any form of therapy (such as the drug Ritalin or special diets) should be reached only after careful consideration. Behavioral techniques will be used in management. A proper therapeutic approach requires coordination between doctor, parents, school, and child.

Bedwetting

Achieving nighttime bladder control has already been discussed on page 82. A number of maturational factors determine when a child will become dry. Many children are dry at night by the age of four, but others are not. With each successive year, many more children will naturally develop control.

Children who have been dry for many months or even years may suddenly wet the bed, often in response to a stressful event. Children may regress because of the arrival of a new brother or sister, a move, or a severe illness. Only occasionally is a urinary tract infection the cause, and then there are usually other symptoms such as increased frequency of urination, burning, abdominal pain, or fever. For such symptoms, the doctor should be seen.

HOME TREATMENT

Most important is your attitude. First, you must expect bedwetting to happen. It should not be considered unusual until beyond the age of six, and it is still normal for children to occasionally lose control during stressful events for the next year or so. A reaction of disgust or anger may make it more difficult for the child to gain control. Second, the child often cannot help it. Bladder control is a complex developmental and neurological task that requires a mature child. Making the child feel guilty about bedwetting will only delay resolution of the problem; you should think in terms of supporting rather than punishing the child.

Because constant sheet changing is often tiresome for parents, placing a short sheet on top of a rubber pad placed on top of the regular sheet will cut down on laundering.

Most of the time, your best course is to ignore the problem and wait it out. After the age of six, you and your child may want to make a "team effort" to solve this problem.

1. Encourage the child to drink in the morning and early afternoon, not in the several hours before bedtime.

2. Have the child void (urinate) just before going to bed. Fluids during the day help ensure that the bladder is big enough, and voiding ensures that it is as empty as possible at bedtime.

3. Develop a chart or use a calendar to indicate when the child has achieved control by awarding a gold star. When you child gets tired of gold stars, you may try drawing smiling faces or sad faces on your calendar. Try these measures for several weeks or months.

4. If progress is not being made, consider getting the child up within the first three hours of sleep for voiding, thus emptying the bladder again. If the child is already wet, awaken for voiding a little sooner.

Older children can participate in the laundering process in order to learn that they are responsible for their actions and some of the consequences.

Nighttime alarms that sound when the child begins to void in bed are effective in the short run. Your doctor may want to include an alarm as *part* of a treatment program. If a child continues to wet the bed after the age of six, a consultation with the doctor can be helpful.

WHAT TO EXPECT AT THE DOCTOR'S OFFICE

The medical history is the most important part of the office visit. The doctor will inquire at what time of night the child wets the bed, as well as how often and under what circumstances. A physical examination with attention to the genitalia will be performed. A urinalysis will be done. You should think of this visit as adding the doctor to the team. The implication that the child is being punished should be avoided.

The doctor will need to get to know your child; this may best be accomplished without your presence. Don't be offended. In some instances, the drug imipramine (Tofranil) will be added to the regimen of home treatment, and there may even be an occasional indication for a drug given nasally called Pitressin.

Extensive tests such as X-rays of the kidneys and bladder are seldom necessary. Bedwetting is almost never due to a surgical condition, and several doctors should be consulted for additional opinions if surgery is suggested.

BEDWETTING

In a previous dry child, are any of the following conditions present?
- Fever
- Abdominal pain
- Increased frequency of urination
- Painful urination
- Blood in the urine

yes → **SEE DOCTOR TODAY**

no ↓

In a previously dry child, has renewed bedwetting lasted longer than a month?

yes → **MAKE APPOINTMENT WITH DOCTOR** (TUESDAY See Doctor)

no ↓

Is the child older than six years and never been dry at night?

yes → **CALL DOCTOR TODAY**

no ↓

APPLY HOME TREATMENT

Constipation

It is not necessary to have a bowel movement every day. Many children have bowel movements only once every three or four days, and they are perfectly normal.

Most parents are concerned about constipation when the stool (feces) is very hard or when a child experiences pain when passing stool. Sometimes the pain is due to a tear in the rectum (rectal fissure). It is often unclear whether the hard fecal matter is responsible for the tear or whether the tear is responsible for the child holding back the bowel movement in order to avoid the pain. In any event, treatment is directed toward softening the stool.

Frequently, infants with a severe diaper rash will withhold bowel movements to avoid pain. If so, work to clear the rash rather than to soften the stool. (See **Diaper Rash**, page 417.)

Infants and older children will frequently be constipated during illnesses. Adequate fluid intake is very important in this case.

Constipation can have emotional causes; for example, it may begin at the time of toilet training. In the struggle of will between child and parent, the child may decide to retain control by holding on to a bowel movement.

If a child retains stool for long enough, liquid material will escape around the hardened bowel plug. This liquid material is apt to leak and soil underpants; hence, the term **soiling**. Soiling in an older child is a sign of longstanding constipation and should be treated by a doctor.

HOME TREATMENT

Dietary changes are usually all that is necessary and are superior to medicines. Prune juice is remarkable in its efficiency. You might eliminate rice cereal for a while because it tends to be constipating. In infants, Karo syrup (one tablespoon to four ounces of water or milk) is helpful. In older children, encouragement of bran products and fiber (celery, whole oranges) often helps. Adequate fluid intake is essential; water is fine.

On rare occasions, a laxative may be needed. Colace, Metamucil, and Maltsupex are acceptable. Glycerin suppositories are safe and effective. Infants should never be given mineral oil because it can cause a serious pneumonia if it finds its way to the lungs. However, in older children it is quite effective. Enemas are virtually never needed and are potentially dangerous.

CONSTIPATION

Are any of the following conditions present?

- Child is in pain when having bowel movement.
- Blood is noticed on the stool.
- No bowel movement for four days.
- A crack or fissure is noticed on child's rectum.

yes → **CALL DOCTOR TODAY**

no ↓

Has the problem persisted for more than four weeks?

yes → **TUESDAY See Doctor — MAKE APPOINTMENT WITH DOCTOR**

no ↓

Has the problem recurred three or more times unaccompanied by an illness?

yes → **TUESDAY See Doctor — MAKE APPOINTMENT WITH DOCTOR**

no ↓

Is your child soiling clothes, or is the problem interfering with school, friends, or family functioning?

yes → **TUESDAY See Doctor — MAKE APPOINTMENT WITH DOCTOR**

no ↓

APPLY HOME TREATMENT

WHAT TO EXPECT AT THE DOCTOR'S OFFICE

An examination of the abdomen and rectum will be done in some instances. In the case of soiling, the doctor will assess whether the child has developed to the point where bowel control can be expected, and, if so, whether stresses might be causing the problem. The child will need your support if a serious soiling problem is to be treated successfully. Rarely, X-rays of the lower bowel are necessary.

Constipation is an example of a case in which the best doctor may well do few or no tests and perform little or no examination, and the less competent doctor may make a non-problem into a big deal. Soiling is a more serious problem and will be investigated more thoroughly.

Overweight

The tendency of infants to be overweight is influenced by parents' genes and behavior. The likelihood of being overweight (or underweight) has a lot to do with inheritance, but parents have tremendous power to influence whether children actually become fat. Although theories about obesity are being constantly developed and debated, one fact remains: A proper balance of diet and exercise will control weight. If the entire family eats and exercises properly, it is unlikely your child will become overweight.

There are a number of ways to detect obesity, but none beats the human eye. Fat is easy to see if you don't try to fool yourself. Weight and height charts often have such a wide range of "normal" weights that there is ample opportunity for self-deception.

Obesity is almost always due to too many calories and not enough exercise. Trying to find a way around this simple truth is one of America's favorite pastimes. The problem is complicated by the peculiar notion that a fat baby is happy and healthy. This, along with our traditional attitudes toward cleaning the plate ("think of the starving orphans in Asia"), can be a tragic combination of concepts and can lead to social problems and premature death as an adult.

There are certain hormone disorders that cause what appears to be obesity, but glands get blamed for a lot more problems than they actually cause. Even to the untrained eye, children with glandular problems do not appear the same as a typical chubby child. For instance, children with thyroid problems usually have retarded growth, meaning they are short. Often weight is gained in a relatively short period of time and is associated with other symptoms. In other words, this type of weight gain is part of an illness and is not an isolated happening. Such illnesses are not common and account for less than 1% of obese children.

Steroid medications can also be responsible for weight gain.

HOME TREATMENT

With an infant, you must overcome the feeling that too much is better than not enough and that a fat baby is a healthy baby. Forcing food is not necessary. Infants have a pretty good idea of how much they need; they will not starve in the presence of food. Remember, a crying infant is not always hungry; comforting may be all that is called for. Also, start at an early age rewarding children with words, not foods.

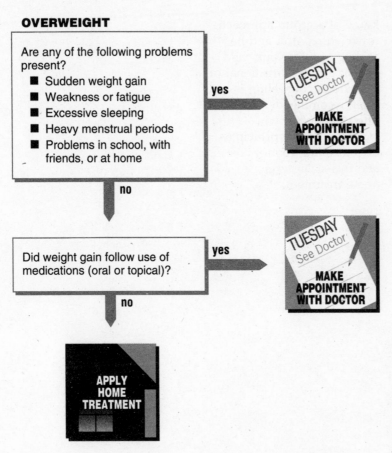

OVERWEIGHT

Are any of the following problems present?
- ■ Sudden weight gain
- ■ Weakness or fatigue
- ■ Excessive sleeping
- ■ Heavy menstrual periods
- ■ Problems in school, with friends, or at home

yes → TUESDAY See Doctor / **MAKE APPOINTMENT WITH DOCTOR**

no ↓

Did weight gain follow use of medications (oral or topical)?

yes → TUESDAY See Doctor / **MAKE APPOINTMENT WITH DOCTOR**

no ↓

APPLY HOME TREATMENT

excuse for not doing anything about obesity when in fact working on the obesity may help with the unhappiness.

To help with the basics of losing weight:

- ■ Keep a diary or chart of calorie intake, exercise, and weight on a daily basis. Charting progress (or lack of it) is central to any method of weight control.
- ■ It is important to eat meals only at the table. The battle of the bulge is most often lost snacking at the refrigerator.
- ■ Rewarding even minor weight loss is important.

Weight reduction is a family problem; children's eating patterns largely are determined by those of their parents. Teenagers can participate in group efforts at weight control; these methods are often helpful. *Whatever method is used, its success will depend on both you and your child.* Do not expect the doctor to solve the problem for you.

Older children and adolescents are a different story. As youngsters become adults, they take on the adult patterns of obesity. Although older children may be initially discouraged in trying to slim down after years of being overweight, they should realize that proper nutrition and exercise at any age will eventually be rewarded.

Because eating is one way of handling stress and is enjoyable, the psychology of obesity is complex. The psychological problems need not be solved before the problem of obesity can be solved. Unhappiness is sometimes used as an

If all else fails, then a visit to the doctor or nutritionist and a fresh look from a different perspective may be needed. However, don't go to the doctor for diet pills; these do not work and can hurt.

WHAT TO EXPECT AT THE DOCTOR'S OFFICE

An extensive dietary history and physical examination will be conducted. Weight charts are more important for infants than for adolescents. Blood tests may also be done but, as in adults, tests for thyroid function as a cause of obesity are usually not needed. The doctor may

know of a group approach to weight control and be willing to refer your child to that group. Nutritional or behavioral counseling is often recommended. Most often the doctor will elaborate on the principles of controlling food intake and increasing exercise as a home treatment.

Underweight

Parents are naturally concerned that their children grow and put on weight properly. Proper growth and weight gain are determined by adequate nutrition as well as genetic factors. Of course, we have control only over nutrition.

Infants

Often parents of a breast-fed baby will be concerned that their child is not as fat as the baby next door or the baby in the food ads; this is because many breast-fed babies gain weight at a slower rate than bottle-fed babies.

There are, of course, feeding problems in both breast- and bottle-fed infants that can cause a child to be underweight. Inadequate intake, excessive vomiting, or long-lasting diarrhea can all lead to inadequate weight gain in infants. Children with underlying heart failure or long-standing urinary tract infections may also gain weight poorly; this may occur with other serious medical problems as well.

Malabsorption is an intestinal disturbance that produces terrible-smelling, greasy-looking bowel movements. Food is absorbed poorly, resulting in insufficient weight gain.

Doctors are as concerned about weight gain as are the parents. They will take height and weight measurements in the office during well-baby visits. A child who is more than two "weight lines" away from the height line is seriously underweight. You can check your child on the charts in Part V. For example, if your daughter is one year old and is 29 inches (74 centimeters) tall, this is on the 50th percentile line. If the weight is only 17 pounds (8 kilograms), this is only the 5th percentile and is more than two lines away from the 50th percentile. This child is underweight and should be seen by a doctor.

Young Children

In general, if nutrition is adequate and weight loss has not been marked, it is far better to be on the lean side than the fat side. Children who are lean (but not markedly underweight) have many health advantages. They often feel better about themselves and find physical activities much easier. In addition, "thin is in" these days.

Some infants and young children are picky eaters and sometimes just refuse to eat. Other children never seem to sit still long enough to finish a meal and are so active that they seem to burn off their few ingested calories instantaneously. These children should have their meals presented as consistently as possible, meals should be at routine times, and feeding allowed for a specified length (for example, 20 minutes). Struggling with children is seldom a useful strategy. If your child is truly underweight, it may be advisable to choose one meal a day in which an extra effort is made to have the child take in adequate calories.

Older Children

Sometimes older children will refuse to eat. While a short diet with a few pounds of weight loss is fine, prolonged or excessive weight loss can be dangerous. Such children frequently will insist that they feel fat and don't wish to eat, even after they have become emaciated and have developed other health problems. If prolonged, this problem, called **anorexia nervosa**, can be life-threatening. Children with this problem have a real aversion to food and sometimes even induce vomiting (**bulimia**) after they do eat. Anorexia nervosa requires medical help.

During periods of rapid growth, children often get gangly and "string out." This is not a cause for concern.

HOME TREATMENT

Children will eat a sufficient amount if it is presented to them. Do not be concerned if your child refuses to eat a meal or two or is a picky eater. Children become underweight only if inadequate amounts of foods are available.

WHAT TO EXPECT AT THE DOCTOR'S OFFICE

A complete history and physical examination will be done. Special attention will be paid to the dietary and bowel patterns. Height and weight will be carefully measured, and further investigations will depend on what is found. If an intestinal problem is suspected, stool analysis and possibly bowel X-rays will be done. If a heart problem is suspected, X-rays or an EKG will be performed; if infection is likely, a blood test and urinalysis will be done.

UNDERWEIGHT

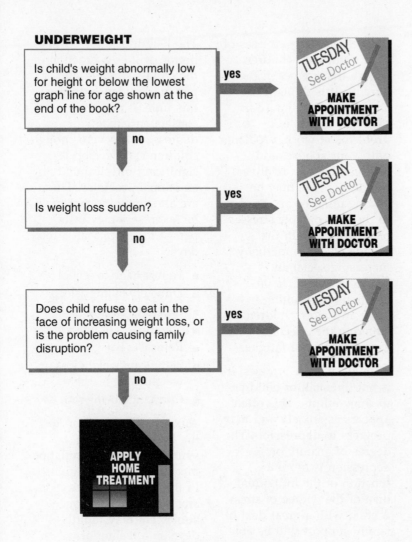

Is child's weight abnormally low for height or below the lowest graph line for age shown at the end of the book?

yes → TUESDAY See Doctor — **MAKE APPOINTMENT WITH DOCTOR**

no ↓

Is weight loss sudden?

yes → TUESDAY See Doctor — **MAKE APPOINTMENT WITH DOCTOR**

no ↓

Does child refuse to eat in the face of increasing weight loss, or is the problem causing family disruption?

yes → TUESDAY See Doctor — **MAKE APPOINTMENT WITH DOCTOR**

no ↓

APPLY HOME TREATMENT

Stress, Anxiety, and Depression

Stress is a normal part of children's lives. Toddlers must learn to communicate. Pre-schoolers need to know how to interact in a socially acceptable fashion. School-age children have peer and teacher pressures. Adolescents must begin to think about future careers; in addition, they will find and often lose meaningful relations with members of the opposite sex. Leaving home is one of the greatest stresses children encounter.

Besides the daily stresses of their own lives, children absorb the stresses of their parents. It is difficult to shield a child from your own concerns about problems at work, problems with your spouse, and serious illness in relatives. In addition, we may be adding unnecessary burdens to the unavoidable stresses of growing up. In the decade of the "superbaby," increasing social and academic pressures on children are resulting in more and more children suffering the effects of excess stress and fatigue.

With all of these stresses, it is not unusual for children to show effects. This often appears as anxiety and may progress to **depression**. The degree of anxiety or depression is more a function of the individual than of the degree of stress. A child with a great deal of family support will be able to deal with minor stresses more easily. Major stresses, such as parental separation or a move, will affect all children.

The length of the recovery period will be a function of the child's strength and parental guidance during this time. If the stress or its consequent anxiety or depression is severe enough, children can develop significant physical and emotional problems. These are all symptoms that must be taken seriously if they persist for more than a few days:

- Failure to go to school
- Difficulty falling asleep
- Excessive sleeping
- Nightmares or night terrors
- Refusal to eat
- Significant weight loss or weight gain
- Constantly being sad

Anxiety can often lead to the **hyperventilation syndrome**. Hyperventilation means excessive breathing. In this condition, an anxious older child or adolescent may rapidly develop a feeling of inability to get enough air into the lungs. Sometimes this is associated with a feeling of chest pain or chest constriction. These sensations lead to further over-breathing, and the lowering of the carbon dioxide level in the blood

STRESS, ANXIETY, DEPRESSION

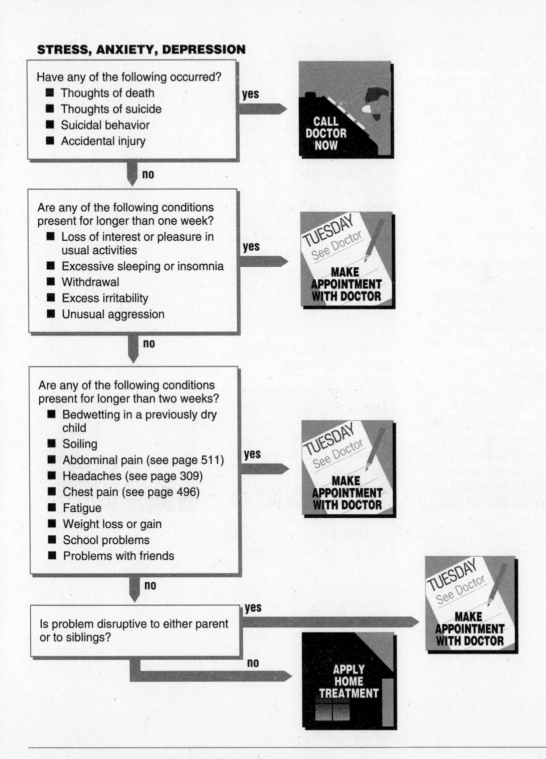

Have any of the following occurred?
- Thoughts of death
- Thoughts of suicide
- Suicidal behavior
- Accidental injury

yes → CALL DOCTOR NOW

no ↓

Are any of the following conditions present for longer than one week?
- Loss of interest or pleasure in usual activities
- Excessive sleeping or insomnia
- Withdrawal
- Excess irritability
- Unusual aggression

yes → TUESDAY See Doctor — MAKE APPOINTMENT WITH DOCTOR

no ↓

Are any of the following conditions present for longer than two weeks?
- Bedwetting in a previously dry child
- Soiling
- Abdominal pain (see page 511)
- Headaches (see page 309)
- Chest pain (see page 496)
- Fatigue
- Weight loss or gain
- School problems
- Problems with friends

yes → TUESDAY See Doctor — MAKE APPOINTMENT WITH DOCTOR

no ↓

Is problem disruptive to either parent or to siblings?

yes → TUESDAY See Doctor — MAKE APPOINTMENT WITH DOCTOR

no → APPLY HOME TREATMENT

(carbon dioxide is present in exhaled air). The lower level of carbon dioxide gives symptoms of numbness and tingling of the hands, feet, and mouth, as well as dizziness. Any of these symptoms can predominate for a particular person.

HOME TREATMENT

The best approach is to be open and to discuss stressful occurrences with your children as they are growing up. Children who feel they can talk with their parents have support in working through their problems when they are young. This hopefully leads to a growing ability to work through stressful situations independently. Identification of the source of anxiety or depression can be difficult and may require the help of a professional; most communities have resources that can help. (See pages 136–137.)

With hyperventilation, recognition that this is a stress reaction is a first step. During an attack, a paper bag can be placed loosely over the nose and mouth so that the child can re-breathe the carbon dioxide in the expelled air. This will raise the blood level of carbon dioxide, and the attack will pass after ten minutes or so. Alternatively, slow the child's breathing by having him or her hold his breath repeatedly.

WHAT TO EXPECT AT THE DOCTOR'S OFFICE

The doctor will attempt to identify the source of the problem. School records and records from other health professionals will help. Seldom can the problem be solved in one visit. Many doctors will rely on social workers, clinical psychologists, or psychiatrists to help with the problem.

Weakness and Fatigue

Children often have periods of being extremely tired or feeling weak; these symptoms may appear quite suddenly. The sudden development of weakness often signifies the beginnings of a cold or other infection. Weakness may precede a fever or occur simultaneously with the fever in the early stages of many infections. In younger children, colds, sore throats, earaches, and stomach flu are the most frequent reasons for weakness. In older children, infectious mononucleosis, a prolonged but seldom serious viral infection, or influenza is often the cause. Weakness is a signal that the body should rest.

The fatigue associated with an infection will usually develop quickly and last only for a short period of time; exceptions are in the case of hepatitis, mononucleosis, and a poorly understood syndrome called *neurasthenia*, which may follow other viral infections on rare occasions. Even in these cases there is nothing to do but wait it out.

Lack of Sleep

Children have an extraordinary amount of energy, and it is unusual for them to complain of being tired for long periods of time. However, we often see young children who seem to have no energy during the day and can trace the problem to late-night watching of television in their rooms. It is not surprising that a young child who is awake until the wee hours of the morning will have little energy for usual daytime activities. Thin walls, noisy neighbors, and late-night talks with brothers or sisters are other factors that can interfere with sleep.

Normally, parents can let nature tell the child when sleep is needed; but if the child is droopy and sleepy late in the day, action may be needed to ensure that adequate amounts of sleep are received.

Prolonged Weakness

Anemia (low blood) is a very occasional cause of chronic weakness or tiredness. The most common anemia is iron-deficiency anemia. Iron is abundant in meat products, cereal, nuts, lima beans, lentils, peas, soybeans, spinach, and many other products. Cow's milk is low in iron and a diet consisting exclusively of cow's milk is dangerous in its ability to produce anemia in infants. Exclusive ingestion of goat's milk can lead to another type of anemia in infants. Restrictive diets, such as the Zen macrobiotic diet, are inadequate for children. Variety of diet, including all the major food groups, is the most important principle, rather than sets of rigid rules. Given enough varied raw materials, the body is very good at selecting out what is needed and eliminating the rest.

Hypoglycemia (low blood sugar) is another commonly discussed but rarely found medical problem. True hypoglycemia causes other symptoms such as sweating, nervousness, headaches, and irritability, as well as fatigue.

Long-standing weakness and tiredness in children may result from **depression**. It is a common mistake to assume that only adults are capable of becoming depressed. Children may react severely—to a move, the loss of a pet or friend, difficulty in school, an inability to be successful with friends, or an inability to compete in sports—by withdrawing. Depressed children often have little energy or self-esteem, feel tired or sad all the time, refuse to eat or eat too much, have trouble falling asleep at night or sleep too much, and may complain of a number of physical ailments. These symptoms should be treated seriously. The longer a child spends away from school or from normal everyday activities, the harder it is to begin functioning normally again. If you are having difficulty dealing with a problem that is disturbing your child, you should seek professional help for the child.

HOME TREATMENT

Watching and waiting is all that is called for in the case of weakness that accompanies a mild infection. As usual, adequate fluid intake and rest will help. If the problem persists for a week or more, then the doctor should be consulted, at least by telephone.

WHAT TO EXPECT AT THE DOCTOR'S OFFICE

A careful history and physical examination with measurement of height and weight will be performed. For some types of infections, specific laboratory tests will be ordered; infectious mononucleosis and hepatitis can be detected by a blood test. Most doctors will check for anemia and hypoglycemia only if the dietary history or symptoms are suggestive of these problems.

WEAKNESS, FATIGUE

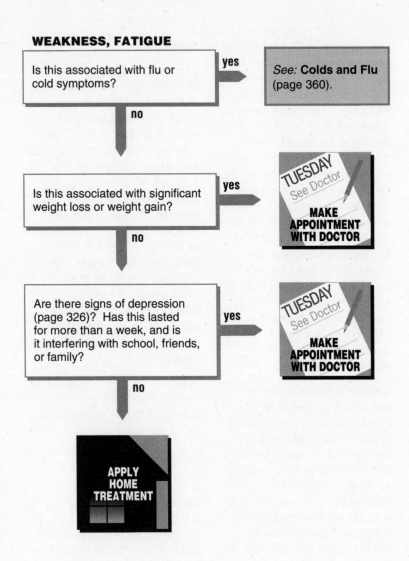

Is this associated with flu or cold symptoms?

yes → *See:* **Colds and Flu** (page 360).

no ↓

Is this associated with significant weight loss or weight gain?

yes → TUESDAY See Doctor — **MAKE APPOINTMENT WITH DOCTOR**

no ↓

Are there signs of depression (page 326)? Has this lasted for more than a week, and is it interfering with school, friends, or family?

yes → TUESDAY See Doctor — **MAKE APPOINTMENT WITH DOCTOR**

no ↓

APPLY HOME TREATMENT

Dizziness and Fainting

These complaints are frustrating to children, parents, and doctors alike. They are common, worrisome, and occasionally frightening, but most often they go unexplained. Some children have a special interest in fainting, it seems, even to the point of inventing games to make themselves pass out. Young children may be subject to the frightening but generally harmless phenomenon of breath-holding spells. Altogether, a problem with dizziness or fainting can be a pretty confusing situation. Defining the terms correctly will help in understanding the problem.

Fainting

The term *fainting* may be used to describe a lightheaded and woozy episode without loss of consciousness, or it may signify a complete collapse with loss of consciousness.

Lightheadedness frequently accompanies viral illnesses. It may also occur if a child suddenly stands up from a reclining or sitting position; it takes a few seconds for the body to adjust to this new position, so that there is temporarily a decrease in flow of blood to the head. Low blood sugar is commonly talked about but hardly ever is responsible. For lightheadedness, the attention of a doctor is needed only if it persists for several weeks. If the child has lost consciousness completely, however, the doctor should be seen without delay.

Breath-holding spells are described on pages 95–96. Breath-holding may lead to complete unconsciousness and even a seizure. Try to prevent injury from falling or from a seizure by observation and support. If you are certain that loss of consciousness occurred because of a breath-holding spell, then apply home treatment and discuss the situation with your doctor on the phone.

Dizziness

Dizziness refers to a situation in which the room seems to spin about the child; this is also called vertigo and may be accompanied by nausea and vomiting. Except for the case when children spin themselves in a circle rapidly, vertigo indicates that the balance mechanism in the inner ear has been disturbed. This most often occurs because of minor viral infections called **labyrinthitis**. Sometimes it occurs with ear infections; if there is an earache, see your doctor.

Some medications can cause vertigo. Telephone your doctor if this is a possibility.

HOME TREATMENT

If the problem is one of lightheadedness upon arising suddenly from a sitting or reclining position, then simple avoidance of such sudden changes in position is all that is necessary. This problem is called **postural hypotension** and does not need the help of the doctor unless it has suddenly become worse. Otherwise, it may wait and be discussed at the next

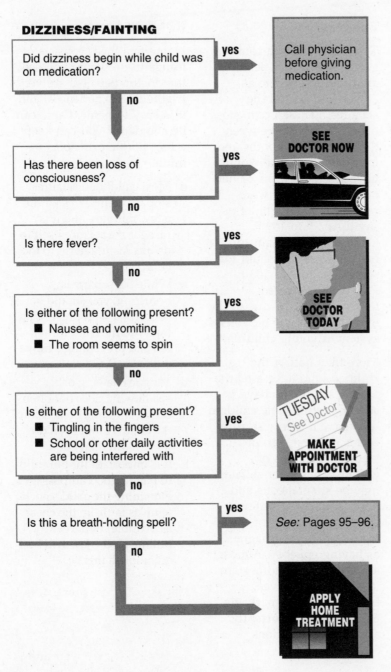

DIZZINESS/FAINTING

Did dizziness begin while child was on medication?
— yes → Call physician before giving medication.
— no ↓

Has there been loss of consciousness?
— yes → **SEE DOCTOR NOW**
— no ↓

Is there fever?
— yes → **SEE DOCTOR TODAY**
— no ↓

Is either of the following present?
- Nausea and vomiting
- The room seems to spin

— yes → **SEE DOCTOR TODAY**
— no ↓

Is either of the following present?
- Tingling in the fingers
- School or other daily activities are being interfered with

— yes → TUESDAY See Doctor / **MAKE APPOINTMENT WITH DOCTOR**
— no ↓

Is this a breath-holding spell?
— yes → *See:* Pages 95–96.
— no ↓

APPLY HOME TREATMENT

routine visit. If it is caused by a virus, it will go away within a few days.

Breath-holding spells are difficult problems. Drugs will not help. In the meantime, one tries to prevent damage from falls or seizures by alerting teachers and friends so that they may be aware of the possibility and ready to help. The child will outgrow them eventually.

WHAT TO EXPECT AT THE DOCTOR'S OFFICE

The history of the episodes provides the most important piece of information in deciding what to do. Physical examination will include examination of the eyes and ears and blood pressure measurements. Sometimes tests on the blood and the urine may be done. X-rays and brain wave tests (EEG) are rarely of use. Tests of the inner ear may be performed. Vertigo may be treated with drugs.

Seizures

About 5% of all children will have a seizure (convulsion, fit, falling-out spell). Seizures occur because of a disruption of the normal electrical impulse pattern of the brain. This disruption can occur spontaneously; be triggered by fevers, poisons, or infection (meningitis); or occur after a breath-holding spell, or an immunization.

There are many types of seizures, but parents will have no trouble recognizing the abnormal condition in their child. Some children will become stiff and roll their eyes backward. Others will exhibit rhythmic jerking of their arms and legs. Some may have staring spells or suddenly collapse. All of these will naturally cause great alarm. Fortunately, most seizures end in a few minutes without permanent damage to the child.

Fever Fits

The most common cause is a high fever; such seizures are often referred to as "fever fits" or "febrile seizures." These occur most often in children between the ages of six months and four years. Anyone would have a seizure if his or her body temperature were raised high enough, let us say to 107°F to 109°F. Children are thought to have a lower threshold for seizing: as low as 103°F. Part of this lowered threshold may be due to the immaturity of the nervous system in young children.

More than half of the children who have a febrile seizure will never have a second. Yet in some children who have a febrile seizure, this is the first sign of recurrent seizures of **epilepsy**, and later seizures may be spontaneous and not associated with fever. Less than one-half of 1% of the population is diagnosed as having epilepsy. Almost all of these individuals are leading normal lives because of the remarkable effectiveness of currently available medications.

HOME TREATMENT

Home treatment consists of things to do before you see the doctor. Seizures are very frightening experiences, and it is easy to panic. Panic can be avoided if you can keep a few simple principles in mind:

- Most childhood seizures stop by themselves within a few minutes, and it is rare for the seizure itself to do any lasting harm to the child.
- While the seizure is occurring, you should be concerned with preventing injury to the child from a fall and preventing a blow to the head.
- Tongue chewing or swallowing does not often occur. Jamming objects into the child's mouth can do more harm than good.
- It is important for the child to have a good air passage. Extending the child's neck (chin as far from the chest as possible) while gently pulling on the jaw will best accomplish this.

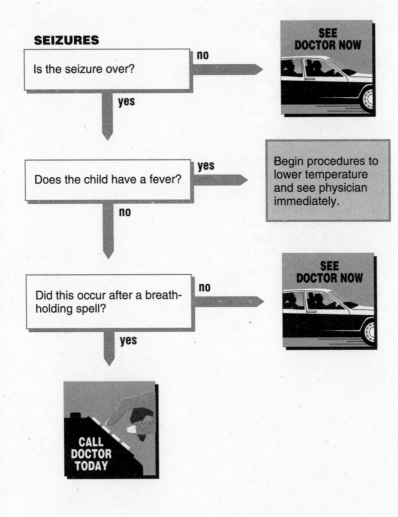

SEIZURES

Is the seizure over? — no → SEE DOCTOR NOW

yes ↓

Does the child have a fever? — yes → Begin procedures to lower temperature and see physician immediately.

no ↓

Did this occur after a breath-holding spell? — no → SEE DOCTOR NOW

yes ↓

CALL DOCTOR TODAY

- If the seizure has stopped and the child has vomited, clear the mouth.
- Following a seizure, the child will be drowsy or may seem to be in a deep sleep. This is called the post-ictal state, and is not harmful. You need not attempt to arouse the child from this state.
- If the seizure has stopped and there is a fever, begin temperature-lowering procedures. You cannot give medicine by mouth, but a rectal aspirin suppository can be used.

Sponging with cool water will help to lower temperature. Beware of giving a bath to a semi-conscious child. Sponging the child can continue during the trip to the doctor. Do not bundle the child up; both you and the child must keep cool. If there is a considerable distance to be traveled, temperature lowering en route becomes important; remember that you can lower a fever as effectively as the doctor can.

Breath-holding spell seizures are seldom serious, but your doctor should be consulted by phone.

WHAT TO EXPECT AT THE DOCTOR'S OFFICE

If the seizure has not stopped, then drugs will be given by injection to bring this about. Following control of the seizure, the doctor will perform a thorough physical examination and will want to talk to you in depth. If a fever is present, then search will be made for the source of the infection. Most often these are the common infections of childhood.

A spinal tap may be done to investigate the possibility of infection in the brain or the spinal cord. Hospitalization is sometimes required for observation, treatment, or further evaluation. A brain wave test (electroencephalogram, or EEG) will usually be done. Often it is better to wait a week before doing this test; brain wave tests do not necessarily require a hospital admission.

Phenobarbital can prevent recurrences of febrile seizures in children. Because phenobarbital has many side effects and must be taken for a long time and because most children will not have a second febrile seizure, we do not recommend its routine use. Phenobarbital will usually be given to children with seizures lasting more than ten minutes or affecting only one side of the body. Discuss the facts fully with your doctor.

Swallowed Objects

B abies and small children delight in swallowing any and all things that they can get their hands on and that are small enough to go down. In this section, we are concerned with things that will not dissolve in the stomach. Anything that *will* dissolve is a potential poison, and **Poison**, page 257, should be consulted.

The children's favorites among non-dissolving objects are coins, buttons, the eyes from teddy bears and dolls, safety pins, and fruit pits. Nature seems to have prepared the digestive tract well because even very sharp objects such as open safety pins, pieces of glass, needles, and straight pins regularly pass through the bowels with the greatest of ease. Therefore, the best strategy is to do nothing unless you're forced to do something.

Of immediate concern is the possibility that an object has become lodged in the windpipe. Violent coughing or difficulty in breathing suggests that the object may be lodged in the windpipe. As long as the child is able to breathe, proceed immediately to the doctor or emergency room. Attempting to dislodge an object that is partially obstructing breathing may lead to complete obstruction. If the windpipe is completely obstructed, a rapidly applied bear hug to the child's chest may force the object loose.

Pain and/or vomiting indicate that help is needed for an abdominal problem. If the object swallowed is very sharp and could possibly puncture the intestine, then you should call your doctor. The purpose of this call is to let the doctor know that assistance could be needed in the next few days on rather short notice. The doctor can make appropriate arrangements. If the object should perforate the intestine, then surgery will be needed, and these prior arrangements can make the surgery go more smoothly. Be prepared, but do not panic. Even razor blades have been passed through the entire digestive system without noticeable effect.

HOME TREATMENT

If the object is smooth, then you may want to look at the bowel movements for the next few days to reassure yourself that it has passed. If you don't see it, it is a far better bet that you missed it than that it did not pass.

If the object is sharp, then the doctor will likely ask you to look for the object to pass and to be on the alert for the symptoms of abdominal pain or vomiting. He or she, in turn, should make sure that surgical help will be available if needed.

SWALLOWED OBJECTS

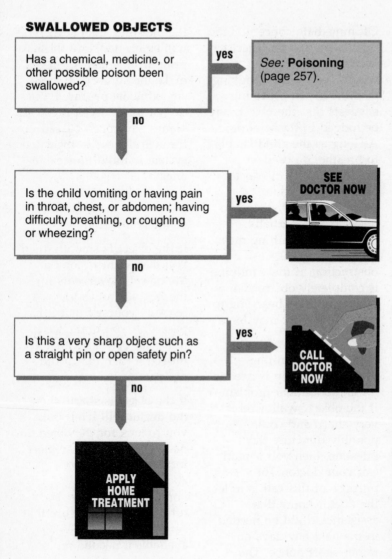

Has a chemical, medicine, or other possible poison been swallowed?

yes → *See:* **Poisoning** (page 257).

no

Is the child vomiting or having pain in throat, chest, or abdomen; having difficulty breathing, or coughing or wheezing?

yes → **SEE DOCTOR NOW**

no

Is this a very sharp object such as a straight pin or open safety pin?

yes → **CALL DOCTOR NOW**

no

APPLY HOME TREATMENT

WHAT TO EXPECT AT THE DOCTOR'S OFFICE

If the object is lodged in the throat, the doctor may be able to remove it in the office. Otherwise, a visit to the hospital may be necessary. If the problem is shortness of breath, wheezing, pain in the chest or abdomen, or vomiting, then a history and physical examination along with X-rays will be performed. Remember that non-metallic objects may not be seen on an X-ray. In this instance, special X-rays (such as a barium swallow) may show the object but may be unwise if puncture of the intestine is suspected.

If puncture of the intestine has occurred, then surgery will be necessary. If the object is in the respiratory tract, it must be removed. This can often be accomplished through an instrument known as a bronchoscope, which enables doctors to visualize and remove foreign objects.

Frequent Illnesses

Frequent illnesses are the rule and not the exception for most children. Recent studies have revealed that the average child has between six and nine viral illnesses per year. In some children, these viral illnesses are so mild that the parent will not notice any symptoms. Other children will have a cough or runny nose or some other minor symptom. We do not think of these viral infections as illnesses but rather as immunizations that serve to protect children against the more serious consequences of these illnesses at a later age.

Besides the frequent minor colds that most children have, the majority of children will also experience one or more ear infections in a lifetime. Similarly, we expect most children to have several bouts of diarrhea while they are young and at least one or two strep throats in the school years. When you start adding up the number of illnesses we expect in children, it is a wonder that they spend so much time free of symptoms.

While the younger children in a family often seem to have more colds than their older brothers and sisters did at the same age, they are just as healthy as their older counterparts. The younger child merely gets many of the common viral illnesses at an earlier age because of exposure to older brothers and sisters. Similarly, while some studies have shown that children in day-care centers or nursery schools seem to

get a few more colds early on, there is no evidence that these children are not every bit as healthy as their peers who remain at home. There are at least 60 different viral strains against which most adults are partly immunized because of childhood illnesses. The sooner the child develops these immunities, the sooner the relatively illness-free years of the adult can begin.

Immune Deficiencies

Abnormalities of the body's immune defense mechanisms, of concern to many parents, virtually never present themselves as frequent minor infections. Children with immune deficiencies often have repeated severe infections of the lungs or skin. These children seldom grow normally and are often quite underweight because of the repeated infections. For any of these problems, a medical evaluation is important. Similarly, if there is history of a family relative dying at a young age because of an overwhelming infection, you should discuss your concern with your doctor.

If your child received a blood transfusion before April 1985, there is a very slight chance of having acquired the human immunodeficiency virus (HIV), which is responsible for AIDS. Children can also acquire HIV while in the uterus. If a child's mother received a transfusion between 1980 and 1985 or used injection drugs or had a sex partner who was using injection drugs, this is also a consideration to be discussed with your doctor.

Allergies

Allergies are often confused with illnesses. Children with allergic rhinitis or hay fever will usually have sneezing, eye itching, and runny noses at a particular time of year, usually in the spring and in the fall. These children will often be rubbing their noses constantly.

HOME TREATMENT

Home treatment of the minor problems of childhood are discussed throughout this book. Frequent illnesses that are troublesome enough to interfere with schoolwork should be evaluated by a doctor.

WHAT TO EXPECT AT THE DOCTOR'S OFFICE

A careful history and physical examination will be performed. Cataloging of the nature and severity of the frequent illnesses will be done. Particular attention will be paid to the child's height and weight development. Depending on the nature of the problem, various laboratory tests may be performed.

True immune deficiency states in children are very rare, and these children need special treatment. Frequent colds or recurrent earaches do *not* warrant the use of gamma globulin shots. Gamma globulin deficiency must be confirmed by blood testing.

FREQUENT ILLNESSES

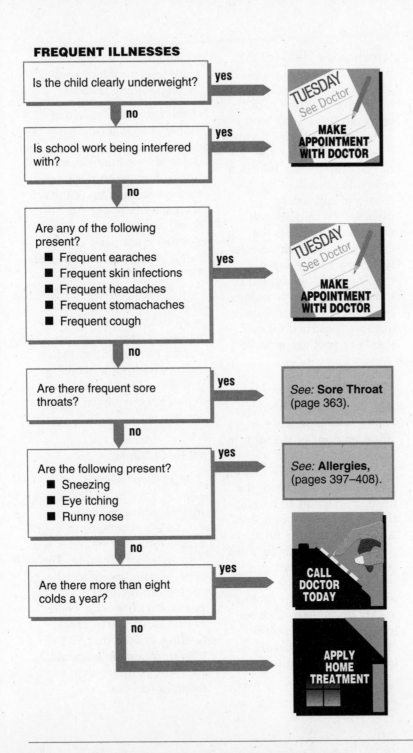

Is the child clearly underweight? — **yes** → **MAKE APPOINTMENT WITH DOCTOR**

no

Is school work being interfered with? — **yes** → **MAKE APPOINTMENT WITH DOCTOR**

no

Are any of the following present?
- Frequent earaches
- Frequent skin infections
- Frequent headaches
- Frequent stomachaches
- Frequent cough

— **yes** → **MAKE APPOINTMENT WITH DOCTOR**

no

Are there frequent sore throats? — **yes** → *See:* **Sore Throat** (page 363).

no

Are the following present?
- Sneezing
- Eye itching
- Runny nose

— **yes** → *See:* **Allergies,** (pages 397–408).

no

Are there more than eight colds a year? — **yes** → **CALL DOCTOR TODAY**

no → **APPLY HOME TREATMENT**

Jaundice

The skin is often a good barometer of what's occurring in your child's body. Parents know this and often speak up with concerns that a child's complexion appears too pale, dusky, flushed, red, and occasionally yellow.

A yellow or orange tinge to a child's skin is not an immediate cause for alarm. Many infants and young children are large consumers of carrots or other foods containing carotenoids that may cause a harmless and reversible discoloration of the skin. In distinction, the yellow coloring of the skin that occurs in jaundice will also be accompanied by a yellowing of the eyes.

Jaundice is a result of the body's inability to properly dispose of a normally occurring chemical called bilirubin. When red blood cells finally collapse because of old age (usually 120 days), they release bilirubin, which is processed by the liver and ultimately disposed of through the intestines. Problems that either cause red blood cells to dissolve more rapidly than normal (hemolysis) or interfere with the workings of the liver (hepatitis) or its outflow (obstruction) can therefore produce jaundice.

Newborns

A majority of newborn babies will experience a slight amount of jaundice in the first few days of life while their livers are getting cranked up to produce the needed enzymes to handle the breakdown of bilirubin. As long as there is no problem with the baby's red cells stemming from incompatibility with the mother's blood (Rh problem or ABO problem), with an inherited blood problem (spherocytosis), or with infection, the baby is considered to have a temporary problem known as **physiologic jaundice**.

If the baby appears quite yellow, blood measurements of the bilirubin may be done to be certain levels are not so high as to pose a risk to the baby's brain. We have seen too many parents terrorized by the bilirubin number game, and even brought to tears if a level of 12 goes to 13. Levels below 20 are generally considered safe, except for very small, very sick, or premature babies. Higher levels are currently being tolerated in infants without a hemolytic problem. Also, infants of Asian ancestry have higher levels.

Older Children

A mild form of jaundice is seen in 1 to 2% of breast-fed infants whose mothers produce a substance in their breast milk that impedes the processing of bilirubin. This type of jaundice generally begins on day four and peaks at two weeks. There is no evidence that breast feeding in the other 98 to 99% of infants is responsible for more jaundice than formula feeding.

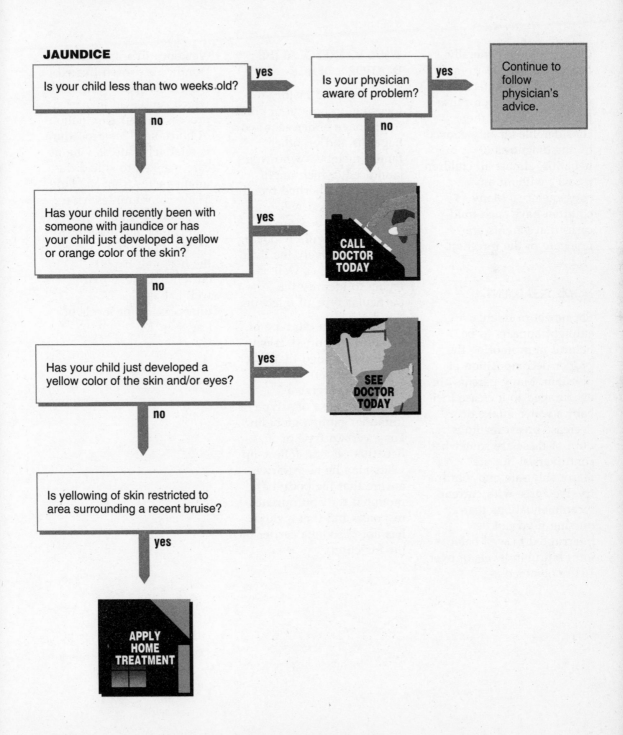

JAUNDICE

Is your child less than two weeks old? — **yes** →

Is your physician aware of problem? — **yes** →

Continue to follow physician's advice.

no ↓

no ↓

Has your child recently been with someone with jaundice or has your child just developed a yellow or orange color of the skin? — **yes** → CALL DOCTOR TODAY

no ↓

Has your child just developed a yellow color of the skin and/or eyes? — **yes** → SEE DOCTOR TODAY

no ↓

Is yellowing of skin restricted to area surrounding a recent bruise?

yes ↓

APPLY HOME TREATMENT

Older children generally become jaundiced from a variety of liver infections, usually viral, which cause hepatitis. Despite the considerable anxiety evoked by the diagnosis of hepatitis, almost all children recover without any consequences. Many children have such mild cases that parents are unaware of the infection.

HOME TREATMENT

For newborn infants, sunlight appears to be helpful in promoting the proper decomposition of bilirubin. Many parents are encouraged to increase their baby's water intake or decrease breast feedings. Both of these are somewhat controversial but are reasonably safe and worth a try. We agree with current recommendations that nursing need not be interrupted in well babies until bilirubin levels of over 16 are obtained.

WHAT TO EXPECT AT THE DOCTOR'S OFFICE

After a careful history focusing on how your child might have either acquired hepatitis and on other family members who were jaundiced (which might suggest an inherited blood problem), blood will undoubtedly be drawn. Your doctor will be looking at how effectively the liver is functioning as well as trying to identify the particular type of infection.

Depending on the type of hepatitis identified, it may be recommended that other family members or playmates receive a protective dose of a type of immune/gamma globulin. For a certain type of hepatitis, several follow-up visits may be necessary to ensure that the body has mounted the appropriate response and that a child has not become a carrier of the infection.

Newborn infants with jaundice are often treated with blue fluorescent lights, which causes an increased processing and elimination of bilirubin. This procedure is relatively safe as long as care is taken to shield the infant's eyes from the light. Many communities have home bilirubin light services that can be far superior to having your infant return to the hospital. Experts currently believe that in well babies this is unnecessary for levels of less than 17.

CHAPTER G

Eye Problems

Eye Burning, Itching, and Discharge

These symptoms usually signal **conjunctivitis**, or "pink eye"; this is an inflammation of the membrane that lines the eye and the inner surface of the eyelids. The inflammation may be due to an irritant in the air, an allergy to something in the air, or an infection (virus, bacteria, or *Chlamydia*).

Irritants

In the newborn, there may be eye discharge in the first two days of life because of irritation if silver nitrate was given to protect against gonococcal infection. This discharge seldom lasts more than two days. Discharges after the third day are usually due to an infection or blocked tear duct (see page 353).

Environmental pollutants in smog can produce burning and itching, which sometimes seem as severe as the symptoms experienced in a tear-gas attack. These symptoms represent a **chemical conjunctivitis** and affect anyone exposed to enough of the chemical. The smoke-filled room, the chlorinated swimming pool, the desert sandstorm, or sun glare on snow can give similar physical or chemical irritation.

In contrast, **allergic conjunctivitis** affects only those people who are allergic. Almost always the allergen is in the air, and grass pollens are probably the most frequent offender. Depending on the season for the offending pollen, this problem may occur in spring, summer, or fall and usually lasts from two to three weeks.

Infections

A minor conjunctivitis frequently accompanies a viral cold. Epidemics of conjunctivitis are most often caused by a virus. The eye discharge is not as thick as in more severe bacterial infections. Often a small swollen lymph gland will be found in front of the ear in **epidemic conjunctivitis**. Conjunctivitis is also common in the first day or so of measles (page 466). Some viruses, such as herpes, may cause deep painful ulcers in the cornea and may interfere with vision.

Bacterial infections cause pus to form, and a thick, plentiful discharge runs from the eye. Often the eyelids are crusted over and "glued" shut upon awakening. *Chlamydia* infections occur in the first few months of life. These infections require antibiotic treatment.

Some major diseases affect the deeper layers of the eye—those layers that control the operation of the lens and the size of the pupillary opening. This condition is termed **iritis** or "uveitis" and may cause irregularity of the pupil or pain when the pupil reacts to light. It is unusual in children. Medical attention is required.

EYE BURNING, ITCHING, AND DISCHARGE

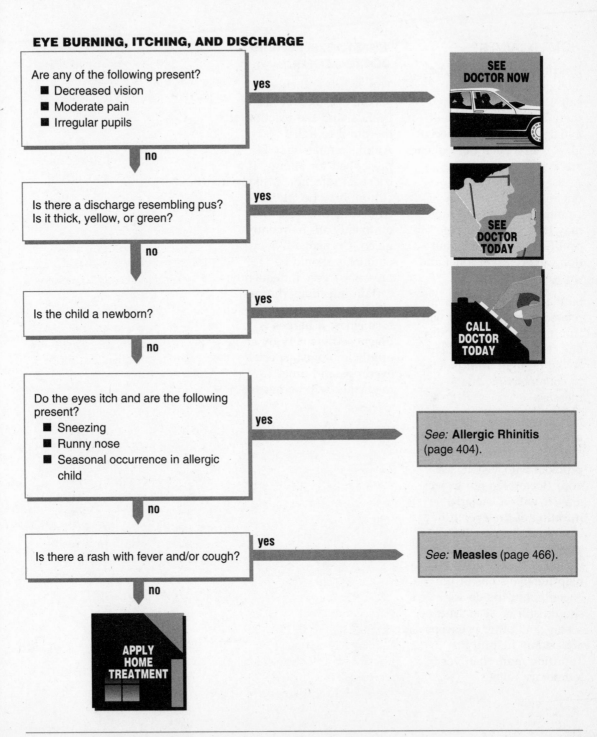

Are any of the following present?
- Decreased vision
- Moderate pain
- Irregular pupils

yes → SEE DOCTOR NOW

no

Is there a discharge resembling pus?
Is it thick, yellow, or green?

yes → SEE DOCTOR TODAY

no

Is the child a newborn?

yes → CALL DOCTOR TODAY

no

Do the eyes itch and are the following present?
- Sneezing
- Runny nose
- Seasonal occurrence in allergic child

yes → *See:* **Allergic Rhinitis** (page 404).

no

Is there a rash with fever and/or cough?

yes → *See:* **Measles** (page 466).

no

APPLY HOME TREATMENT

HOME TREATMENT

If a physical, chemical, or allergic exposure is the cause of the symptoms, there is nothing to do but avoid the exposure. Special glasses that keep out pollen are available for severe cases. Antihistamines obtained either over the counter or by prescription may help slightly if the problem is an allergy, but don't expect total relief and drowsiness may occur.

Similarly, a viral infection related to a cold or flu will run its course in a few days, and it is best to be patient. Avoiding light affords relief in conjunctivitis due to measles.

If it doesn't clear up, if the discharge gets thicker, or if there is eye pain or a problem with vision, see your doctor. Do not expect a fever with a bacterial infection of the eye; it may be absent. Because the infection is superficial, washing the eye gently will help remove some of the bacteria, but the doctor should still be seen. Murine, Visine, and other eyedrops are seldom helpful in affording more than very temporary relief.

WHAT TO EXPECT AT THE DOCTOR'S OFFICE

The doctor will check vision, eye motion, the eyelids, and the reaction of the pupil to light. Antihistamines may be prescribed for allergy. If a bacterial infection is likely, a culture may be taken. Antibiotic eyedrops or ointments are frequently given. Cortisone-like eye ointments should be prescribed very infrequently; certain infections (herpes) will get worse with these medicines. If herpes is diagnosed, usually by an ophthalmologist, special eyedrops and other medicines will be needed.

Swollen Eyes

It is easy and quite common for eyelids to become swollen. Not surprisingly, the most common reasons for swelling are linked to everyday events.

Irritants

Insect bites near the eye are notorious for causing eyelid swelling. Search carefully for a tell-tale bite near the affected eye. Usually, only one eyelid is swollen unless there has been a mosquito onslaught on your child's face.

Young children manage to rub a variety of irritating substances in their eyes, which can simultaneously cause redness of the eye and lid swelling. Irritants can range from dirt to household cleansers to spices. Allergies also contribute to swollen eyelids; hayfever and allergic conjunctivitis usually include eye itching and redness along with eyelid swelling. Allergic problems tend to occur repeatedly, and parents can learn to effectively begin home treatment. If eyelid swelling accompanies swelling elsewhere in the body (for example, in the mouth, hands or feet), this suggests a more widespread allergic reaction or accumulation of fluid characteristic of certain kidney disorders. Obviously, both of these should receive immediate medical attention.

Infections

Certain infections such as **sinusitis** (in older children) and **conjunctivitis** also characteristically produce eyelid swelling. However, certain serious infections may also be responsible for swollen eyes. **Periorbital cellulitis** is a bacterial infection carried by the blood and generally produces swelling of an area around the eyes. Usually, more than the eyelids are involved, the skin may have a red or purple color, and fever is present. Even more unusual is **infection of the eye cavity** (orbit) that will, in addition to the above features, have a bulging (sometimes painful) eye. This condition, often following trauma, is quite serious and merits immediate medical attention.

HOME TREATMENT

For insect bites and other minor irritations, washing the area around the eye will be both cleansing and soothing. Allergic eye reactions can be helped with a variety of antihistamines (for example, chlorpheniramine). Wrap-around sunglasses may also offer some protection in heavy pollen season. If your child has had allergic reactions in the past, your doctor may also have recommended Benadryl, now available as an over-the-counter preparation.

SWOLLEN EYES

Are any of the following present?
- Protruding eye
- Fever
- Skin color change
- Eye pain
- Swelling after injury

yes → SEE DOCTOR NOW

no ↓

Is there redness and discharge that is not clear, or are eyelids stuck together (see pages 346–348)?

yes → SEE DOCTOR TODAY

no ↓

Is swelling increasing or are both eyelids swollen?

yes → CALL DOCTOR TODAY

no ↓

APPLY HOME TREATMENT

WHAT TO EXPECT AT THE DOCTOR'S OFFICE

A careful history and physical examination will be performed with focus on the eye and face. Blood tests may be taken to check for indicators of serious infection or to evaluate for bacteria in the blood. If sinusitis is suspected, X-rays may be taken. More extensive radiological evaluations will be performed for suspected orbital cellulitis. If a kidney problem is suspected, a urinalysis will be performed.

Antibiotics are commonly prescribed for milder infections, whereas daily injections or hospitalization for intra-venous antibiotics are appropriate for serious infections.

Object in Eye/Eye Pain

All eye injuries should be taken seriously; if there is any question, visit the doctor. The stakes are too high.

A foreign body in the eye must be removed, or the threat of infection and loss of sight in that eye is present. Be particularly careful if the foreign body was caused by the striking of metal on metal; this can cause a small metal particle to strike the eye with great force and to penetrate the eyeball.

Under a few circumstances, you can treat the injury at home. If the foreign body is minor, such as sand, and did not strike the eye with great velocity, it may feel as though it is still in the eye even when it is not. Small, round particles like sand rarely stick behind the upper lid for long.

If it feels as though a foreign body is present but it is not, then the covering of the eye (cornea) has been scraped or cut. A minor corneal injury will usually heal quickly without problems; a major one requires medical attention.

Even if you think the injury to be minor, run through the questions on the decision chart daily. If any symptoms at all are present after 48 hours and are not clearly resolving, see the doctor. Minor problems will heal within 48 hours; the eye repairs injury quickly.

Eye pain is an unusual symptom that may signify a serious inflammation of the eye (iritis) or a serious infection with herpes virus. A doctor should be consulted.

HOME TREATMENT

Be gentle. Wash the eye out; water is fine. Inspect the eye and have someone else check it as well. Use a good light and shine it from both the front and the side. Pay particular attention to the cornea—this is a clear membrane that covers the colored portion of the eye. Do not rub the eye; if a foreign body is present you will abrade or scratch the cornea.

An eye patch will relieve pain and usually it is needed for 24 hours or less. Make the patch with several layers of gauze and tape firmly in place. Check vision each day and compare the two eyes. If you are not sure that all is going well, see the doctor.

WHAT TO EXPECT AT THE DOCTOR'S OFFICE

The doctor will perform a vision check, inspection of the eye, and inspection under the upper lid; this is not painful. Usually, a fluorescent stain will be dropped into the eye, and the eye then examined under ultra-violet light; this too is not painful or hazardous.

A foreign body, if found, will be removed. In the office, this may be done with a cotton swab, an

OBJECT IN EYE/EYE PAIN

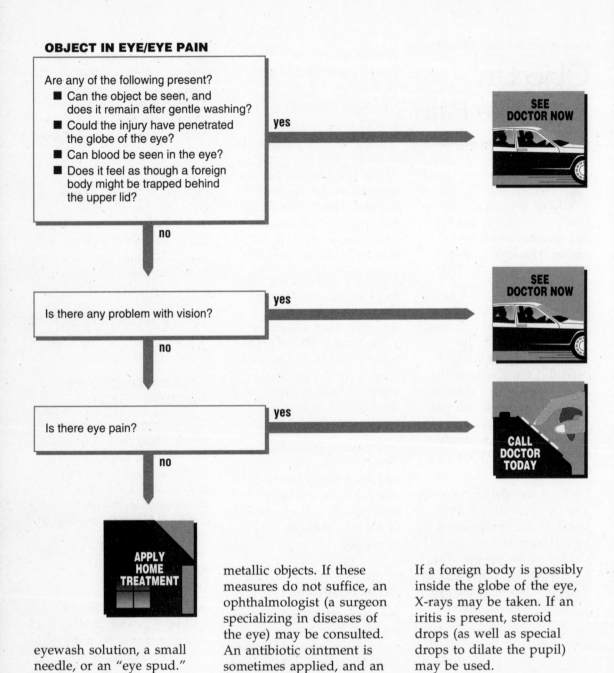

Are any of the following present?
- Can the object be seen, and does it remain after gentle washing?
- Could the injury have penetrated the globe of the eye?
- Can blood be seen in the eye?
- Does it feel as though a foreign body might be trapped behind the upper lid?

yes → SEE DOCTOR NOW

no

Is there any problem with vision?

yes → SEE DOCTOR NOW

no

Is there eye pain?

yes → CALL DOCTOR TODAY

no

APPLY HOME TREATMENT

eyewash solution, a small needle, or an "eye spud." Strong magnets are sometimes used to remove metallic objects. If these measures do not suffice, an ophthalmologist (a surgeon specializing in diseases of the eye) may be consulted. An antibiotic ointment is sometimes applied, and an eye patch may be provided.

If a foreign body is possibly inside the globe of the eye, X-rays may be taken. If an iritis is present, steroid drops (as well as special drops to dilate the pupil) may be used.

Styes and Blocked Tear Ducts

We might have called this problem "bumps around the eyes" because that is how they appear. Styes are infections (usually with staphylococcal bacteria) of the tiny glands in the eyelids. They are really small abscesses, and the bumps are red and tender. They grow to full size over a day or so.

Another type of bump in the eyelid called a **chalazion** appears more slowly over many days or even weeks and is not red or tender. A chalazion often requires drainage by a doctor, whereas most styes will respond to home treatment alone. However, there is no urgency to the treatment of a chalazion, and home treatment will not cause any harm.

Tear Ducts

Tears are the lubricating system of the eye. They are continually produced by the tear glands and then drained away into the nose by the tear ducts. These tear ducts are often incompletely developed at birth so that the drainage of tears is blocked. When this happens, the tears may collect in the tear duct and cause it to swell, appearing as a bump along the side of the nose just below the inner corner of the eye (see figure on page 354). This bump is not red or tender unless it has become infected. Most blocked tear ducts will open by themselves in the first month of life, and most of the remainder will respond to home treatment. Tears running down the cheek are seldom noted in the first month of life because the infant produces only a small volume of tears.

The eyeball itself is *not* involved in a stye or a blocked tear duct. Problems with the eyeball, and especially with vision, should not be attributed to these two relatively minor problems.

HOME TREATMENT

Stye

Apply warm, moist compresses for 10 to 15 minutes at least three times a day. The compresses help the abscess to "point." This means that the tissue over the abscess becomes quite thin, and the pus in the abscess is very close to the surface. After an abscess points, it often will drain spontaneously. If this does not happen, then it is ready to be lanced by the doctor. Sometimes the stye goes away without coming to a point and draining.

Chalazions usually do not respond to warm compresses but neither will they be harmed. If no improvement is noted with home treatment after 48 hours, then see the doctor.

Blockage of the Tear Ducts

Simply massage the bump downward with warm, moist compresses several times a day. If the bump is not red and tender (indicating infection), this may be continued for up to

…

EYE PROBLEMS

STYES AND BLOCKED TEAR DUCTS

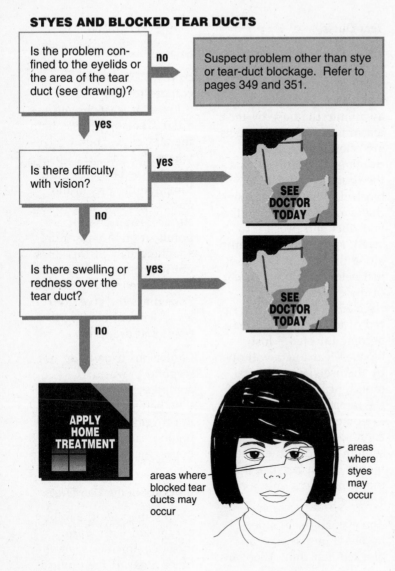

Is the problem confined to the eyelids or the area of the tear duct (see drawing)? **no** → Suspect problem other than stye or tear-duct blockage. Refer to pages 349 and 351.

yes ↓

Is there difficulty with vision? **yes** → **SEE DOCTOR TODAY**

no ↓

Is there swelling or redness over the tear duct? **yes** → **SEE DOCTOR TODAY**

no ↓

APPLY HOME TREATMENT

areas where blocked tear ducts may occur

areas where styes may occur

several months. If the problem exists for this long, discuss it with your doctor. If the bump becomes red and swollen, antibiotic drops will be needed.

WHAT TO EXPECT AT THE DOCTOR'S OFFICE

If the stye is pointing and ready to be drained, the doctor will open it with a small needle. If it is not pointing, then compresses usually will be advised and antibiotic eyedrops sometimes will be added. Attempting to drain a stye that is not pointing is usually not very satisfactory.

If the doctor feels that the problem is a chalazion, then it may be removed with minor surgery. Whether to have the surgery will be up to you; chalazions are not dangerous, and the operation may be disturbing to the child.

If the child is over six months of age and is still having problems with blocked tear ducts, they can be opened in almost all cases with a very fine probe. This probing is successful on the first try in about 75% of all cases and on subsequent attempts in the remainder. Only rarely is a surgical procedure necessary to establish an open tear duct. For red and swollen ducts, antibiotic drops as well as warm compresses usually will be recommended.

Vision Problems

Although children can see at birth, the eye is not fully developed. The eye completes its development by about six months of age, and coordination between the two eyes is complete at about one year of age. It is difficult to determine how clearly the child sees (visual acuity) before the age of three or four. Problems with vision may be suggested by the child not reaching for objects or failing to follow a moving object with his or her eyes. Remember that the child's vision is still developing in the first months of life; do not be too quick in your judgment. Even in older children, significant problems with vision may not be detected without the use of eye tests, and eye tests are important for the school-age child.

Crossed eyes may be striking in the newborn child. This occurs simply because muscle coordination between the eyes is not yet fully developed; the problem usually corrects itself by the sixth month of life. If crossed eyes persist beyond this age, then you should discuss this at a regular checkup.

Strabismus is a problem of the eye muscles also known as "lazy eye"; one or both eyes may be involved. If the lazy eye is allowed to remain lazy, vision may be lost in that eye. Strabismus should be suspected in a child in which one or both eyes turn in or out after the age of four months. Ordinarily, a light shining from several feet in front of the child's eyes should reflect in the same location in each eye.

Fortunately, sudden loss of vision is less frequent in children than in adults. When it does occur, a quick trip to the doctor is obviously warranted.

An optometrist and ophthalmologist are often involved in the care of these problems. Before referral, however, you should have the opportunity to discuss the problem with your doctor; this may help sort out the problem and will result in a referral if necessary.

HOME TREATMENT

Home treatment is reserved for children under the age of six months who have crossed eyes. This problem is usually present at birth and gets better as the child gets older. Full correction may take as long as a year. Home treatment consists merely of observation to make sure that the problem corrects itself. Crossed eyes occurring at any other age should be discussed with your doctor.

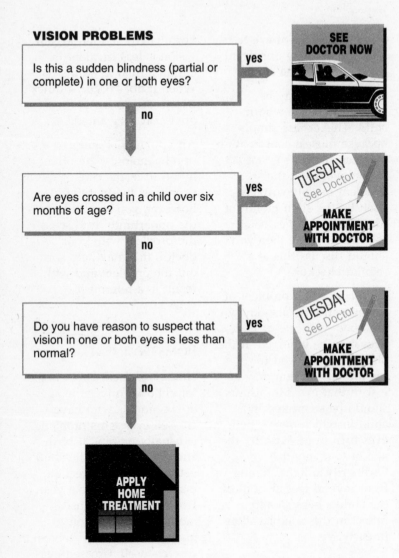

VISION PROBLEMS

| Is this a sudden blindness (partial or complete) in one or both eyes? | **yes** → | **SEE DOCTOR NOW** |

no ↓

| Are eyes crossed in a child over six months of age? | **yes** → | **TUESDAY** See Doctor / **MAKE APPOINTMENT WITH DOCTOR** |

no ↓

| Do you have reason to suspect that vision in one or both eyes is less than normal? | **yes** → | **TUESDAY** See Doctor / **MAKE APPOINTMENT WITH DOCTOR** |

no ↓

APPLY HOME TREATMENT

use of both eyes. Eye movement in a suddenly uncovered eye is a sign of strabismus.

The inside of the eye will be examined with a hand-held instrument called an ophthalmoscope. A special kind of microscope, called a slit lamp, may be used to examine the front portion of the eye.

If necessary, corrective lenses will be prescribed (this can be done even in very small children). Sometimes surgery will be required to correct crossed eyes and will be necessary in occasional cases of decreased vision. Most often, strabismus will be treated with an eye patch.

The younger the age at which this condition is detected, the better the chances for saving the vision in the eye. Make sure that your doctor always checks for visual acuity while your child is young.

WHAT TO EXPECT AT THE DOCTOR'S OFFICE

Visual acuity, eye movements, and pupils will be examined. With children too young to read, eyecharts with pictures are used. The doctor will alternately cover each eye while asking the child to look at a distant object. Charts may be used to check for simultaneous

CHAPTER H

Ear, Nose, and Throat Problems

Is It a Virus, Bacteria, or an Allergy?

The following sections discuss upper respiratory problems including colds and flu, sore throats, ear pain or stuffiness, runny nose, cough, hoarseness, swollen glands, and nosebleeds. A central question is important to each of these complaints: Is it caused by a virus, bacteria, or an allergic reaction? In general, only for bacterial infection does the doctor have more effective treatment than is available at home. Remember that viral infections and allergies do *not* improve with treatment with penicillin or other antibiotics. To demand a "penicillin shot" for a cold or allergy is to ask for a drug reaction, to risk a more serious "super-infection," and to waste time and money. Among common problems well treated at home are the following:

- The common cold—often termed "viral URI (upper respiratory infection)" by doctors
- The flu when uncomplicated
- Hay fever
- Mononucleosis—infectious mononucleosis, or "mono"

Medical treatment *is* commonly required for:

- Strep throat
- Ear infection

How can you tell these conditions apart? Table 5 and the charts for the following problems will usually suffice. Here are some brief descriptions.

VIRAL SYNDROMES

Viruses usually involve several portions of the body and cause many different symptoms. Three basic patterns (or syndromes) are common in viral illnesses. Overlap between these three syndromes is not unusual. Your child's illness may sometimes have features of each.

Viral URI. This is the "common cold." It includes some combination of the following: sore throat, runny nose, stuffy or congested ears, hoarseness, swollen glands, and fever. One symptom usually precedes the others, and another (usually hoarseness or cough) may remain after the others have disappeared.

The flu. Fever may be quite high. Headache can be severe; muscle aches and pain (especially low back and eye muscles) are equally troublesome.

Viral gastroenteritis. This is the "stomach flu" with nausea, vomiting, diarrhea, and crampy abdominal pain. It may be incapacitating and can mimic a variety of other more serious conditions including appendicitis.

STREP THROAT

Most strep throats will be accompanied by soreness of the throat. However, symptoms outside the respiratory tract can occur, most commonly fever and swollen lymph glands in the neck (from draining the infected material). The rash of scarlet fever sometimes may help to distinguish a streptococcal (strep) from a viral infection. Abdominal pain or headaches may be associated with a strep throat. This disorder should be diagnosed and treated because serious heart and kidney complications can follow if adequate antibiotic therapy is not given. (See **Sore Throat**, page 363.)

Remember, viral infections and allergies do not improve with treatment by penicillin or other antibiotics.

HAY FEVER

The seasonal runny nose, sneezing, and itchy eyes are well known. As with viruses, this disorder is treated simply to relieve symptoms; given enough time, the condition runs its course without doing any permanent harm. Allergies tend to recur whenever the pollen or other allergic substance is encountered. (See Chapter I on hay fever.)

TABLE 5   *Is It a Virus, Bacteria, or an Allergy?*

	Virus	*Bacteria*	*Allergy*
Runny nose?	Often	Rarely	Often
Aching muscles?	Usually	Rarely	Never
Headache (non-sinus)?	Often	Rarely	Never
Fever?	Often	Often	Never
Cough?	Often	Sometimes	Rare
Croup?	Usually	Rarely	Never
Recurs at a particular season?	Never	Never	Often
Do antibiotics help?	Never	Yes	Never
Can the doctor help?	Seldom	Yes	Sometimes
Dizziness?	Often	Rarely	Rarely
Dry cough?	Often	Rarely	Sometimes
Raising sputum?	Rarely	Often	Rarely
Hoarseness?	Often	Rarely	Sometimes
Only a single complaint (sore throat, earache, sinus pain, or cough)?	Rarely	Usually	Rarely

Colds and Flu

All children have "colds," and most parents become experts in the treatment of their children's common viral upper respiratory infections rather quickly. Colds are almost invariably caused by viruses and do not require or respond to antibiotic treatment. An uncomplicated cold can hardly be considered an illness. In fact, a great many viral infections in children are so mild that no symptoms at all are noticed. Viruses can cause **Headache** (page 309), **Runny Nose** (page 373), **Sore Throat** (page 363), **Muscle Aches** (page 477), **Cough** (page 376), **Vomiting** (page 504), **Diarrhea** (page 507), and **Eye Discharges** (page 346) among other symptoms. These individual problems may be consulted for further discussion.

Parents are also frequently concerned that their children may have too many colds. Most healthy children have between six and nine viral infections each year. Some of these colds will be very short and mild, whereas others may last a week or longer. Children with frequent colds seldom have anything seriously wrong with them. These children do *not* need gamma globulin shots. Children who have real problems with their immune defense system most often have extremely serious illnesses rather than frequent mild illnesses. The child's immune system is being made stronger with each cold, so even the sniffles have their positive side.

Complications

Most parents are worried about complications. These do, of course, occur, but are fortunately much less common than the uncomplicated illness. The most frequent is a blockage in the tube that drains the middle ear and, consequently, an ear infection. Other parents worry about colds progressing to a pneumonia. The observant parent is often able to tell when a child is developing a complication of a cold.

- Children with a developing complication are often more fussy and may be taking their food poorly.
- Younger children may tug at their ears or cry frequently if an ear infection is starting. Older children may complain of pain in their ears.
- A breathing rate greater than 40 per minute is also an indication that a cold may be becoming complicated by a developing pneumonia.

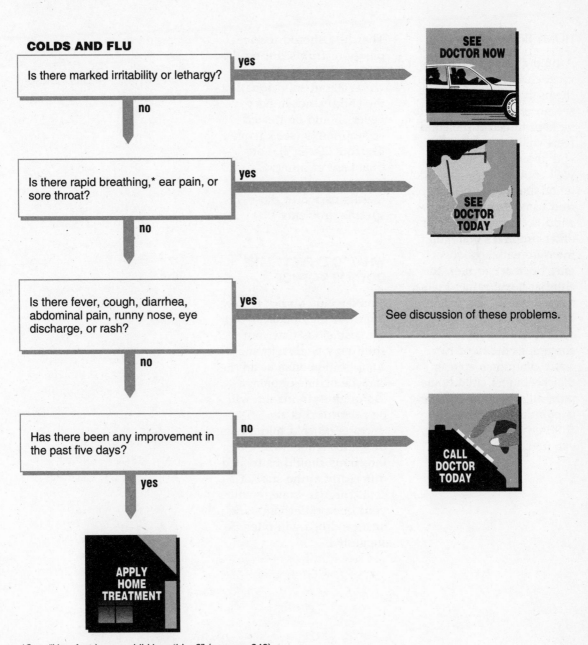

COLDS AND FLU

Is there marked irritability or lethargy?

yes → SEE DOCTOR NOW

no ↓

Is there rapid breathing,* ear pain, or sore throat?

yes → SEE DOCTOR TODAY

no ↓

Is there fever, cough, diarrhea, abdominal pain, runny nose, eye discharge, or rash?

yes → See discussion of these problems.

no ↓

Has there been any improvement in the past five days?

no → CALL DOCTOR TODAY

yes ↓

APPLY HOME TREATMENT

*See: "How fast is your child breathing?" (on page 249)

HOME TREATMENT

A tired child needs **rest**, but most children will restrict themselves adequately. We see no reason for a child to be kept in bed if the child feels well enough to be up and about. If a child feels well enough to go to school, he or she should be allowed to do so. Of course, if the child is coughing or having other problems that will interfere with schoolwork, it may be better to keep the child at home rather than to have the child sent home in the middle of the day. Colds are usually the most contagious a day or two *before* symptoms appear, so that keeping a child home until all symptoms are gone is pointless; the other children have already been exposed.

The child should receive plenty of **fluids** during the time of the cold. Don't worry about solid food if the child is not hungry. Fever should be treated appropriately (see Chapter E, pages 299–307). For treatment of other symptoms, please refer to the discussion of those specific problems.

WHAT TO EXPECT AT THE DOCTOR'S OFFICE

Simple colds do not require a visit to the doctor. If you are suspicious that your child may be developing a complication such as an earache or pneumonia, a thorough examination will be performed at the doctor's office. Children prone to frequent ear infections should be treated differently at the start of a cold. Discuss strategy with your doctor; decongestants or nose drops will often be suggested.

Sore Throat

Sore throat is one of the most common complaints of childhood. Very often a sore throat is accompanied by a fever, a headache, or even abdominal pain. Infants seldom have a sore throat; but as children approach school age, sore throats become more frequent.

Sore throats can be caused by either viruses, bacteria, or special types of organisms such as *Mycoplasma* and *Chlamydia*. Often, especially in the winter, sleeping with an open mouth or mouth breathing can cause drying and irritation of the throat. This type of irritation always subsides quickly after the pharynx becomes moist again.

Mono

Older children and adolescents frequently develop a viral sore throat known as infectious **mononucleosis**, or "mono." Despite the formidable sounding name of this illness, complications seldom occur. The sore throat is often more severe and is often prolonged beyond a week, and the child may feel particularly weak. Occasionally, the spleen, one of the internal organs in the abdomen, may enlarge during mononucleosis, and resting will be important. A viral sore throat that does not resolve within a week might be caused by the virus responsible for mononucleosis.

Strep Throat

Virtually all sore throats caused by bacteria are due to the streptococcal bacteria. These sore throats are commonly referred to as **strep throat**. A strep throat should be treated with an antibiotic because of complications.

First, an **abscess** may form in the throat. This is an extremely rare complication but should be suspected if the child has extreme difficulty in swallowing, has an excess of salivation, or has difficulty opening his or her mouth.

The more significant complications of strep throat occur from one to four weeks after the pain in the throat has disappeared. **Acute glomerulonephritis** causes an inflammation of the kidney. Although antibiotics will not prevent this complication, the antibiotics will prevent the strep from spreading to other family members or friends.

Of greatest concern is the complication of **rheumatic fever**, less common today than in the past but still a problem in many parts of the country. Rheumatic fever is a complicated disease that causes painful swollen joints, unusual skin rashes, and heart damage in half of its victims. It can be prevented by antibiotic treatment of a strep throat.

Unfortunately, it is not possible to definitely distinguish a viral sore throat from a strep throat on the basis of symptoms. Neither the height of the fever, the appearance of the throat, nor the amount of pain present will prove whether a sore throat is due to a virus or strep. The only method of distinguishing viral from strep throat is through a **throat culture**.

In many parts of the country, a throat culture can be obtained without a doctor's office visit fee. If you have no other concerns about your child and merely need to know whether the child has a strep throat, inquire about obtaining a throat culture without a full office visit.

HOME TREATMENT

Cold liquids and acetaminophen are effective for the pain and fever. Many home remedies include saltwater gargles and honey or lemon in tea. Time is the most important healer for pain; a vaporizer makes the waiting more comfortable for some.

WHAT TO EXPECT AT THE DOCTOR'S OFFICE

A throat culture will usually be taken. New tests are available to detect the streptococcal bacteria in just a few minutes. However, these tests are not quite as accurate as throat cultures, so your doctor may use either the rapid method or prefer the culture technique; some will use both.

Most doctors will delay treating a sore throat until the test results are known; delaying treatment by one or two days does not increase the risk of developing rheumatic fever. Furthermore, acetaminophen is as effective as antibiotic treatment in reducing the discomfort of a sore throat. Since the majority of sore throats are due to viruses,

> **F**requent and recurrent sore throats are common, especially in children between the ages of five and ten. There is no evidence that removing the tonsils decreases this frequency.

treating all sore throats with antibiotics would expose children needlessly to the risks of allergic reactions to the antibiotics. Doctors often will begin antibiotics immediately if there is a family history of rheumatic fever, if the child has scarlet fever (a rash accompanying a sore throat), if rheumatic fever is commonly occurring in the community at the time, or if there are other compelling reasons.

If one child has a strep throat, the chances are very good that other family members will also have a strep throat, and it is common for doctors to take cultures from brothers and sisters.

SORE THROAT

Is there severe difficulty swallowing or breathing, or is there excessive drooling? — **yes** → 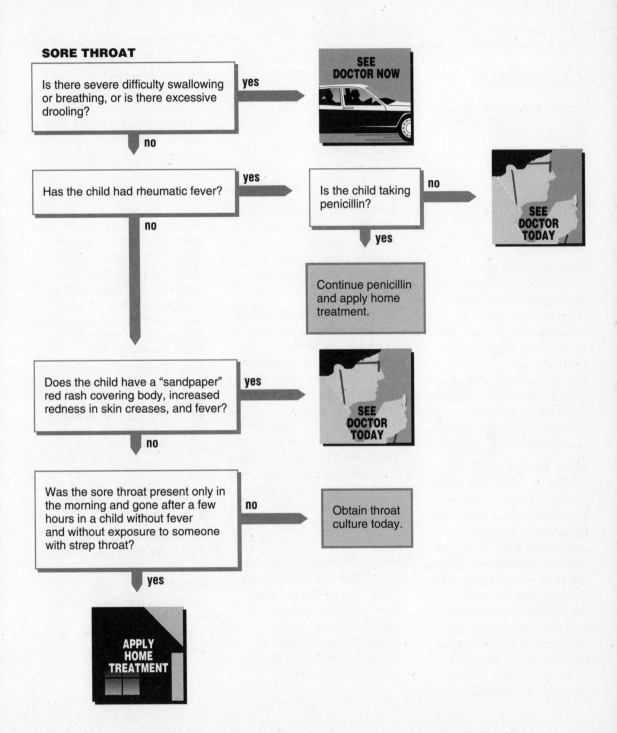 **SEE DOCTOR NOW**

no ↓

Has the child had rheumatic fever? — **yes** → Is the child taking penicillin? — **no** → **SEE DOCTOR TODAY**

no ↓

yes ↓ Continue penicillin and apply home treatment.

Does the child have a "sandpaper" red rash covering body, increased redness in skin creases, and fever? — **yes** → **SEE DOCTOR TODAY**

no ↓

Was the sore throat present only in the morning and gone after a few hours in a child without fever and without exposure to someone with strep throat? — **no** → Obtain throat culture today.

yes ↓

APPLY HOME TREATMENT

Earaches

Most children will have at least one earache while growing up, and many will have frequent earaches. Ear pain is caused by a buildup of fluid and pressure in the child's middle ear. Under normal circumstances, the middle ear is drained by a short narrow tube (the eustachian tube) into the nasal passages. Often during a cold, the eustachian tube will become swollen shut; this occurs most easily in small children in whom the tube is smaller.

Many infants are given bottles of milk while lying in bed; drinking milk while lying down may, in some instances, cause the eustachian tube to become irritated and closed. When the tube closes, the normal flow of fluid from the middle ear is prevented, and the fluid begins to accumulate. Bacteria grow rapidly in this stagnant fluid, and hence a bacterial infection often results.

The symptoms of an ear infection may include:

- Fever
- Ear pain
- Fussiness
- Increased crying
- Irritability
- Pulling at the ears

Because infants cannot tell you that their ears hurt, increased irritability or ear pulling should make a parent suspicious of ear infection.

Ear pain and ear stuffiness can also result from high altitudes, as when descending in an airplane. Here again, the mechanism for the stuffiness or pain is obstruction of the eustachian tube. Swallowing frequently relieves this pressure. Closing the mouth and holding the nose closed while pretending to blow one's nose is another method of opening the eustachian tube.

Parents are often concerned about hearing impairment after ear infections. While most children will have a temporary and minor hearing loss during and immediately following an ear infection, there is seldom any permanent hearing loss with adequate medical management.

HOME TREATMENT

Ear infections generally require antibiotic treatment and hence a doctor visit. In some countries, such as The Netherlands, ear infections are treated with antibiotics only if fever has lasted more than two days. This approach has been neither investigated nor adopted in the United States. However, parents can begin pain and fever treatment with acetaminophen immediately on suspecting an ear infection. Some ear infections may be avoided by keeping bottles out of baby's bed.

WHAT TO EXPECT AT THE DOCTOR'S OFFICE

An examination of the ear, nose, and throat as well as the bony portion of the skull behind the ears, known as the mastoid, will be performed. Pain, tenderness, or redness of the mastoid signifies a serious infection.

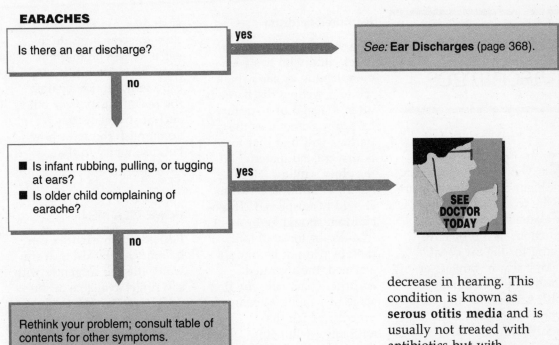

EARACHES

Is there an ear discharge? → **yes** → *See:* **Ear Discharges** (page 368).

no ↓

- Is infant rubbing, pulling, or tugging at ears?
- Is older child complaining of earache?

→ **yes** → **SEE DOCTOR TODAY**

no ↓

Rethink your problem; consult table of contents for other symptoms.

Therapy will generally consist of an antibiotic. Decongestants do not appear to be of value. Antihistamine preparations are seldom useful except in a child known to be allergic. Antibiotic therapy generally will be prescribed for at least a week. Be sure to give all of the antibiotic prescribed, and on schedule.

If your child still has a fever after two days of treatment, a different antibiotic may be needed. Be sure to call your doctor. Children who have frequent ear infections may benefit from taking antibiotics daily for a period of several months in order to prevent new bouts.

Occasionally, fluid in the middle ear will persist for a long period without infection. In this event, there may be a slight decrease in hearing. This condition is known as **serous otitis media** and is usually not treated with antibiotics but with attempts to open the eustachian tube and allow drainage. If this condition persists, the doctor may resort to placement of ear tubes in order to establish proper functioning of the middle ear once again. Placing ear tubes sounds frightening, but this is actually a simple and very effective surgical procedure.

Most doctors will wish to re-examine your child's ears to make sure the infection has completely cleared and to make sure that the child's hearing has returned to normal.

Ear Discharges

Ear wax is almost never a problem unless attempts are made to "clean" the child's ear canals. Ear wax functions as a protective lining for the ear canal. Taking warm showers or washing the external ears with washcloths dipped in warm water provides enough vapor to prevent the buildup of wax that is thick and caked. Children often like to push things in their ear canal, and they may pack the wax tightly enough to prevent vibration of the eardrum and hence interfere with hearing. Well-meaning parents armed with a cotton swab on a stick often accomplish the same awkward result.

Ruptured Eardrum

In a young child or in an older child who has been complaining of ear pain, a white or yellow discharge is often the sign of a ruptured eardrum. Sometimes the parents will find that there is dry crusted material on the child's pillow. Here again, a ruptured eardrum should be suspected. These children should be brought to a doctor for antibiotic therapy. Do not be unduly alarmed; the ruptured eardrum is actually the first stage of a natural healing process, which the antibiotics will help. Children have remarkable healing powers and most eardrums will heal completely within a matter of weeks.

Swimmer's Ear

In the summertime, ear discharges are commonly caused by swimmer's ear, an irritation of the ear canal and not a problem of the middle ear or eardrum. Children will often complain that their ears are itchy. In addition, tugging on the ear will often cause pain; this can be a helpful clue to an inflammation of the outer ear and canal, such as swimmer's ear. The urge to scratch inside the ear is very tempting but must be resisted. We especially caution against the use of hairpins or other such instruments to accomplish the scratching because injury to the eardrum can result.

HOME TREATMENT

Packed-down ear wax can be removed by using warm water flushed in gently with a syringe, available at the drugstore. A water jet, *set at the very lowest setting*, can also be useful but can be frightening to young children and is dangerous at higher settings. We do not advise that parents attempt to remove impacted ear wax unless they are dealing with an older child and can see the impacted, blackened ear wax.

Wax softeners, such as Cerumenex, or ordinary olive oil are useful; all commercial products can be irritating if not used properly. Cerumenex, for example, must be flushed out of the ear within 30 minutes.

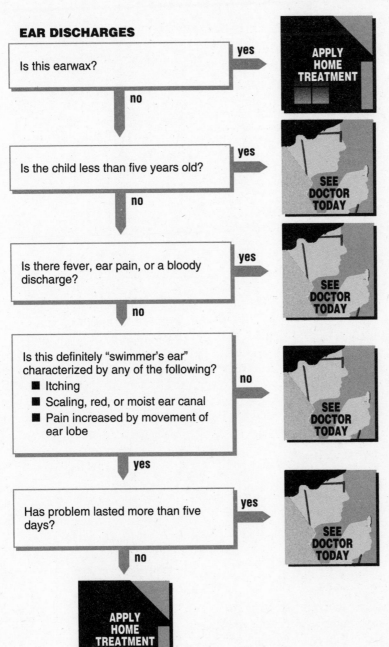

EAR DISCHARGES

Is this earwax? — yes → **APPLY HOME TREATMENT**

no ↓

Is the child less than five years old? — yes → **SEE DOCTOR TODAY**

no ↓

Is there fever, ear pain, or a bloody discharge? — yes → **SEE DOCTOR TODAY**

no ↓

Is this definitely "swimmer's ear" characterized by any of the following?
- Itching
- Scaling, red, or moist ear canal
- Pain increased by movement of ear lobe

no → **SEE DOCTOR TODAY**

yes ↓

Has problem lasted more than five days? — yes → **SEE DOCTOR TODAY**

no ↓

APPLY HOME TREATMENT

Two warnings:

- The water must be close to body temperature; the use of cold water may result in dizziness and vomiting.
- Washing should never be attempted if there is any question about the condition of the eardrum.

Although swimmer's ear (or other causes of similar *otitis externa*) is often caused by a bacterial infection, this infection does not often require antibiotic treatment because the infection is very shallow. Effective treatment is to place a cotton wick soaked in Burrow's solution in the ear canal overnight, followed by a brief irrigation with 3% hydrogen peroxide, followed by warm water. Success has also been reported with Merthiolate mixed with mineral oil (enough to make it pink), followed by the hydrogen peroxide and warm water rinse. For particularly severe or itching cases or persistence beyond five days, a doctor's visit is advisable.

WHAT TO EXPECT AT THE DOCTOR'S OFFICE

A thorough examination of the ear will be performed. In severe cases, a culture for bacteria may be taken. Corticosteroid and antibiotic preparations to be placed in the ear canal may be prescribed, or one of the regimens described above under Home Treatment may be advised. Oral antibiotics will usually be given if a perforated eardrum is causing the discharge.

Hearing Loss

Problems with hearing can be divided into two broad categories: sudden and slow onset. When a child of five or older complains of difficulty in hearing developing over a short period of time, then the problem is usually a blockage in the ear of one type or another. On the outside of the eardrum, such blockage may be due to the accumulation of wax, a foreign object that the child has put in the ear canal, or an infection of the ear canal. On the inside of the eardrum, fluid may accumulate and cause blockage because of an ear infection caused by a virus or bacteria.

In the other category are hearing problems that are slow in developing or are present from birth and become evident over a long period of time. Many parents become concerned that their infant or small child is not hearing normally. Hearing can now be tested in a child of any age through the use of computers to analyze changes in brain waves in response to sounds.

More simply, a child with normal hearing will react in a characteristic manner to a noise. A hand clap, horn, or whistle can be used to produce the sound. From birth up to three months, the infant will blink or open the eyes, move arms or legs, turn the head, or begin sucking in response to sound. If a child is moving or vocalizing before the sound is made, these activities may stop. At three months, children begin to attempt to find a sound by moving the head and looking for it. The ability to find the sound no matter where it is (below, behind, or above the child) may not be fully developed until the age of two.

Normal speech development relies on hearing. A child whose speech is developing slowly or not at all may in fact have difficulty in hearing. Children who babble continually beyond a year without forming words should also be suspected of hearing difficulties.

HOME TREATMENT

The need for an accurate ear examination usually necessitates a trip to the doctor. However, if the problem is known with certainty to be due to wax accumulation, this may be effectively treated at home. The ear is simply flushed gently with tepid or warm water, and the wax washed out.

Ear syringes or other devices for squirting water into the ear canal are available in drugstores. A water jet set *at the very lowest setting* can be used with considerable success, but we do not recommend it for young children because of the frightening noise. Wax softeners, such as Cerumenex, may be needed when the wax is hard and impacted; follow the instructions on the label.

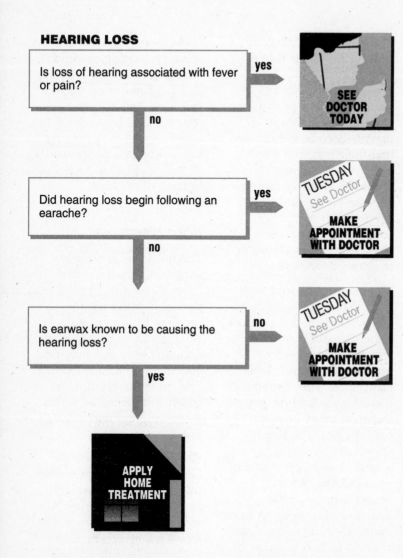

HEARING LOSS

Is loss of hearing associated with fever or pain?

yes → SEE DOCTOR TODAY

no ↓

Did hearing loss begin following an earache?

yes → TUESDAY See Doctor / MAKE APPOINTMENT WITH DOCTOR

no ↓

Is earwax known to be causing the hearing loss?

no → TUESDAY See Doctor / MAKE APPOINTMENT WITH DOCTOR

yes ↓

APPLY HOME TREATMENT

- Washing should never be attempted if there is any doubt about the eardrums being intact and undamaged.

Caution must be advised with respect to removing foreign bodies. Unless the object is easily accessible and removing it clearly poses no threat of damage to ear structures, do not try to remove. Sharp instruments should never be used in an attempt to remove foreign bodies. Many times efforts to remove objects at home lead to pushing the object further into the ear or to damage of the eardrum itself.

WHAT TO EXPECT AT THE DOCTOR'S OFFICE

A thorough examination of both ears often reveals the cause of the hearing loss. If it does not, then the doctor may test the hearing in the manner described above or, if the child is older, recommend audiometry (an electronic hearing test).

(Debrox has gained a reputation for irritating ear canals, perhaps unjustly.) A few words of warning:

- The water must be as close to body temperature as possible; the use of cold water may result in dizziness and vomiting.

Runny Nose

A runny nose is one of the most frequent occurrences of childhood. Many infants will have a runny nose and sneeze during the first two weeks of life. The cause is unknown; it is certainly a natural phenomenon. Children will also have runny noses during episodes of crying and sometimes after exercising. Runny noses in these instances are temporary and of no concern.

The hallmark of the common cold is the runny nose. It is intended by nature to help the body fight the virus infection. Nasal secretions contain antibodies, which act against the viruses. The profuse outpouring of fluid carries the virus outside the body.

Allergy

Allergy is another common cause of runny noses. Children whose runny noses are due to an allergy are deemed to have **allergic rhinitis**; hay fever is one kind of allergic rhinitis caused by the ragweed pollen. The nasal secretions in this instance are often clear and very thin. Children with allergic rhinitis will often have other symptoms simultaneously, including sneezing, and itching, watery eyes. They will rub their noses so often that a crease in the nose may appear. This problem lasts longer than a viral infection, often for weeks or months, and occurs most commonly during the season when pollen particles or other allergens are in the air. A great many other substances may aggravate allergic rhinitis, including house dusts, molds, and animal danders.

Other Causes

The runny nose may also be due to a small **object** that a young child has pushed into the nose. Usually, this will produce a discharge from only one nostril. Often the discharge will be foul-smelling and yellow or green.

Another common cause of runny noses as well as stuffy noses is prolonged use of nose drops. This problem of excess medication is known as **rhinitis medicamentosum**. Nose drops should never be used for longer than three days.

Post-nasal Drip

Complications from the runny nose are due to the excess mucus. The mucus may cause a post-nasal drip and a cough that is most prominent at night. The mucous drip may plug the eustachian tube between the nasal passages and the ear, resulting in ear infection and pain. It may plug the sinus passages, resulting in secondary sinus infection and sinus pain.

HOME TREATMENT

Two major types of drugs are used to control a runny nose:

- **Decongestants** such as pseudo-ephedrine and ephedrine act to shrink the mucous membranes and to open the nasal passages.

- **Antihistamines** act to block allergic reactions and decrease the amount of secretion.

Decongestants make some children overly active, and antihistamines may cause drowsiness as well as interfere with sleep. Because of the complications of the medications, runny noses should be treated only when they are severely impairing the child's comfort, Often a tissue is the best approach; it has no side effects, costs less, and helps get the virus outside the body!

Should you choose to treat a runny nose, topical decongestants (nose drops) are suitable. Saline nose drops are fine for young infants. As children get older, they can graduate from ¼% to ½% Neo-synephrine nose drops, or may take Afrin nose drops. These nose drops should never be used for longer than three consecutive days. Oral decongestants that can be purchased over the counter contain phenylephrine, phenylpropanolamine, or pseudoephedrine. Often they are combined with antihistamines such as chlorpheniramine or brompheniramine.

Complications such as ear and sinus infection may be prevented by ensuring that the mucus is thin rather than thick and sticky, although this has not been proven. This may prevent plugging of the nasal passages. Increased humidity in the air with a vaporizer or a humidifier helps liquefy the mucus. Inside a house, heated air is often very dry; cooler air contains more moisture and is preferable. Drinking a great deal of liquid also helps liquefy the secretions.

If symptoms persist beyond three weeks, your doctor should be contacted.

Sneezes are healthy, for they remove germs, allergens, and dust from the child's nose. The only danger is that of infecting others with the same germ or virus. Teach children to cover their mouths with tissues or handkerchiefs before sneezing. And teach them to wash their hands afterwards.

WHAT TO EXPECT AT THE DOCTOR'S OFFICE

If a visit is made, a thorough examination of the ears, nose, and throat will be made. A check for tenderness over the sinuses will be performed. A doctor may take a swab of the nasal secretions to examine under a microscope. The presence of certain types of cells, known as eosinophils, will indicate the presence of allergic rhinitis. If allergic rhinitis is found, antihistamines or nasal chromolyn may be prescribed, and a plan of dust, mold, dander, and pollen avoidance explained (see Chapter I, "Allergies," page 397).

RUNNY NOSE

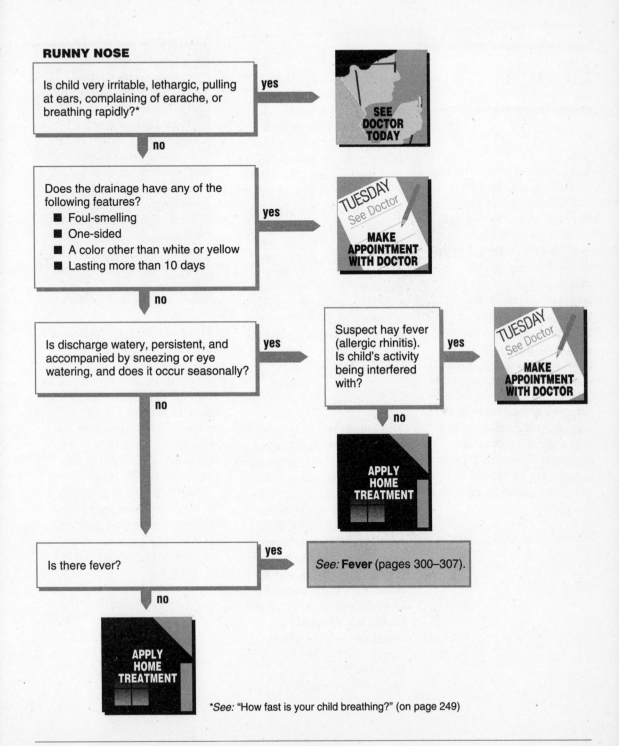

Is child very irritable, lethargic, pulling at ears, complaining of earache, or breathing rapidly?*

yes → **SEE DOCTOR TODAY**

no ↓

Does the drainage have any of the following features?
- Foul-smelling
- One-sided
- A color other than white or yellow
- Lasting more than 10 days

yes → **MAKE APPOINTMENT WITH DOCTOR** (TUESDAY See Doctor)

no ↓

Is discharge watery, persistent, and accompanied by sneezing or eye watering, and does it occur seasonally?

yes → **Suspect hay fever (allergic rhinitis). Is child's activity being interfered with?**

yes → **MAKE APPOINTMENT WITH DOCTOR** (TUESDAY See Doctor)

no ↓

APPLY HOME TREATMENT

no ↓

Is there fever?

yes → *See:* **Fever** (pages 300–307).

no ↓

APPLY HOME TREATMENT

*See: "How fast is your child breathing?" (on page 249)

Cough

Coughing, of course, has several causes. In very young infants, coughing is unusual and may indicate a serious lung problem. In older infants who are prone to swallowing foreign objects, an object may become lodged in the windpipe and cause coughing. Young children also tend to inhale bits of peanut and popcorn, which can produce coughing and serious problems in the lung. In some children a chronic cough may be the first sign of asthma or sinusitis. However, most coughs are produced by infections, usually viral.

Pneumonia

Is it pneumonia? This question worries parents more than any other. Pneumonia is a serious infection of the lung that often requires antibiotics and hence needs the doctor. Fortunately, it is rare when the only symptom is a cough. A rapid breathing rate is often the best indicator of a pneumonia. This problem may follow an ordinary upper respiratory infection (such as a cold) by a few days. So, if the fever from a simple cold doesn't go down after a few days, see the doctor.

Cough Reflex

The cough reflex is one of the body's best defense mechanisms. An irritation of the breathing tubes will trigger this reflex, and a violent rush of air helps clear material from the breathing tubes. Hence, material in the lungs that should not be present is removed by the coughing. Consequently, much of the treatment directed at coughing is directed at increasing the ability of the lungs to clear out unnecessary material.

Hiccups, which are caused by an irregularity in contractions of the diaphragm, may occasionally prove troublesome. Home remedies include drinking large amounts of water, or drinking water in rhythmic sips. For older children, research indicates the most effective treatment is ½ teaspoon of dry sugar placed on the back of the tongue.

Often a minor irritation in the breathing tubes will trigger a cough reflex even when there is no material to be expelled. At other times, mucus from the nasal passages will drip into the breathing tubes at night (post-nasal drip) and will start the cough reflex. Coughing that is interfering with a child's sleep can be counterproductive and this is the only type of cough that should be stopped.

COUGH

Is child less than three months of age?
yes → SEE DOCTOR TODAY

no ↓

Did violent cough begin suddenly without signs of a cold in a child who might have inhaled a small object?
yes → SEE DOCTOR NOW

no ↓

Are any of the following present?
- Rapid breathing*
- Difficulty breathing
- Wheezing
- Fever lasting more than four days

yes → SEE DOCTOR TODAY

no ↓

Has cough persisted for more than ten days?
yes → TUESDAY See Doctor / MAKE APPOINTMENT WITH DOCTOR

no ↓

APPLY HOME TREATMENT

*See: "How fast is your child breathing?" (on page 249)

HOME TREATMENT

The reason for home treatment is to liquefy the secretions in the breathing tubes in order to enhance clearing unwanted materials from the lungs. The mucus in the breathing tubes may be made thinner by several means. Increased humidity in the air may help, and a cool mist vaporizer can often provide this. This is especially important in the severe "croupy" cough of small children (**Croup**, page 379). Drinking large quantities of fluid may be helpful for cough, particularly if fever is present to dry out and dehydrate the body.

There is serious doubt about whether any of the oral cough medications help very much in liquefying secretions, although guaifenesin may be of use. Decongestants or nose drops may help nighttime coughs that are caused by post-nasal drip.

Children were often given cough suppressants containing dextromethorphan or codeine to allow them to get some rest. Unfortunately, recent studies have not documented the effectiveness of these preparations for children.

WHAT TO EXPECT AT THE DOCTOR'S OFFICE

An examination of the ears, nose, throat, and chest will be made. If a child is suspected of having inhaled a foreign body or if pneumonia is suspected, a chest X-ray will be taken. Antibiotics will be prescribed only for those few cases suspected of being caused by a bacterial infection (such as pneumonia or sinusitis). Other evaluations and therapies may be instituted if asthma is suspected.

Croup

Croup is one of the most frightening illnesses that parents will ever encounter. It generally occurs in children under the age of three or four. In the middle of the night, a child may sit up in bed gasping for air. Often there will be an accompanying cough that sounds like the barking of a seal. The child's symptoms are so frightening that panic is often the response. However, the most severe problems with croup usually can be relieved safely, simply, and efficiently at home.

Croup is usually caused by one of several different viruses. The viral infection causes a swelling and outpouring of secretions in the larynx (voice box), trachea (windpipe), and the larger airways going to the lungs. The air passages of the young child are made narrower because of the swelling. This is further aggravated by the secretions, which may become dried out and caked. This combination of swelling and thickened, dried secretions makes it extremely difficult to breathe. There may also be a considerable amount of spasm of the airway passages, further complicating the problem. Treatment is designed to dissolve the dried secretions.

In some children, croup is a recurring problem; these children may have three or four bouts of croup. Seldom does this represent a serious underlying problem, but a doctor's advice should be sought. Croup will be outgrown as the airway passages grow larger; it is unusual after the age of seven.

Epiglottitis

Occasionally, a more serious obstruction caused by a bacterial infection and known as **epiglottitis** can be confused with croup. Epiglottitis is more common in children over the age of three, but there is considerable overlap in the ages of children affected by these two conditions. Children with epiglottitis often have more serious difficulty breathing. They may have an extremely hard time handling their saliva and will be found to be drooling. Often they assume a characteristic position, with their head tilted forward and their jaw pointed out, and will gasp for air.

Epiglottitis will *not* be relieved by the simple measures that bring prompt relief of croup. It must be brought to medical attention immediately. Epiglottitis is caused by the *Hemophilus influenza* bacteria and is now preventable with immunization.

HOME TREATMENT

Mist is the backbone of therapy for croup and is supplied efficiently by a cold-steam vaporizer. We prefer cold-steam vaporizers to hot-steam ones because of the possibility of scalding from the hot water.

If the breathing is very hard, you can get faster results by taking the child to the bathroom and turning up the hot shower to make thick clouds of steam. (*Do not put the child in the hot shower!*) Steam can be created more efficiently if there is some cold air in the room. Remember that steam rises, so the child will not benefit from sitting on the floor.

Relief usually occurs promptly and should be noticeable within the first 20 minutes. Not becoming alarmed and keeping the child calm is also important; holding the child may comfort him or her and may help relieve some of the airway spasm. If the child is not showing significant improvement within 20 minutes, you should contact your doctor or the local emergency room immediately. They will want to see the child and will make arrangements in advance while you are in transit. Unfortunately, few emergency rooms can provide steam as well as the home shower can.

WHAT TO EXPECT AT THE DOCTOR'S OFFICE

If the doctor feels confident that this is croup, a further trial of mist or inhaled epinephrine may be offered. On occasion, children with severe croup may need hospitalization and treatment with inhaled medicine, including epinephrine and steroids.

In the case of croup, the trip to the doctor often cures the problem that was resistant to steam at home; keep the car windows open a bit to let in the cool night air.

Differentiating croup from epiglottitis can be difficult. X-rays of the neck are a reliable way of differentiating croup from epiglottitis. A swollen epiglottis often can be seen in the back of the throat, but this examination has its risks and should not be tried at home. If epiglottitis is found, the child will be admitted to the hospital, an airway will be placed in the child's trachea to enable the child to breathe, and intra-venous antibiotics directed at curing the bacterial infection will be started.

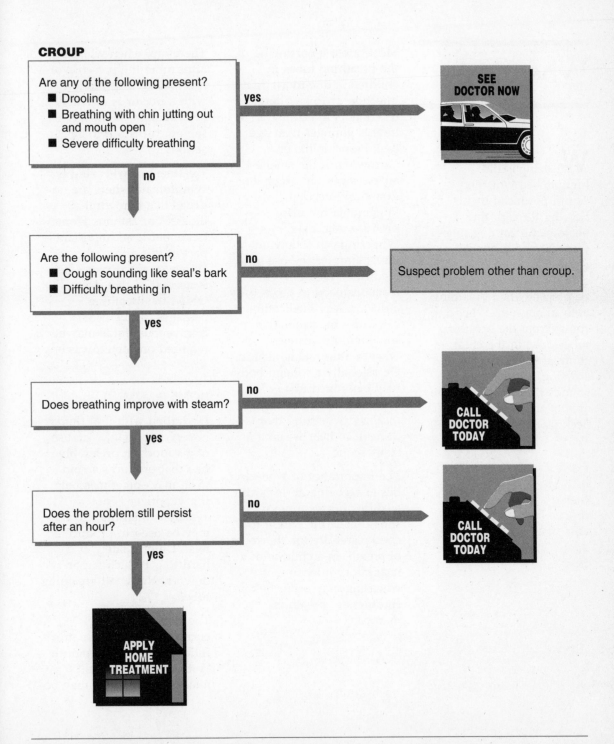

CROUP

Are any of the following present?
- Drooling
- Breathing with chin jutting out and mouth open
- Severe difficulty breathing

yes → SEE DOCTOR NOW

no ↓

Are the following present?
- Cough sounding like seal's bark
- Difficulty breathing in

no → Suspect problem other than croup.

yes ↓

Does breathing improve with steam?

no → CALL DOCTOR TODAY

yes ↓

Does the problem still persist after an hour?

no → CALL DOCTOR TODAY

yes ↓

APPLY HOME TREATMENT

Wheezing

Wheezing is the highpitched whistling sound produced by air flowing through the narrowed breathing tubes (bronchi and bronchioles). It is most obvious when the child breathes out but may be present when breathing both in and out. Wheezing comes from the breathing tubes deep in the chest, in contrast to the croupy, crowing, or whooping sounds that come from the area of the voice box in the neck (see **Croup**, page 379).

Most often, a narrowing of the breathing tubes in children is due to a viral infection or to an allergic reaction as in asthma. In infants younger than age two, bronchiolitis or narrowing of the smallest air passages can occur due to a viral infection. Pneumonia can also produce wheezing. Wheezing can follow an insect sting or the use of a medicine; these allergic reactions need to be seen by a doctor. Any medication can cause the problem; some individuals even wheeze after taking aspirin. Occasionally a foreign body may be lodged in a breathing tube, causing a localized wheezing that is difficult to hear without a stethoscope.

The importance of wheezing lies in its being an indicator of difficult breathing; it should alert the parent to check carefully for shortness of breath. In a child with a respiratory infection, wheezing may occur before shortness of breath is marked.

Therefore, when wheezing appears in the presence of a fever, early consultation with a doctor is advisable, even though the illness seldom turns out to be serious.

Treatment of wheezing is symptomatic; there are no drugs that cure viral illnesses or asthma. Home treatment is an important part of this approach. However, the doctor's help is needed so that drugs that widen the breathing passages can be used. Intravenous fluids may be required on some occasions.

HOME TREATMENT

Hydration with oral fluids is very important. The use of a vaporizer, preferably one that produces a cold mist, may sometimes help. If a vaporizer is not available, then the shower may be used to produce a mist. Unfortunately, it is hard to get much vapor down to the small breathing tubes.

The child should be encouraged to take as much fluid as possible by mouth. Water is best, but fruit juices or soft drinks may be

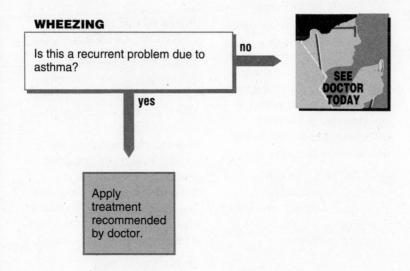

WHEEZING

Is this a recurrent problem due to asthma?

no → **SEE DOCTOR TODAY**

yes ↓

Apply treatment recommended by doctor.

used if this will increase the amount taken. These measures will be part of the therapy that the doctor recommends and may be begun immediately, even though a visit to the doctor will be necessary.

WHAT TO EXPECT AT THE DOCTOR'S OFFICE

Physical examination will focus on the chest and neck. Questions will be asked not only about the current illness but also about a past history of allergies either in the child or in the family. The possibility that a foreign body has been swallowed may also be investigated.

Drugs to open up the breathing tubes may be given by injection, by mouth, or by inhalation. (See **Asthma**, pages 401–403.) Occasionally, hospitalization will be necessary in order to permit medications to be given through a vein and other medications and oxygen given by inhalation. Most important, the child can be closely watched; the hospital is used as a precautionary measure against things getting worse before they get better.

Hoarseness

Hoarseness is usually caused by a problem in the vocal cords. In infants under three months of age, this can be due to a serious problem such as a birth defect or thyroid disorder. In young children, hoarseness is more often due to prolonged or excessive crying, which puts a strain on the vocal cords.

In older children, viral infections are the most common cause of hoarseness. If the hoarseness is accompanied by either difficulty in breathing or a cough that sounds like a barking seal, the hoarseness is considered a symptom of **croup** (see page 379). Croup is characteristic in children under the age of four, whereas the symptom of hoarseness by itself is more common in older children.

If hoarseness is accompanied by difficulty breathing, difficulty swallowing, drooling, gasping for air, or breathing with the mouth wide open and the chin jutting forward, a doctor must be seen immediately because this is a medical emergency. This problem is known as **epiglottitis**, a bacterial infection that involves the entrance to the airway.

In older children who develop hoarseness or laryngitis without any other symptoms, a virus is most often responsible.

HOME TREATMENT

Hoarseness, unassociated with other symptoms, is very resistant to medical therapy. Nature must heal the inflamed area. Humidifying the air with a vaporizer or taking in fluids can offer some relief. However, the child must wait for healing to occur, and this may take several days. Resting the vocal cords makes sense; crying or shouting makes the situation worse. For the treatment of hoarseness associated with coughs, see **Cough** (page 376).

WHAT TO EXPECT AT THE DOCTOR'S OFFICE

If there is severe difficulty in breathing, the first order of business is to ensure that the child has an adequate air passage. This may require placement, in the emergency room, hospital, or doctor's office, of a breathing tube. If X-rays of the neck are taken, a doctor should accompany the child at all times.

In uncomplicated hoarseness that has persisted for a long period of time, a doctor will look at the vocal cords with the aid of a small mirror. Occasionally, hormonal disorders caused by the thyroid or adrenal gland will be found by more extensive physical examination and confirmed by blood tests.

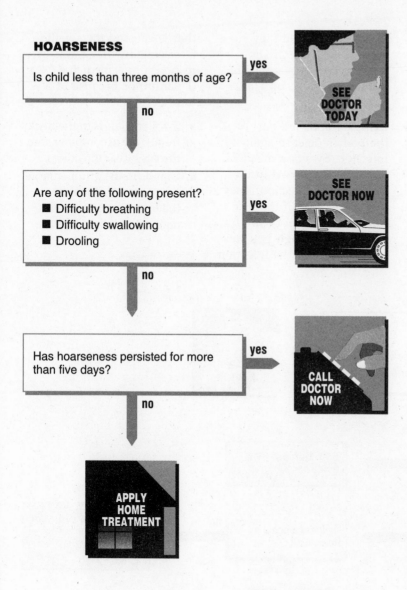

HOARSENESS

Is child less than three months of age? — **yes** → SEE DOCTOR TODAY

no

Are any of the following present?
- Difficulty breathing
- Difficulty swallowing
- Drooling

yes → SEE DOCTOR NOW

no

Has hoarseness persisted for more than five days? — **yes** → CALL DOCTOR NOW

no

APPLY HOME TREATMENT

Swollen Glands

The most common types of swollen glands are lymph and salivary glands. The biggest salivary glands are located below and in front of the ears. When they swell, the characteristic swollen jaw appearance of **mumps** is the result (see page 461).

Lymph glands are part of the body's defense against infection. Swelling means a gland is taking part in the fight against infection. They may swell even if the infection is trivial or not apparent, although you can usually identify the infection that is causing the swelling. Occasionally, they swell in response to a medicine (Dilantin) or because of a tumor.

- Swollen glands in the **neck** frequently accompany sore throats or ear infections.
- Glands in the **groin** enlarge when there is infection in the feet, legs, or genital region. Sometimes the

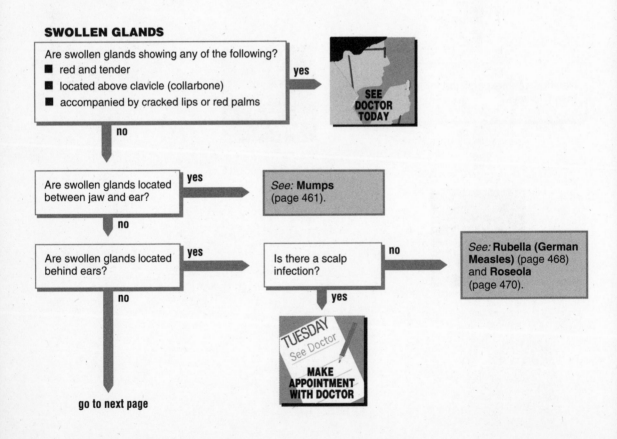

SWOLLEN GLANDS

Are swollen glands showing any of the following?
- red and tender
- located above clavicle (collarbone)
- accompanied by cracked lips or red palms

yes → SEE DOCTOR TODAY

no ↓

Are swollen glands located between jaw and ear? — **yes** → *See:* **Mumps** (page 461).

no ↓

Are swollen glands located behind ears? — **yes** → Is there a scalp infection? — **no** → *See:* **Rubella (German Measles)** (page 468) and **Roseola** (page 470).

no ↓   **yes** ↓

go to next page   TUESDAY See Doctor — **MAKE APPOINTMENT WITH DOCTOR**

basic problem may be so minor as to be overlooked (as with athlete's foot).

- Glands that swell in the **elbow** or **armpit** are usually due to an infection of the hand or chest.
- Swollen glands **behind the ears** are often the result of an infection in the scalp, or German measles (see page 468). Infectious

mononucleosis (mono) can also cause swelling of the glands behind the ears.

If a swollen gland is red and tender, there may be a bacterial infection within the gland that requires antibiotic treatment. Swollen glands otherwise require no treatment because they are merely fighting infections elsewhere. If there is an accompanying sore throat or

earache, these should be treated as described on pages 363 and 366, respectively.

However, swollen glands are usually the result of viral infections that require no treatment. Another bacterial infection is responsible for swollen red nodes that are part of cat scratch disease. Glands located above the clavicle

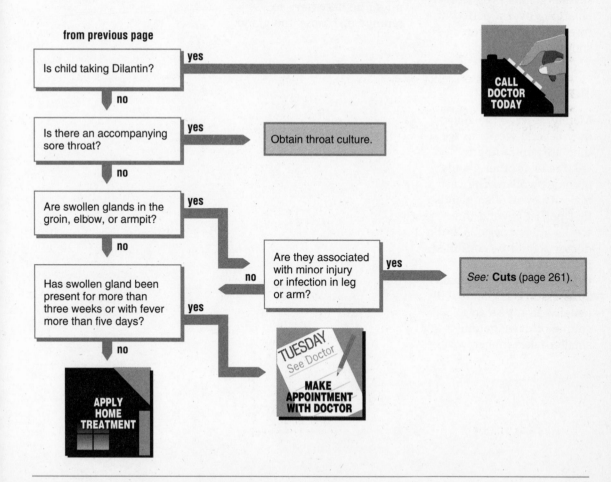

from previous page

Is child taking Dilantin? — **yes** → CALL DOCTOR TODAY

no ↓

Is there an accompanying sore throat? — **yes** → Obtain throat culture.

no ↓

Are swollen glands in the groin, elbow, or armpit? — **yes** → Are they associated with minor injury or infection in leg or arm? — **yes** → *See:* **Cuts** (page 261).

no ↓

Has swollen gland been present for more than three weeks or with fever more than five days? — **no** (to injury box) / **yes** → MAKE APPOINTMENT WITH DOCTOR

no ↓

APPLY HOME TREATMENT

(collarbone) may represent a serious infection (tuberculosis) or other serious problem (cancer) and should prompt a doctor visit.

Cracked lips with red or peeling palms may indicate a rare but treatable illness known as **Kawasaki Disease**.

If you have noticed one or several glands progressively enlarging over a period of three weeks, a doctor should be consulted.

HOME TREATMENT

Merely observe the glands over several weeks to see if they are continuing to enlarge or if other glands become swollen. The vast majority of swollen glands that persist beyond three weeks are not serious, but a doctor should be consulted if the glands show no tendency to become smaller. Soreness in the glands will usually disappear in a couple of days; getting smaller takes longer.

WHAT TO EXPECT AT THE DOCTOR'S OFFICE

The doctor will examine the glands and search for infections or other causes of swelling. Inquiry will be made into fever, weight loss, or other associated symptoms. The doctor may decide to observe the glands for a period of time, or decide that blood tests are indicated. Eventually, it might be necessary to remove or biopsy the gland.

Nosebleeds

The blood vessels within the nose lie very near the surface, and bleeding may occur with the slightest injury. In children, picking the nose is a common cause. Keeping fingernails cut and discouraging the habit is good preventive medicine. Occasionally, a foreign body in the nose may be the cause of bleeding. Accidents and fights produce their share of nosebleeds, but more often than not, the onset is spontaneous.

Nosebleeds are frequently due to irritation by a virus or to vigorous nose blowing. The main problem in this case is the cold, and treatment of cold symptoms will reduce the probability of the nosebleed. If the mucous membrane of the nose is dry, cracking and bleeding is more likely.

These key points should be remembered:

- You can almost always stop the child's nose bleeding yourself.
- The great majority of nosebleeds are associated with colds or minor injury to the nose.
- Treatment such as packing the nose with gauze has significant drawbacks and should be avoided if possible.
- Investigation into the cause of recurrent nosebleeds is not urgent and is best accomplished when the nose is *not* bleeding.

HOME TREATMENT

The nose consists of a bony part and a cartilaginous part: a "hard" portion and a "soft" portion. The area of the nose that usually bleeds lies within the "soft" portion, and compression will control the nosebleed. Simply squeeze the nose between thumb and forefinger just below the hard portion of the nose. Pressure should be applied for at least five minutes. The child should be seated. Holding the head back is not necessary. It merely directs the blood flow backward rather than forward. Cold compresses or ice applied across the bridge of the nose may help. Almost all nosebleeds can be controlled in this manner if *sufficient time* is allowed for the bleeding to stop.

NOSEBLEEDS

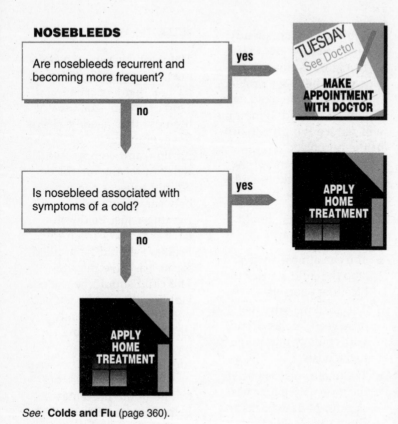

Are nosebleeds recurrent and becoming more frequent?

yes → **MAKE APPOINTMENT WITH DOCTOR**

no ↓

Is nosebleed associated with symptoms of a cold?

yes → **APPLY HOME TREATMENT**

no ↓

APPLY HOME TREATMENT

See: **Colds and Flu** (page 360).

WHAT TO EXPECT AT THE DOCTOR'S OFFICE

The child will be seated with head back and nostrils compressed. This will be done even if the child has been doing this at home, and it will usually work. Packing the nose or attempting to cauterize a bleeding point is less desirable. If the nosebleed cannot be stopped, the nose will be examined to see if a bleeding point can be identified. If a bleeding point is seen, coagulation by either electrical or chemical cauterization may be attempted. If this is not successful, then packing of the nose may be unavoidable. Such packing is uncomfortable and may lead to infection; thus, the child must be carefully observed.

If a doctor is seen because of recurrent nosebleeds, questions about events preceding the bleeds and a careful examination of the nose itself should be expected. Depending on the history and the physical examination, blood-clotting tests may on rare occasions be ordered.

Nosebleeds are more common in the winter, when both viruses and dry, heated air indoors are common. A cooler house and a vaporizer to return humidity to the air help many children.

If nosebleeds are a recurrent problem, are becoming more frequent, and are not associated with a cold or other minor irritation, then a doctor should be consulted on a non-urgent basis. A doctor need not be seen immediately after the nosebleed because examination at that time may simply restart the nosebleed.

Bad Breath

Children seldom have the problem with bad breath in the morning that is so common with adults, and regular tooth brushing should eliminate this problem. Other causes may be more important to identify.

Infections of the mouth as well as sore throats may be the cause of bad breath; because of the possibility of bacterial infection, the doctor should be consulted.

A common cause of prolonged bad breath in our experience is a foreign body in the child's nose. This is especially common in toddlers who have inserted, unnoticed, some small object. Often, but not always, there is a white, yellowish, or bloody discharge from one or both nostrils. Sinus infection or even tonsillitis can also be responsible. A severely decaying tooth may cause bad breath. Finally, unusual problems such as abscesses of the lung or heavy worm infestations have been reported to cause bad breath, although we have not seen these in our practices.

HOME TREATMENT

Proper dental hygiene will prevent most cases of bad breath. If this does not eliminate the odor, then a foreign body in a young child's nose is likely, and a trip to the doctor will be necessary. Although some foreign bodies may be seen very close to the child's nostril, most are located very deep within the nasal cavity where they are extremely difficult to remove.

Do not use mouthwashes to perfume the breath; these cover up, but do not treat, the underlying problem.

WHAT TO EXPECT AT THE DOCTOR'S OFFICE

A thorough examination of the mouth and the nose will be done. If there is a sore throat or mouth sores, a culture may be taken; subsequently, antibiotics may be prescribed. If there is an object in the nose, the doctor will use a special forceps to remove the object. Older children should be evaluated for sinusitis and pharyngitis.

BAD BREATH

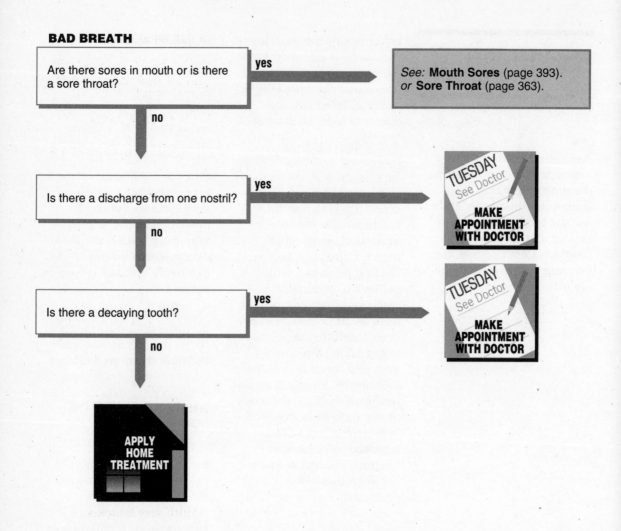

Are there sores in mouth or is there a sore throat?

yes → *See:* **Mouth Sores** (page 393). *or* **Sore Throat** (page 363).

no ↓

Is there a discharge from one nostril?

yes → TUESDAY See Doctor — **MAKE APPOINTMENT WITH DOCTOR**

no ↓

Is there a decaying tooth?

yes → TUESDAY See Doctor — **MAKE APPOINTMENT WITH DOCTOR**

no ↓

APPLY HOME TREATMENT

Mouth Sores

Problems in the mouth are very common in children. Doctors frequently use the term *lesion* to describe anything that may be wrong, be it a sore, a patch, or a pimple. In young children, large, white spots may appear on the roof of the mouth, due to a monilial yeast infection commonly referred to as **thrush**. Thrush can be treated effectively by medication, but it often disappears by itself.

Bacteria and viruses can also be responsible for mouth lesions. A bacterial infection more common in older children and adults is commonly known as **trench mouth**; lesions in trench mouth often occur on the gums.

Lesions of the gum are more likely to be caused by a **herpes virus**. *Herpes lesions* often start as blisters and then change to small spots with white, ulcerous centers surrounded by redness. With the first infection, they may be found on the gums, inner parts of the lips, cheeks, and even tongue. In repeat infections, it is more usual for the virus to involve only the lips. Because these herpes infections are almost always accompanied by a fever, they are known as fever blisters. The blisters have usually ruptured, and generally parents only observe the remaining underlying sore.

A **canker sore** often follows an injury, such as accidentally biting the inside of the lip or the tongue, or it may appear without obvious cause. These problems are minor and disappear in a short time.

Another virus that can cause mouth lesions is the **Coxsackie** virus. These lesions are often accompanied by spots on the hands and feet; hence the name "hand–foot–mouth syndrome." Again, this problem will go away by itself.

Allergic reactions to drugs may cause mouth ulcers. In such cases, a skin rash may be present on other parts of the body as well, and the doctor must be contacted.

HOME TREATMENT

Mouth sores caused by viruses heal by themselves. The goal of treatment is to reduce fever, relieve pain, and maintain adequate fluid intake.

Children will seldom want to eat when they have painful mouth lesions. Although children can go

MOUTH SORES

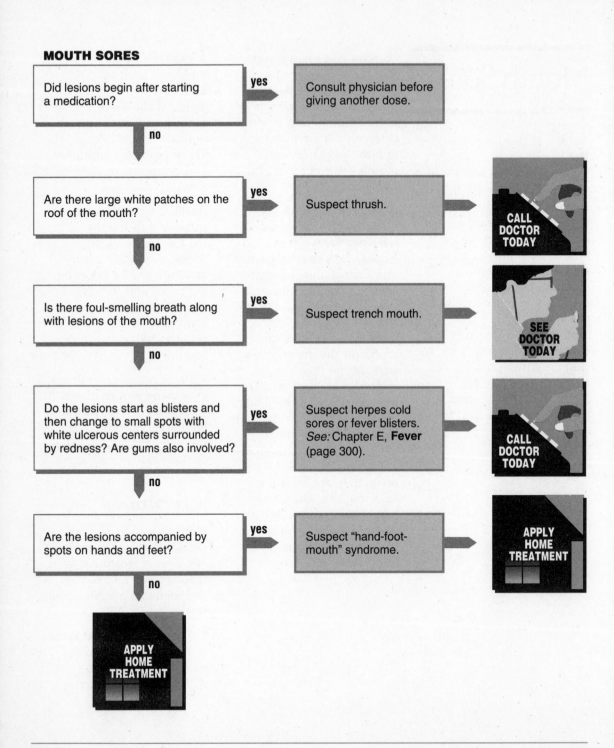

| Did lesions begin after starting a medication? | **yes** → | Consult physician before giving another dose. |

↓ **no**

| Are there large white patches on the roof of the mouth? | **yes** → | Suspect thrush. | → **CALL DOCTOR TODAY** |

↓ **no**

| Is there foul-smelling breath along with lesions of the mouth? | **yes** → | Suspect trench mouth. | → **SEE DOCTOR TODAY** |

↓ **no**

| Do the lesions start as blisters and then change to small spots with white ulcerous centers surrounded by redness? Are gums also involved? | **yes** → | Suspect herpes cold sores or fever blisters. *See:* Chapter E, **Fever** (page 300). | → **CALL DOCTOR TODAY** |

↓ **no**

| Are the lesions accompanied by spots on hands and feet? | **yes** → | Suspect "hand-foot-mouth" syndrome. | → **APPLY HOME TREATMENT** |

↓ **no**

APPLY HOME TREATMENT

several days without taking solid foods, it is imperative that they maintain an adequate liquid diet. Cold liquids are the most soothing and Popsicles or iced frozen juices often are helpful.

For sores inside the lip and on the gums, a preparation called Orabase, available over the counter, may be applied for protection. For cold sores and fever blisters on the outside of the lips, one of the phenol and camphor preparations (Blistex, Campho-Phenique) may provide relief, especially if applied early. If one of these preparations appears to cause further irritation, then discontinue its use. If the external sores have crusted over, then cool compresses may be applied to remove the crusts.

Mouth sores usually resolve in one to two weeks; any sore that persists beyond three weeks should be seen by the doctor.

WHAT TO EXPECT AT THE DOCTOR'S OFFICE

A thorough examination of the mouth will be done. A drug called Nystatin will usually be prescribed for thrush. If a bacterial cause of trench mouth is suspected, an antibiotic may be prescribed. For most viral infections, doctors have no more to offer than home remedies. Under certain circumstances, doctors may use Acyclovir to treat herpes infections.

We caution against the use of oral anesthetics such as viscous Xylocaine. This anesthetic can interfere with proper swallowing and can lead to inhalation of food into the lungs.

Toothaches

A toothache is the sad result of a poor program of dental hygiene. Although resistance to tooth decay is partly inherited, the majority of dental problems are preventable. (See the discussion of dental care on pages 208–212.)

Dental Problems

If you can see a decayed tooth or an area of redness surrounding a tooth, a toothache is most likely. Tapping on the teeth with a wooden Popsicle stick will often accentuate the pain in an affected tooth, even though it appears normal.

If your child appears ill, has a fever, and has swelling of the jaw or redness surrounding the tooth, a tooth abscess is likely, and antibiotics will be necessary in addition to proper dental care. In such circumstances a visit to the doctor may be in order before seeing the dentist. Alternatively, see the dentist.

Other Causes

Occasionally, it is difficult to distinguish a toothache from other sources of pain. Earaches, sore throats, mumps, sinusitis, and injury to the joint that attaches the jaw to the skull may all be confused with a toothache. If a pain occurs every time your child opens his or her mouth widely, it is likely that the joint of the jaw has been injured; this can occur from a blow or just by trying to eat too big a sandwich.

HOME TREATMENT

Acetaminophen can be used for pain when a toothache is suspected and while a dental appointment is being arranged. Acetaminophen is also helpful for problems in the joint of the jaw.

WHAT TO EXPECT AT THE DENTIST'S OFFICE

At the dentist's office, fillings or extractions will be performed. Often in baby teeth, an extraction will be most likely. Root canals as opposed to extraction are generally performed on permanent teeth if the problem is severe. If there is fever or swelling of the jaw, an antibiotic usually will be prescribed.

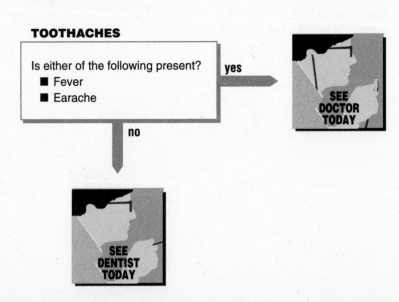

TOOTHACHES

Is either of the following present?
■ Fever
■ Earache

yes → SEE DOCTOR TODAY

no → SEE DENTIST TODAY

CHAPTER I

Allergies

Allergy was first described at the turn of the century by a pediatrician named Clemens von Pirquet. The term "allergy" meant "changed activity" and described changes that occurred after contacting a foreign substance. Two types of change were noticed.

- One change is beneficial; developing protection against a foreign substance after having been once exposed to it. This response prevents us from developing many infectious diseases for a second time and provides the scientific basis for most immunizations.
- The other type of response was known as a hypersensitivity response, is generally not beneficial, and is the response for which the term "allergy" is generally used.

Allergy is now known to be possible even without previous exposure to the substance. All persons are capable of allergic responses; for example, anyone given a transfusion with the wrong type of blood will have an allergic reaction.

However, the term "allergy" is overused. When your eyes smart in a smoggy city, they are not allergic to the air but are experiencing a direct chemical irritation from the pollutants. Similarly, skin coming in contact with some plants or chemicals experiences direct damage and not an allergic response. Doctors often blame milk or food allergy

for vomiting, diarrhea, colic, crying, irritability, fretfulness, or sneezing in infants. Although allergy can cause these symptoms, so can countless other things.

Over the next few pages, we discuss common allergies (such as those to food, insects, drugs, pets, pollen, and dust), the common allergic problems (such as asthma, hay fever, hives, and other skin problems), and the medical treatments available.

Food Allergy

Almost any food can produce an allergic response; only breast milk appears to be incapable of causing an allergy. Even in this case, certain foods in a mother's diet may be detected in the breast milk and cause an allergy in an infant.

Food allergy is not the only cause of digestive upsets and gets blamed for much that it does not cause. For example, some children are born without an important digestive enzyme, known as **lactase**, which is necessary to digest the sugars present in milk. Other children lose the ability to make lactase after the age of three or four. The absence of lactase can produce diarrhea, abdominal pain, bloating, and vomiting after the drinking of milk. This is only one example of a digestive problem that can be confused with food allergy; there are many others.

SYMPTOMS OF FOOD ALLERGY

Food allergy may produce swelling of the mouth and lips, hives, skin rashes, vomiting, diarrhea, blood loss from the intestinal tract, asthma, runny nose, and shock (extremely rarely). Of course, many other allergens besides food can cause these problems, making it difficult to prove that a particular food is the culprit. The best approach for detecting food allergy is for the parent to think like Sherlock Holmes. If your child's lips swell only after eating strawberries, you have your suspect!

FOODS RESPONSIBLE FOR ALLERGY

Cow's milk is frequently blamed for food allergy by parents, pediatricians, and allergists. Because almost any symptom can be blamed on allergy and because infants consume so much cow's milk, it is easy to see why milk is so quickly blamed. Infant intestines are capable of absorbing proteins that older children's digestive tracts would not absorb. These proteins may set up altered reactions or allergies.

Controversy exists over the relationship of early exposure to cow's milk and the later development of asthma. Some doctors maintain that avoidance of cow's milk will delay or eliminate the development of asthma. Cow's milk may have some effects on children likely to develop allergies but probably should not be of particular concern in children with no family history of allergy.

Cow's milk does cause allergic responses in some infants, with diarrhea and even blood loss through the intestines. This situation is a clear indication for removal of cow's milk, if the severe diarrhea is documented by a doctor. The necessary tests are simple and require analysis of the stools (feces). Stool analysis should be repeated after the child has been taken off cow's milk.

Other foods that have been associated with allergic reactions in children include wheat, eggs, citrus fruits, beef and veal, fish, and nuts. Severe reactions are very rare in children; parents need not be anxious about giving their children new foods. Families with a strong history of allergy can introduce new foods to an infant once every few days so that if an allergy develops, the cause is obvious.

SOYBEAN SUBSTITUTES FOR MILK

The amount of soybean formula produced in this country exceeds the amount necessary to provide for children with a cow's milk allergy or intolerance to cow's milk because of lack of the enzyme lactase.

There are two possible explanations for the purchase of soybean preparations. First, parents may buy them because they like them. They are nutritious, an excellent source of iron, and children tolerate them well. However, they tend to be more expensive than cow's milk.

The other reason for the high consumption of soybean formula is that parents have been instructed to substitute for cow's milk formulae at the slightest suspicion of an allergy. Every childhood symptom known has been attributed to cow's milk allergy—but the condition is rare.

Many stories of children getting better on a soybean preparation result from the child's spontaneous recovery from whatever was formerly producing the troublesome symptom. If you wish, you can switch back to a cow's milk formula to check if symptoms recur. If they do, a soybean formula is probably well warranted.

Asthma

Asthma is a potentially severe disorder, which is often due to an allergic response and is discussed further under its most prominent symptom— **wheezing** (see page 382). The wheezing in asthma is caused by spasm of the muscles in the walls of the smaller air passages in the lungs. An excess amount of swelling and mucous production further narrows the air passages and can aggravate the difficulty in getting the air out, as well as cause a troublesome cough.

Infections and foreign bodies in the air passages can mimic asthma. All wheezing in children is potentially serious and should be evaluated by a medical professional, at least for the first few occurrences. Asthma tends to occur in families where other members have either asthma, hay fever, or eczema.

An attack can be triggered by an infection, a change in weather, an emotionally upsetting event, exercise, or exposure to an allergen. Common allergens include house dust, pollen, mold, food, and shed animal materials, or "animal danders." It is sometimes easy to identify airborne allergens to which a person is susceptible. Some children will wheeze only around cats, others only during a particular pollen season. (Pollens most often cause seasonal hay fever or allergic rhinitis, rather than asthma.) Most often, there is no clear reason for a particular asthmatic attack. If asthma is severe, it is desirable to identify the offending allergens if possible.

Some doctors maintain that children never truly outgrow asthma, but the evidence is otherwise. More than half of the children diagnosed as having asthma will never have an asthmatic attack as an adult. Another 10% will have only occasional attacks during adult life.

TREATMENT OF ASTHMA

The treatment of asthma varies according to the severity of the problem. Some children have only one or two episodes of asthma and are never troubled again. We wonder if we should even call these asthma attacks. Other children will have daily attacks, which severely compromise their growth, development, and ability to function well in school and at home.

Therapy provides relief of symptoms, often dramatically so, but must also work to remove the cause, be it allergic, infectious, or emotional. Our approach to asthma involves three components: avoidance, physical control, and medication.

Allergen Avoidance

Occasionally, an offending allergen is easy to identify. More often, no offender is found, and a general approach to reducing allergen exposure is advisable. A relatively clean and dust-free house is all but essential for the allergic. An asthmatic child's room should be particularly allergen-free, because sleeping requires eight to ten hours in the room. Except in very severe cases, we do not recommend changing the entire household furnishings to reduce potential allergen exposure. Even then, removal of items should progress on a rational basis after suspected allergens have been identified.

- Rugs, furniture, drapes, bedspreads, and other items that are particular dustcatchers should be vacuumed regularly.
- Clothing attracts dust, pollen, and other troublesome particles. Keeping non-seasonal clothing outside your child's room or in plastic storage containers may be helpful.
- Children and pets may do fine together, although it is best not to allow pets to sleep in an allergic child's room.
- Toy animals should be kept clean; washable ones are the best. A home without stuffed animals seems to some like a morning without sunshine and orange juice, but avoid products that may be stuffed with animal hair.
- Change heating filters and air-conditioner filters regularly.

Physical Control

Exercise can cause breathing problems, especially sports that call for sudden bursts of energy. Warming up slowly before playing and staying in good physical condition can minimize the wheezing caused by exercise.

Asthmatics can participate in athletics; athletes with asthma have been Olympic gold medal winners in sports ranging from swimming to the heptathalon. Swimming appears to be far and away the best exercise and best sport for the asthmatic child. Exercise programs with long and steady energy requirements seem to work the best, and swimmers come into contact with allergens rather less often than in most other sports.

Medication

Medications are available to prevent an attack or to treat an attack by opening up the air passages. Drugs that open the air passages are known as **bronchodilators**; some have been in use for over 75 years. They are available in forms for inhalation, oral consumption, and injection. Most asthmatics will have encountered an injection of epinephrine, which should not be used as an inhalant, although it can be purchased over the counter.

The latest advance in asthma is the availability of highly effective **inhalants**, including terbutaline, metaproterenol (Alupent), albuterol (Ventolin), isoetharine (Bronkosol, Bronkometer), and isoproterenol (Isuprel, Medihaler-Iso). Older children can be trained to use hand-held metered dose inhalers; younger children may need to use plastic inhaling chambers or nebulization machines.

Oral drugs are of two types: **sympathomimetic drugs**, such as metaproterenol (Alupent, Metaprel, etc.), terbutaline (Brethine, Brethaire, Bricanyl, etc.), and albuterol (Proventil, Ventolin, etc.); and **theophylline**. These drugs are excellent, although accompanied by more side effects than inhalants. In particular, theophylline is now used as a second-line drug if inhalants and other oral agents are not effective.

An effective drug currently used to prevent attacks (if taken daily) is **cromolyn**, recently available as a metered dose inhalant that is far more acceptable than the inhaled powder that many children found objectionable. It is also effective in reducing exercise-induced wheezing if taken 15 minutes before the exercise.

Cortiocosteroid drugs (prednisone, beclomethasone) are quite powerful and used if the above medications are not successful. If used, older children would be given the inhaled form (beclomethasone).

Antihistamines are not useful in asthma, and their drying effect may cause airway plugging. We also advise strongly against any combination of drugs for asthma (Asbron G, Brondecon, Bronkolixir, Marax, Quadrinal, Quibron Plus, Tedral, and so on).

Infection Control

Because infections can trigger asthma, a physical examination is important during any first or frequently recurring attacks.

Antibiotics should not be given unless a definite infection is present.

Hydration Therapy

Water and other fluids taken by mouth are very important. Water can help loosen the mucus in the lungs and make breathing easier. Mist is not too helpful during asthmatic attacks because the affected airway passages are beyond the reach of the mist. Vaporizers are most useful for problems of the upper air passages of ears, nose, sinuses, mouth, and throat.

Supportive Therapy

Severe asthma is strenuous for parents and family, as well as for the affected child. Assistance is often required to manage the emotional consequences of asthma for the whole family. Do not hesitate to seek this assistance; social workers and other counselors can be invaluable here.

Allergic Rhinitis

Allergic rhinitis is the most common allergic problem. A stuffy runny nose, watering itchy eyes, headache, and sneezing are all common. The cause in infants is often dust or food, and in adults, dust or pollens. Most individuals are troubled only in pollen season; ragweed, the cause of **hay fever**, is particularly troublesome. The problem seems to run in families.

Treatment is directed toward both symptomatic relief and avoidance of the offending allergen. The first line in symptomatic relief is the use of tissues or handkerchiefs; often this is not enough. Drugs that will reduce symptoms may be prescribed or purchased over the counter; all have some side effects.

Antihistamines block the action of histamine, a substance released during allergic reactions. They also have a drying effect and improve nasal stuffiness. They also may be useful in reducing itching, helping motion sickness, or decreasing vomiting. The antihistamines used most often in allergic rhinitis are diphenhydramine (Benadryl), chlorpheniramine maleate (Chlor-Trimeton), tripelennamine (Pyribenzamine), brompheniramine (Dimetane) and terfenadine (Seldane). These five drugs are from four different classes of antihistamine compounds. Individuals respond differently to different drugs, and a trial of the different types of antihistamines may be necessary to determine the most effective type.

The most common side effect of antihistamines is drowsiness, and this may interfere with a child's schoolwork. Seldane causes less drowsiness, but it may also be less effective. Antihistamines should not be used as sleeping pills, because the drowsiness they produce *decreases* the amount of deep sleep necessary for normal rest.

For severe cases, nasal sprays containing steroids or cromolyn may be recommended.

Eczema

Atopic dermatitis, known commonly as eczema, is an allergic skin condition characterized by dry, itching skin. This itching often leads to scratching. The scratching then produces weeping, infected skin. Dried weepings lead to crusting. This weeping and crusting condition is what doctors refer to as eczema. Sufficient scratching will produce a thickened, rough skin, which is characteristic of long-standing atopic dermatitis.

Atopic dermatitis runs in families with asthma and allergic rhinitis. Like asthma, a variety of conditions can aggravate it.

These conditions include infection, emotional stress, food allergy, and sweating.

Infants seldom exhibit any signs of this problem at birth. The first signs may be red, chapped cheeks. Often infants can rub these itchy areas and cause secondary infections.

As the child grows older, the atopic dermatitis can spread. It may be found on the back of the legs and front of the arms. Adults often have problems with their hands; this is especially true of people whose hands are in frequent contact with water. Water tends to have a drying effect on the skin and tends to aggravate the dry skin–itch–scratch–weep–crust cycle.

Therapy is based on avoidance of allergens and maintenance of good skin care.

- Avoid wool, which tends to aggravate itching.
- Avoid excessively warm clothing, which will cause sweat retention and aggravate itching.
- Keep the child's fingernails clipped short.
- Avoid bathing with soap and water in moderate to severe cases because these tend to dry the skin. Instead use non-lipid containing cleansers. Some cleansers with cetyl alcohol aid in preventing drying of the skin (Cetaphil lotion, etc.).

- When inflammation is prominent (the skin is red and oozing), a wet dressing of Burrow's solution and total-body oatmeal baths can be applied.
- Avoid all oil or grease preparations. They occlude the skin and increase sweat retention and itching.
- For dry skin, lubricating lotions can be applied. You can make your own combining Crisco and glycerine or purchase Keri or Lubriderm (lotions), Keri or Nivea (creams), or Eucerin (creamy paste).

- Avoiding cow's milk is often suggested; make sure this really works for your child before permanently changing to more expensive feedings. When trying your child on any milk-avoidance diet, make *no* other changes in food or other care for a full two weeks unless absolutely necessary.
- Itching is often worse at bedtime. Aspirin is an effective and inexpensive medication for reducing itching. Antihistamines also reduce itching but should be used only if necessary.
- Steroid creams are useful. When possible, steroids should be used only as long as needed. Weaker preparations are used on the face.

- Antibiotics are often necessary to clear up badly infected skin.
- Emotional factors may need attention; they may be the key to successful therapy.
- There has been no benefit demonstrated from either skin testing or hyposensitization.

For more information on treating eczema, see page 433.

Allergy Testing and Shots

The purpose of allergy testing is to help decide what is causing the allergy; it is not a treatment, and as a test it is not always accurate. Once an allergy test is positive, there are two treatment approaches: avoidance and hyposensitization (desensitization).

Avoidance is sometimes, though not usually, possible. Seldom is a child allergic to cats and nothing else, and usually such an isolated allergy is noted by alert parents and children. Avoiding dusts, pollens, trees, and flowers is next to impossible, so hyposensitization is sometimes reasonable if the problem is severe.

Hyposensitization involves injecting a tiny amount of the offending allergen. Gradually larger and larger amounts are injected until the child is able to tolerate exposure to the allergen with only mild symptoms.

Hyposensitization works in many cases, but there are many problems. Local reactions at the site of the injection are common but can be minimized by injecting through a different needle from the one used to withdraw the material from the bottle.

Hyposensitization requires weekly injections for months or years. It may be considered for children with moderate or severe asthma or severe hay fever but appears unwarranted, as does the preliminary skin testing, for children with mild asthma or with mild allergic rhinitis.

ADDITIONAL READING

M. Eric Gershwin and E.L. Klingelhofer, *Asthma: Stop Suffering, Start Living*, 2nd ed. Reading, Mass.: Addison-Wesley, 1992.

M. Eric Gershwin and E.L. Klingelhofer, *Conquering Your Child's Allergies*. Reading, Mass.: Addison-Wesley, 1989.

Thomas F. Plaut, *Children with Asthma: A Manual for Parents*, 2nd ed. Amherst, Mass.: Pedipress, 1988.

CHAPTER J

Skin Problems

Skin problems must be approached somewhat differently from other medical problems. Decision charts that proceed from complaints such as "red bumps" can be developed, but the charts are complicated and somewhat unsatisfactory. This is because most people, including doctors, identify skin diseases by recognizing a particular pattern. This pattern is composed of not only what the skin problem looks like at a particular time but also how it began, where it spread, and whether it is associated with other symptoms such as itching or fever. Also important are elements of the medical history that may suggest an illness to which the child has been exposed. Fortunately, many times you already have a good idea of the problem, and it is possible to proceed immediately with the question of whether this is or is not poison ivy or ringworm or whatever.

If you are confused about what skin problem is present, we have provided two tables that will help you find a place to start, the accompanying decision chart, and Table J (page 412). Each decision chart in this section begins with the question of whether the problem is compatible with the essentials of the pattern for that skin disease. (Note that a more complete description of the pattern is given in the text that accompanies the chart.) If it is not, you are directed to reconsider the problem and consult the tables.

Most cases of a particular skin disease do not look exactly as a textbook says they should, so we have not provided you with pictures. We have tried to allow for a reasonable amount of variation in the descriptions, and you will have to exercise your common sense a good deal. Don't be afraid to ask for other opinions; grandmother or your friends have seen a lot of skin problems over the years and know what the problems we describe in print look like in the flesh.

We have listed some of the more common problems, but by no means all. If your problem doesn't seem to fit anything, then common sense—about whether the problem is serious and a call to the doctor if it is—will take you a long way.

Finally, because every case is at least a little bit different, even the best doctors will not be able to immediately identify all skin problems. Simple office laboratory methods can help sort these out. The vast majority of skin problems, fortunately, are minor, self-limited, and pose no major threat to health.

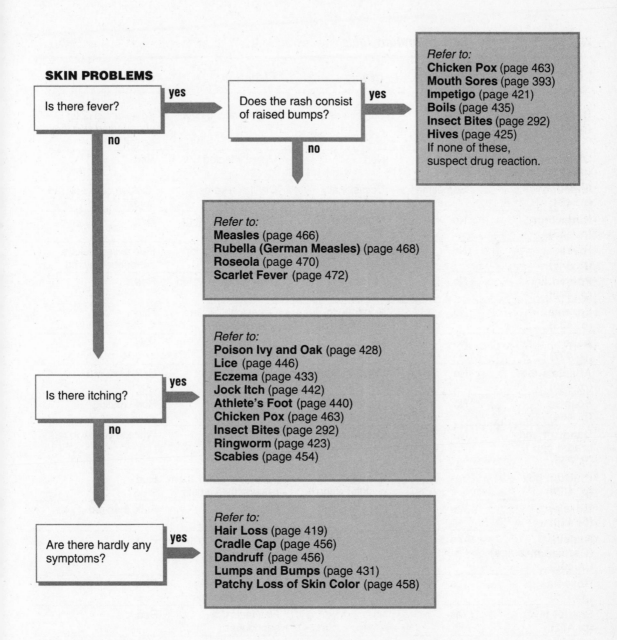

SKIN PROBLEMS

Is there fever? — yes → Does the rash consist of raised bumps? — yes →

Refer to:
Chicken Pox (page 463)
Mouth Sores (page 393)
Impetigo (page 421)
Boils (page 435)
Insect Bites (page 292)
Hives (page 425)
If none of these, suspect drug reaction.

no ↓

Refer to:
Measles (page 466)
Rubella (German Measles) (page 468)
Roseola (page 470)
Scarlet Fever (page 472)

Is there itching? — yes →

Refer to:
Poison Ivy and Oak (page 428)
Lice (page 446)
Eczema (page 433)
Jock Itch (page 442)
Athlete's Foot (page 440)
Chicken Pox (page 463)
Insect Bites (page 292)
Ringworm (page 423)
Scabies (page 454)

Are there hardly any symptoms? — yes →

Refer to:
Hair Loss (page 419)
Cradle Cap (page 456)
Dandruff (page 456)
Lumps and Bumps (page 431)
Patchy Loss of Skin Color (page 458)

TABLE J   *Skin Symptom Table*

	Fever	Itching	Elevation	Color
Baby rashes (p. 414)	No	Sometimes	Slightly raised dots	White or red dots; surrounding skin may be red
Diaper rash (p. 417)	No	No	Only if infected	Red
Impetigo (p. 421)	Sometimes	Occasionally	Crusts on sores	"Golden crusts on red sores"
Ringworm (p. 423)	No	Occasionally	Slightly raised rings	Red
Hives (p. 425)	No	Intense	Raised with flat tops	Pale raised lesions surrounded by red
Poison ivy (p. 428)	No	Intense	Blisters are elevated	Red
Eczema (p. 433)	No	Moderate to intense	Occasional blisters when infected	Red
Acne (p. 437)	No	No	Pimples, cysts	Red
Athlete's foot (p. 440)	No	Mild to intense	No	Colorless–red
Scabies (p. 454)	No	Intense	Slight	Red-crusting
Dandruff and cradle cap (P. 456)	No	Occasionally	Some crusting	White to yellow to red
Chicken pox (p. 463)	Yes	Intense during pustular stage	Flat, then raised, then blisters, then crusts	Red
Measles (p. 466)	Yes	None to mild	Flat	Pink; then red
Rubella (German measles) (p. 468)	Yes	No	Flat or slightly raised	Red
Roseola (p. 470)	Yes	No	Flat, occasionally with a few bumps	Pink
Scarlet fever (p. 472)	Yes	No	Flat; feels like sandpaper	Red
Fifth disease (p. 474)	No	No	Flat; lacy appearance	Red

Location	Duration of Problem	Other Symptoms
Trunk, neck, skin folds on arms and legs	Until controlled	
Under diaper	Until controlled	
Arms, legs, face first; then most of body	Until controlled	
Anywhere, including scalp and nails	Until controlled	Flaking or scaling
Anywhere	Minutes to days	
Exposed areas	7–14 days	Oozing; some swelling
Elbows, wrists, knees, cheeks	Until controlled	Moist; oozing
Face, back, chest	Until controlled	Blackheads
Between toes	Until controlled	Cracks; scaling; oozing blisters
Arms, legs, trunk; Infants: head, neck, hands, feet	Until controlled	
Scalp, eyebrows, behind ears, groin	Until controlled	Fine, oily scales
May start anywhere; most prominent on trunk and face	4–10 days	Lesions progress from flat to tiny blisters, then become crusted
First face; then chest and abdomen; then arms and legs	4–7 days	Preceded by fever, cough, red eyes
First face; then trunk; then extremities	2–4 days	Swollen glands behind ears; occasional joint pains in older children and adults
First trunk; then arms and neck; very little on face and legs	1–2 days	High fever for 3 days that disappears with rash
First face; then elbows; spreads rapidly to entire body in 24 hours	5–7 days	Sore throat; skin peeling afterwards, especially palms
First face; then arms and legs; then rest of body	3–7 days	"Slapped-cheek" appearance, rash comes and goes

Baby Rashes

The skin of the newborn child may exhibit a wide variety of bumps and blotches. Fortunately, almost all of these are harmless and clear up by themselves. The most common of these conditions are addressed in this section, and only one, heat rash, requires any treatment. If the baby was delivered in a hospital, many of these conditions may occur before discharge so that advice will be readily available from nurses or doctors.

Heat Rash

Heat rash is caused by blockage of the pores that lead to the sweat glands. It actually can occur at any age but is most common in the very young child in which the sweat glands are still developing. When heat and humidity rise, these glands attempt to provide sweat as they would normally; but because of the blockage, this sweat is held within the skin and forms little red bumps. It is also known as "prickly heat" or "miliaria."

Milia

On the other hand, the "little white bumps of milia" are composed of normal skin cells that have overaccumulated in some spots. As many as 40% of children have these bumps at birth. Eventually, the bumps break open, the trapped material escapes, and the bumps disappear without requiring any treatment.

Erythema Toxicum

Erythema toxicum is an unnecessarily long and frightening term for the flat, red splotches that appear in up to 50 percent of all babies. These seldom appear after five days of age and have usually disappeared by seven days. The children involved are perfectly normal, and whether or not any real toxin is involved is not clear.

Acne

Because the baby is exposed to the mother's adult hormones, a mild case of acne may develop, just as may occur when a child begins to produce adult hormones during adolescence. (The little white dots often seen on a newborn's nose represent an excess amount of normal skin oil, *sebaceous gland hyperplasia,* that has been produced by the hormones.) Acne usually becomes evident at between two and four weeks of age and clears up spontaneously within six months to a year. It virtually never requires treatment.

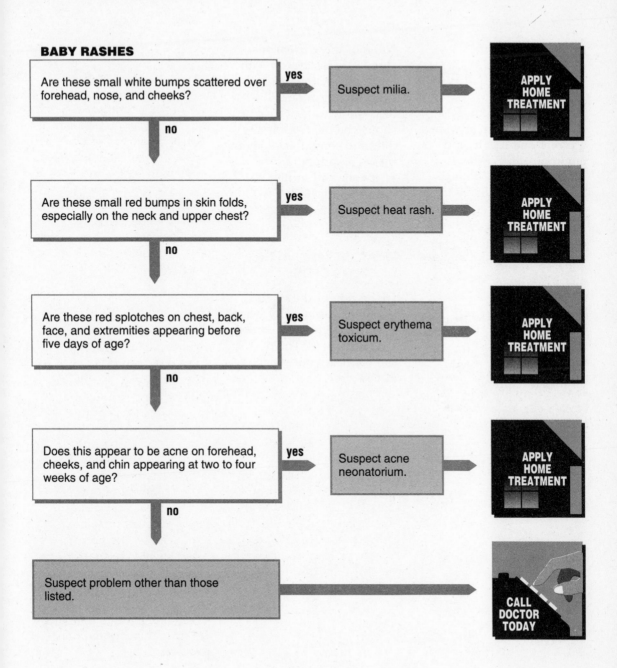

BABY RASHES

Are these small white bumps scattered over forehead, nose, and cheeks? — **yes** → Suspect milia. → **APPLY HOME TREATMENT**

no ↓

Are these small red bumps in skin folds, especially on the neck and upper chest? — **yes** → Suspect heat rash. → **APPLY HOME TREATMENT**

no ↓

Are these red splotches on chest, back, face, and extremities appearing before five days of age? — **yes** → Suspect erythema toxicum. → **APPLY HOME TREATMENT**

no ↓

Does this appear to be acne on forehead, cheeks, and chin appearing at two to four weeks of age? — **yes** → Suspect acne neonatorium. → **APPLY HOME TREATMENT**

no ↓

Suspect problem other than those listed. → **CALL DOCTOR TODAY**

HOME TREATMENT

Heat Rash

Heat rash is effectively treated simply by providing a cooler and less humid environment. Powders carefully applied do no harm but are unlikely to help. Ointments and creams should be avoided because they tend to keep the skin warmer and block the pores.

Acne

Acne should *not* be treated with the medicines used by adolescents and adults. Normal washing usually is all that is required.

None of these problems should be associated with fever and, with the exception of minor discomfort in heat rash, should be painless. If any question should arise about these conditions, a telephone call to the doctor's office often will answer your questions.

WHAT TO EXPECT AT THE DOCTOR'S OFFICE

Discussion of these problems can usually wait until the regular scheduled well-baby visit. The doctor can confirm your diagnosis at that time.

Diaper Rash

The only children who never have diaper rash are those who never wear diapers. An infant's skin is particularly sensitive and likely to develop diaper rash, an irritation caused by dampness and the interaction of urine and skin. An additional irritant is thought to be the ammonia in urine.

Two factors that tend to keep the baby's skin wet and exposed to the irritant promote diaper rash:

- Continuously wet diapers
- The use of plastic pants

Treatment consists of reversing these factors.

The irritation of simple diaper rash may become complicated by an infection due to **yeast** (*Candida*) or bacteria. When yeast is the culprit, small red spots may be seen. Also, small patches of the rash may appear outside the area covered by the diaper, as far away as the chest. Infection with bacteria leads to development of large fluid-filled blisters. If the rash is worse in the skin creases, a mild underlying skin problem known as **seborrhea** may be present. This skin condition is also responsible for cradle cap and dandruff.

Occasionally, parents will notice blood or what appear to be blood spots when boys have diaper rash. This is due to an irritating rash at the urinary opening at the end of the penis.

HOME TREATMENT

Treatment of diaper rash is aimed at keeping the skin dry. The first things to do are to change the diapers more frequently and to discontinue the use of plastic pants. Leaving the diapers off altogether for as long as possible will also help. Cloth diapers should be washed in a mild soap and rinsed thoroughly; occasionally, the soap residues left in the diapers will act as an irritant. Adding a half-cup of vinegar to the last rinse cycle may help counter the irritating ammonia. Cloth and paper diapers are equivalent insofar as causing diaper rash. Moisture and not material is the culprit.

Complete clearing of the rash will take several days, but definite improvement should be noted within the first 48 to 72 hours. If this is not the case or if the rash is severe, the doctor should be consulted.

To prevent diaper rash, some parents use zinc oxide ointment (Desitin) or A + D ointment. Others use baby powders. (**Caution**: talc dust can injure lungs.) Caldesene powder is helpful in preventing seborrhea and monilial rashes. Always place powder in your hand first, and then pat on baby's bottom. We do not feel that all babies need powders and creams. If a rash has begun, ointments should *not* be used because they delay healing.

Sometimes a rash will not heal until a precipitating problem, such as diarrhea, has resolved. Dealing with the diarrhea is the initial approach to healing the diaper rash.

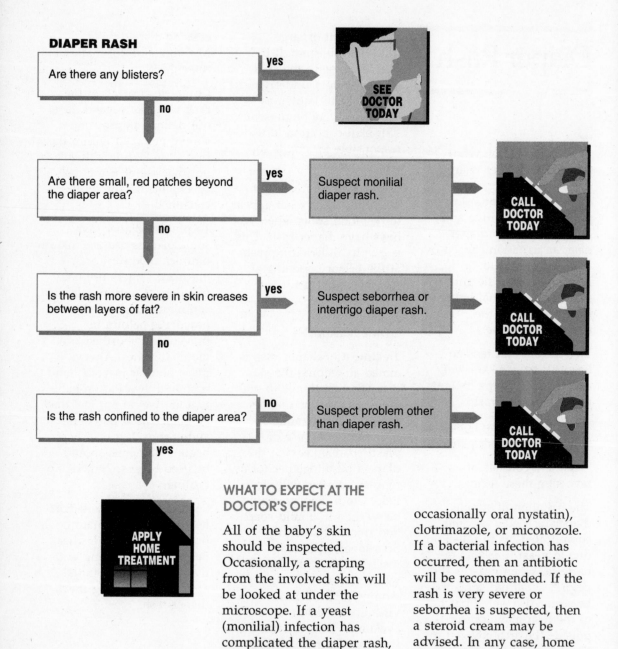

DIAPER RASH

Are there any blisters?

yes → **SEE DOCTOR TODAY**

no ↓

Are there small, red patches beyond the diaper area?

yes → Suspect monilial diaper rash. → **CALL DOCTOR TODAY**

no ↓

Is the rash more severe in skin creases between layers of fat?

yes → Suspect seborrhea or intertrigo diaper rash. → **CALL DOCTOR TODAY**

no ↓

Is the rash confined to the diaper area?

no → Suspect problem other than diaper rash. → **CALL DOCTOR TODAY**

yes ↓

APPLY HOME TREATMENT

WHAT TO EXPECT AT THE DOCTOR'S OFFICE

All of the baby's skin should be inspected. Occasionally, a scraping from the involved skin will be looked at under the microscope. If a yeast (monilial) infection has complicated the diaper rash, the doctor will prescribe a medication to kill the yeast (nystatin cream and occasionally oral nystatin), clotrimazole, or miconozole. If a bacterial infection has occurred, then an antibiotic will be recommended. If the rash is very severe or seborrhea is suspected, then a steroid cream may be advised. In any case, home therapy may be begun safely before seeing the doctor.

Hair Loss

Hair loss may cause concern in childhood. Often all the hair in one small area will be completely lost, but the scalp underneath will be normal. This problem is called **alopecia areata**, and its cause is unknown. Usually the hair will be completely re-grown within 12 months, although about 40% of children will have a similar loss within the next four to five years. This problem resolves by itself. Cortisone creams may make the hair grow back faster, but the new hair falls out again when the treatment is stopped, so these creams are of little use.

Types of hair loss that may need treatment by a doctor are characterized by abnormalities in the scalp skin or the hairs themselves. The most frequent problem in this category is **ringworm** (see page 423). Ringworm may be red and scaly, or there may be pustules with oozing. The ringworm fungus infects the hairs so that they become thickened and break easily. Whenever the scalp skin or the hairs themselves appear abnormal, the doctor may be able to help.

Hair pulling by the children themselves, or occasionally a friend, often is responsible for hair loss. Tight braids or ponytails may also cause some hair loss. If your child constantly pulls out his or her hair, you should consider this unusual behavior and discuss it with a doctor.

HOME TREATMENT

In this instance, home treatment is reserved for presumed alopecia areata and consists of watchful waiting. The skin in the involved area must be completely normal to make a diagnosis of alopecia areata. Should the appearance of scalp or hairs become abnormal, then the doctor should be consulted.

HAIR LOSS

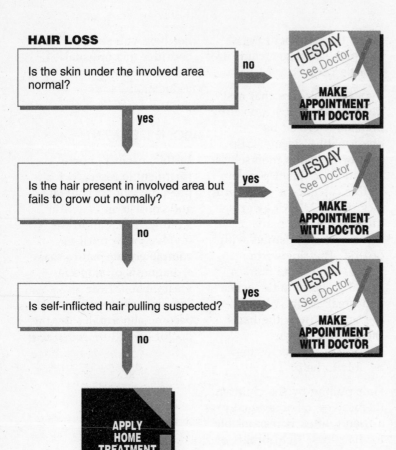

Is the skin under the involved area normal? — **no** → **TUESDAY** See Doctor / **MAKE APPOINTMENT WITH DOCTOR**

yes ↓

Is the hair present in involved area but fails to grow out normally? — **yes** → **TUESDAY** See Doctor / **MAKE APPOINTMENT WITH DOCTOR**

no ↓

Is self-inflicted hair pulling suspected? — **yes** → **TUESDAY** See Doctor / **MAKE APPOINTMENT WITH DOCTOR**

no ↓

APPLY HOME TREATMENT

WHAT TO EXPECT AT THE DOCTOR'S OFFICE

An examination of the hair and scalp is usually sufficient to determine the nature of the problem. Occasionally, the hairs themselves may be examined under the microscope. Certain types of ringworm of the scalp can be identified because they fluoresce (glow) under a Wood's lamp. Ringworm of the scalp will require the use of an oral drug, griseofulvin, because creams and lotions applied to the affected area will not penetrate into the hair follicles to kill the fungus. Simultaneously, topical therapy may be given to prevent spread to other parts of the scalp or other family members.

We hope that no doctor would recommend the use of X-rays today as some did 15 or 20 years ago. If it is offered, it should be flatly rejected and you should find yourself another doctor.

Impetigo

Impetigo is particularly troublesome in the summer and especially in warm, moist climates. It can be recognized by the characteristic appearance of the lesions. These lesions begin as small red spots and progress to tiny blisters that eventually rupture, producing an oozing, sticky, honey-colored crust. These lesions are usually spread very quickly by scratching fingers. Another characteristic of impetigo is that it is extremely contagious, and children pass it on to their brothers, sisters, and playmates very easily.

Impetigo is a skin infection caused by streptococcal bacteria; occasionally, other bacteria may also be found. Impetigo, if it spreads, can be a very uncomfortable problem. There is usually a great deal of itching, and scratching hastens spreading of the lesions. After the sores heal, there may be a slight decrease in skin color at the site. Skin color usually returns to normal, so this need not concern parents.

Complication in the Kidneys

Of greatest concern is a rare, complicating kidney problem known as **glomerulonephritis**, which occasionally occurs in epidemics. Glomerulonephritis will cause the urine to turn a dark brown (cola) color and is often accompanied by headache and elevated blood pressure. Although this problem has a formidable name, the kidney problem is short-lived and heals completely in most children.

Unfortunately, antibiotics will not prevent glomerulonephritis but can prevent the impetigo from spreading to other children, thus protecting them from both impetigo and glomerulonephritis. They are effective in healing the impetigo.

Although there is some debate on this matter, many doctors believe that if only one or two lesions are present and the lesions are not progressing, home treatment may be used for impetigo. The exception to this rule is if an epidemic of glomerulonephritis is occurring within your community.

HOME TREATMENT

Crusts may be soaked off with either warm water or Burrow's solution (Domeboro, Bluboro, etc.). Some antibiotic ointments are no more effective than soap and water, although Bacitracin is often used and may prevent some spread to others. The lesions should be scrubbed with soap and water after the crusts have been soaked off. If lesions do not show prompt improvement or if they seem to be spreading, the child should be seen by a doctor without delay.

WHAT TO EXPECT AT THE DOCTOR'S OFFICE

After examining the sores and taking an appropriate medical history, the doctor usually will either prescribe an effective topical agent (mupirocin) or an antibiotic

IMPETIGO

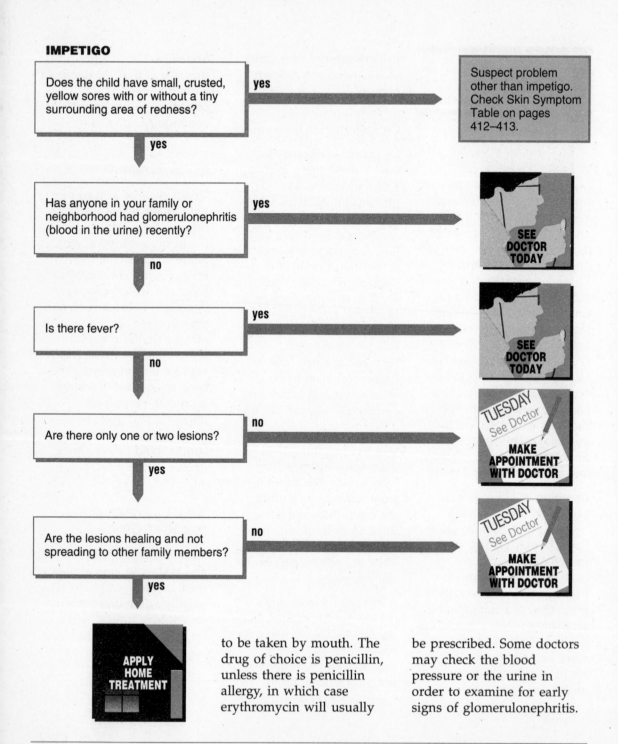

Does the child have small, crusted, yellow sores with or without a tiny surrounding area of redness? — **yes** → Suspect problem other than impetigo. Check Skin Symptom Table on pages 412–413.

yes ↓

Has anyone in your family or neighborhood had glomerulonephritis (blood in the urine) recently? — **yes** → SEE DOCTOR TODAY

no ↓

Is there fever? — **yes** → SEE DOCTOR TODAY

no ↓

Are there only one or two lesions? — **no** → TUESDAY See Doctor — MAKE APPOINTMENT WITH DOCTOR

yes ↓

Are the lesions healing and not spreading to other family members? — **no** → TUESDAY See Doctor — MAKE APPOINTMENT WITH DOCTOR

yes ↓

APPLY HOME TREATMENT

to be taken by mouth. The drug of choice is penicillin, unless there is penicillin allergy, in which case erythromycin will usually be prescribed. Some doctors may check the blood pressure or the urine in order to examine for early signs of glomerulonephritis.

Ringworm

Ringworm is a shallow fungus infection of the skin. Worms have nothing whatsoever to do with this condition; the designation "ringworm" is derived from the characteristic red ring that appears on the skin.

Ringworm can generally be recognized by its pattern of development. The lesions begin as small, round, red spots and get progressively larger. When they are about the size of a pea, the center begins to clear. When the lesions are about the size of a dime, they have the appearance of a ring. The border of the ring is red, elevated, and scaly. Often there are groups of infections so close to one another that it is difficult to recognize them as individual rings.

Ringworm may also affect the scalp or the nails. These infections are more difficult to treat but fortunately are not seen very often. Epidemics of ringworm of the scalp were common many years ago.

HOME TREATMENT

Tolnaftate (Tinactin) applied to the skin is an effective treatment for ringworm. It is available in cream, solution, and powder and can be purchased over the counter. Either the cream or the solution should be applied two or three times a day. Only a small amount is required for each application. Resolution of the problem may require several weeks of therapy, but improvement should be noted within a week. Selsun Blue shampoo, applied as a cream several times a day, will often do the job just as well and is less expensive. Ringworm that shows no improvement after a week of therapy or that continues to spread should be checked by a doctor.

WHAT TO EXPECT AT THE DOCTOR'S OFFICE

The diagnosis of ringworm can be confirmed by scraping the scales, soaking them in a potassium hydroxide solution, and viewing them under the microscope. Some doctors may culture the scrapings. One of four agents will be prescribed if Tinactin has failed: haloprogin (Halotex), clotrimazole (Lotrimin), miconazole (MicaTin), or ciclopirox.

RINGWORM

Are all of the following conditions present?

- Rash begins as a small, red, color-less, or depigmented circle that becomes progressively larger.
- The circular border is elevated and perhaps scaly.
- The center of the circle begins healing as the circle becomes larger.

no → Suspect problem other than ringworm. Check Skin Symptom Table on pages 412–413.

yes ↓

Is the scalp infected?

yes → TUESDAY See Doctor **MAKE APPOINTMENT WITH DOCTOR**

no ↓

APPLY HOME TREATMENT

In infections involving the scalp, an ultra-violet light (called a Wood's lamp) will cause affected hairs to become fluorescent (glowing). The Wood's lamp is used to make the diagnosis; it does not treat the ringworm. Ringworm of the scalp must be treated by griseofulvin, taken by mouth, usually for at least a month; this medication is also effective for fungal infections of the nails but must be taken for 6 to 18 months. Ringworm of the scalp should never be treated with X-rays.

Hives

Hives are an allergic reaction. Unfortunately, the reaction can be to almost anything, including cold or heat and even emotional tension. Unless you already have a good idea what is causing the hives or if a new drug has just been taken, the doctor is unlikely to be able to determine the cause; most often, a search for a cause is fruitless.

Here is a list of some of the things that are frequently mentioned as causes:

- Drugs
- Eggs
- Milk
- Wheat
- Chocolate
- Pork
- Shellfish
- Freshwater fish
- Berries
- Cheese
- Nuts
- Pollens
- Insect bites

The only sure way to know whether one of these is the culprit is to expose the child to it purposely. The problem with this approach is that if an allergy does exist, then the allergic reaction may include not only hives but dangerous reactions causing difficulty with breathing or circulation.

As indicated by the decision chart, a **systemic reaction** along with the hives is a potentially dangerous situation, and the doctor should be consulted immediately. Avoid exposure to a suspected cause to see if the attacks will cease. Such "tests" are difficult to interpret because attacks of hives are often separated by long periods of time. Actually, most people have only one attack, lasting from a period of hours to weeks.

HOME TREATMENT

Determine whether there has been any pattern to the appearance of the hives. Do they appear after meals? After exposure to the cold? During a particular season of the year? If these seem to be likely possibilities, eliminate these and see what happens.

If the reactions seem to be related to foods, an alternative is available. Lamb and rice virtually never cause allergic reactions. The child may be placed on a diet consisting only of lamb and rice until completely free of hives. Foods are then added back to the diet one at a time, and the child is observed for the development of hives.

Itching may be relieved by the application of cold compresses, oatmeal baths, calamine lotion, the use of aspirin, or the use of antihistamines such as diphenhydramine (Benadryl) (see "The Home Pharmacy," page 223).

WHAT TO EXPECT AT THE DOCTOR'S OFFICE

If the child is suffering a systemic reaction with difficulty breathing or dizziness, then injections of adrenaline and other drugs may be given. In the more usual case of hives alone, the doctor may do two things for your child. First, he or she can prescribe an antihistamine or use adrenaline injections to relieve swelling and itching. Second, the doctor can

review the history of the reaction to try to find an offending agent and advise you as above. Remember that most often the cause of hives goes undetected.

If severe hives are accompanied by serious symptoms such as breathing problems following an insect sting, several approaches are available, including:

- Desensitization to the insect through injections
- Instructions in the use of an emergency epinephrine-injection kit

HIVES

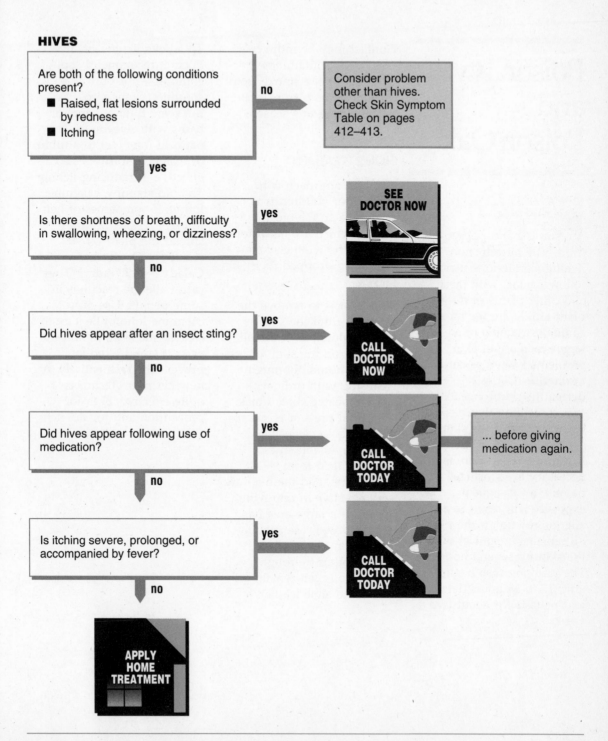

Are both of the following conditions present?
- Raised, flat lesions surrounded by redness
- Itching

no → Consider problem other than hives. Check Skin Symptom Table on pages 412–413.

yes ↓

Is there shortness of breath, difficulty in swallowing, wheezing, or dizziness?

yes → SEE DOCTOR NOW

no ↓

Did hives appear after an insect sting?

yes → CALL DOCTOR NOW

no ↓

Did hives appear following use of medication?

yes → CALL DOCTOR TODAY → ... before giving medication again.

no ↓

Is itching severe, prolonged, or accompanied by fever?

yes → CALL DOCTOR TODAY

no ↓

APPLY HOME TREATMENT

Poison Ivy and Poison Oak

Poison ivy and poison oak need little introduction. The itching skin lesions that follow contact with these and other plants of the *Rhus* plant family are the most common example of a larger category of skin problems known as **contact dermatitis**. Contact dermatitis simply means that something that has been applied to the skin has caused the skin to react to it. An initial exposure is necessary to "sensitize" the patient; a subsequent exposure will result in an allergic reaction if the plant oil remains in contact with the skin for several hours. The resulting rash begins after 12 to 48 hours delay and persists for about two weeks.

Contact may be indirect, from pets, contaminated clothing, or the smoke from burning *Rhus* plants. It can occur during any season.

HOME TREATMENT

The best approach is to teach your children to recognize and avoid the plants, which are hazardous even in the winter when they have dropped their leaves.

Next best is to remove the plant oil from the skin as soon as possible. If the oil has been on the skin for less than six hours, thorough cleansing with ordinary soap, repeated three times, will often prevent reaction. Alcohol-base cleansing tissues, available in pre-packaged form (Alco-wipe, and the like) are very effective in removing the oil. If a large area has been exposed, use rubbing alcohol on a washcloth. Wash clothes thoroughly because oils can remain on fabric and shoe leather.

To relieve itching, many doctors recommend cool compresses of Burrow's solution (Domeboro, BurVeen, Bluboro, etc.) or baths with Aveeno or oatmeal (one cup to a tubful of water). Aspirin is also effective in reducing itching. The old standby, calamine lotion, is sometimes of help in early lesions but may spread the plant oil. (**Caution**: Do not use Caladryl or Zyradryl. They cause allergic reactions in some people. Use only calamine lotion.) Be sure to cleanse the skin, as above, even if you are too late to prevent the rash entirely. A new cleanser (Tecnu) is quite effective, both for prevention and for the itch.

POISON IVY AND POISON OAK

Are all of the following conditions present?
- Itching
- Redness, minor swelling, blisters, or oozing
- Probable exposure to poison ivy, poison oak, or poison sumac

no →

Suspect problem other than poison ivy or poison oak. Check the Skin Symptom Table on pages 412–413.

yes ↓

Have any of the following occurred?
- Extensive involvement of the face
- Involvement of the genitals
- Pus or swelling around lesions
- Large, fluid-filled blisters

yes →

CALL DOCTOR TODAY

no ↓

APPLY HOME TREATMENT

Another useful method of obtaining symptomatic relief for older children is the use of a hot bath or hot shower. Heat releases histamine, the substance in the cells of the skin that causes the intense itching. Therefore, a hot shower or bath will cause intense itching as the histamine is released. The heat should be gradually increased to the maximum tolerable and continued until the itching has subsided. This process will deplete the cells of histamine, and the child will obtain up to eight hours of relief from the itching. This method has the advantage of not requiring frequent applications of ointments to the lesions and is a good way to get some sleep at night.

Poison ivy or poison oak will persist for the same length of time despite any medication. If secondary bacterial infection occurs, healing will be delayed; hence scratching is not helpful. Cut the nails to avoid damage to the skin through scratching.

Poison ivy is not contagious; it cannot be spread once the oil has been absorbed by the skin or removed.

If the lesions are too extensive to be easily treated, if home treatment is ineffective, or if the itching is so severe that the child can't tolerate it, a visit to the doctor may be necessary.

WHAT TO EXPECT AT THE DOCTOR'S OFFICE

After a history and physical examination, the doctor may prescribe a steroid cream to be applied four to six times a day to the lesions. This is often of moderate help.

An alternative is to give a steroid (such as prednisone) by mouth for short periods of time. A rather large dose is given the first day, and the dose is then gradually reduced. We are reluctant to recommend oral steroids except in children with previous severe reactions to poison ivy or poison oak or extensive exposure.

The itching may be treated symptomatically with either an antihistamine (Benadryl, Vistaril), an anti-inflammatory agent such as aspirin or a cleanser like Tecnu. The antihistamines may cause drowsiness and interfere with sleep.

Skin Lumps, Bumps, and Warts

Lumps and bumps are common at all ages. The lump may be:

- Above the skin, as with warts, moles, and insect bites
- Within the skin, as with boils and certain kinds of moles
- Under the skin, as in the case of swollen lymph glands or small collections of fat called lipomas.

If there is only one lump and it is red, hot, tender, and swollen (inflamed), then it should be considered a boil until proven otherwise. (See **Boils**, page 435).

A dark mole that appears to be enlarging or changing color might be a melanoma, a kind of skin cancer, although melanomas are extremely unusual in children.

The most commonly found lumps in children are swollen lymph glands. These are discussed in more detail in **Swollen Glands**, page 386.

Warts are caused by viral infections and often spontaneously resolve by themselves. However, warts can occasionally be troublesome, especially if they are on the fingers where they may interfere with writing or on the face where they are cosmetically disturbing.

HOME TREATMENT

Treatment of many lumps and bumps is discussed under **Insect Bites** (page 292), **Boils** (page 435), **Swollen Glands** (page 386), and **Arm and Leg Lumps** (page 492).

Warts can be removed with over-the-counter medicines used consistently and carefully. Available preparations include salicylic acid plasters, Compound W, Vergo, and preparations with salicylic and lactic acid (Duofilm). On your child's next visit to the doctor, you can ask about any skin lumps or warts that are of concern to you; these are seldom worth a special trip.

WHAT TO EXPECT AT THE DOCTOR'S OFFICE

The doctor may be able to make the diagnosis simply by inspecting the lump or wart and asking you a few questions. A wart may be removed by freezing it with liquid nitrogen or by using an electric needle or by chemical cauterization. It is seldom treated surgically because warts are caused by viruses; cutting them out may leave the virus to cause a later recurrence of the wart. A recurrence of warts is common, and repeated treatments may be necessary.

SKIN BUMPS, LUMPS, AND WARTS

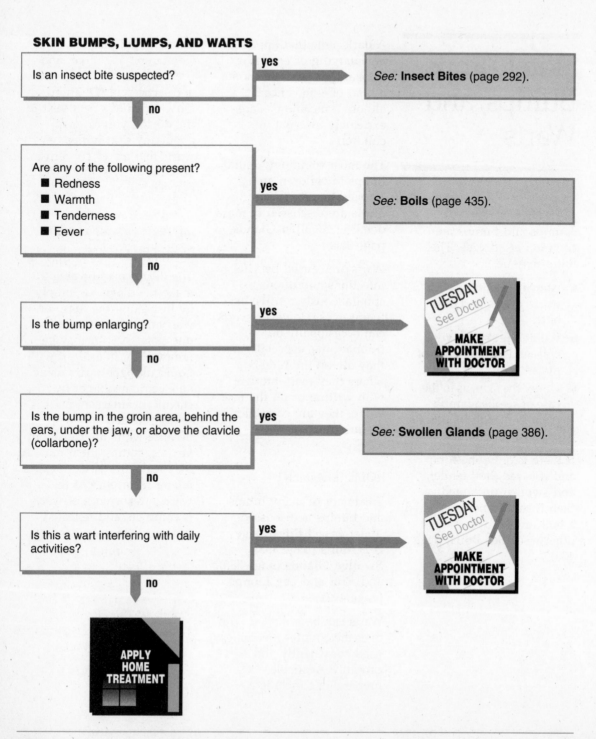

Is an insect bite suspected?

yes → *See:* **Insect Bites** (page 292).

no ↓

Are any of the following present?
- Redness
- Warmth
- Tenderness
- Fever

yes → *See:* **Boils** (page 435).

no ↓

Is the bump enlarging?

yes → TUESDAY See Doctor. **MAKE APPOINTMENT WITH DOCTOR**

no ↓

Is the bump in the groin area, behind the ears, under the jaw, or above the clavicle (collarbone)?

yes → *See:* **Swollen Glands** (page 386).

no ↓

Is this a wart interfering with daily activities?

yes → TUESDAY See Doctor **MAKE APPOINTMENT WITH DOCTOR**

no ↓

APPLY HOME TREATMENT

Eczema

Eczema is commonly found in children with a family history of either eczema, hay fever, or asthma. The underlying problem is the inability of the skin to retain adequate amounts of water; the skin of children with eczema is consequently very dry, which causes the skin to itch.

Children with atopic dermatitis have skin that responds to changes in the environment, allergic substances in food, and emotional stress, with intense itching. Most of the manifestations of eczema are a consequence of scratching. Atopic skin also has an increased tendency to swell and ooze when scratched.

In young infants who are unable to scratch, the most common manifestation is red, dry, mildly scaling cheeks. Although the infant cannot scratch his or her cheeks, the cheeks can be rubbed by moving against the sheets and thus become red. In infants, eczema may also be found in the area where the plastic pants meet the skin. The tightness of the elastic produces the characteristic red scaling lesion. In older children, it is very common for eczema to involve the area behind the knees and in front of the elbows.

If there is a fair amount of weeping or crusting, the eczema has become infected with bacteria, and a visit to the doctor will most likely be required.

The course of eczema is quite variable. Some children only have a brief mild problem; others have manifestations throughout life.

HOME TREATMENT

Attempts must be made to prevent the skin from becoming too dry, and frequent bathings make the skin even dryer. Although the child will feel comfortable in the bath, itching will become more intense after the bath because of the drying effect. Lubricating lotions help keep the skin moist and include Keri and Lubriderm lotions, Keri and Nivea creams, and Eucerin paste.

Sweating aggravates eczema; consequently, children should not be overdressed so that they perspire. Light night clothing is important. Contact with wool or silk seems to aggravate eczema in some children and should be avoided.

When inflammation is prominent—the skin is red, swollen, oozing, and itching—wet dressings of Burrow's solution and oatmeal baths can be applied. Nails should be kept trimmed short to minimize the effects of scratching. In older children who are helping with household chores, rubber

ECZEMA

Are the following conditions present (at least two of three)?
■ Itching
■ Flat red areas on cheeks, behind ears, along edge of plastic pants, behind knees, or in front of elbows
■ Family history of allergy

no →

Suspect problem other than eczema. Check Skin Symptom Table on pages 412–413.

yes ↓

Is there any crusting of lesions?

yes →

CALL DOCTOR TODAY

no ↓

APPLY HOME TREATMENT

WHAT TO EXPECT AT THE DOCTOR'S OFFICE

By history and examination of the lesions, the doctor can determine whether the problem is eczema. If home treatment has not improved the problem, steroid creams and lotions may be prescribed. While these are effective, they are not curative; eczema is characterized by repeated occurrences. If crusted or weeping lesions are present, bacterial infection is likely, and an oral antibiotic will be prescribed. Hydroxyzine (Atarax, Vistaril) is an antihistamine that is often used to reduce itching.

gloves can help prevent drying of the hands after washing dishes or the car.

Washing is best accomplished with a cleansing and moisturizing agent such as Cetaphil lotion. Many soaps aggravate eczema. Dove soap is milder and less drying than most.

While freshwater or pool swimming can aggravate eczema by causing loss of skin moisture, ocean swimming does not do so and can be freely undertaken. (Also see "Allergies," page 397.)

Boils

The familiar phrase "painful as a boil" emphasizes the severe discomfort that can arise from this common skin problem. A boil is a localized infection due to the staphylococcus germ; usually, a particularly savage strain of the germ is responsible. When this particular germ inhabits the child's skin, recurrent problems with boils may persist for months or years. Often several family members will be affected at about the same time.

Boils may be single or multiple, and they may occur anywhere on the body. They range from the size of a pea to the size of a walnut or larger. The surrounding red, thickened, and tender tissue increases the problem even further. The infection begins in the tissues beneath the skin and develops into an abscess pocket filled with pus. Eventually, the pus pocket "points" toward the skin surface and finally ruptures and drains. Then it heals.

Boils often begin as infections around hair follicles; hence the term **folliculitis** for minor infections. Often areas under pressure (such as the buttocks) are likely spots for boils to begin. A boil that extends into the deeper layers of the skin is called a **carbuncle**.

Special consideration should be given to boils on the face because they are more likely to lead to severe complicating infections.

HOME TREATMENT

Boils are handled gently because rough treatment can force the infection deeper inside the body. Warm, moist soaks are applied gently several times each day to speed the development of the pocket of pus and to soften the skin for the eventual rupture and drainage. Once drainage begins, the soaks will help keep the opening in the skin clear. The more drainage, the better. Frequent thorough soaping of the entire skin helps prevent re-infection. *Ignore all temptation to squeeze the boil.*

WHAT TO EXPECT AT THE DOCTOR'S OFFICE

If there is fever or a facial boil, the doctor will usually prescribe an antibiotic. Otherwise, antibiotics may not be used; they are of limited help in abscess-like infections.

If the boil feels as if fluid is contained in a pocket but has not yet drained, the doctor may lance the boil. In this procedure, a small incision is made to allow the pus to drain. After drainage, the pain is reduced, and healing is quite prompt. Although "incision and drainage" is not a complicated procedure, it is tricky enough that you should not attempt it yourself.

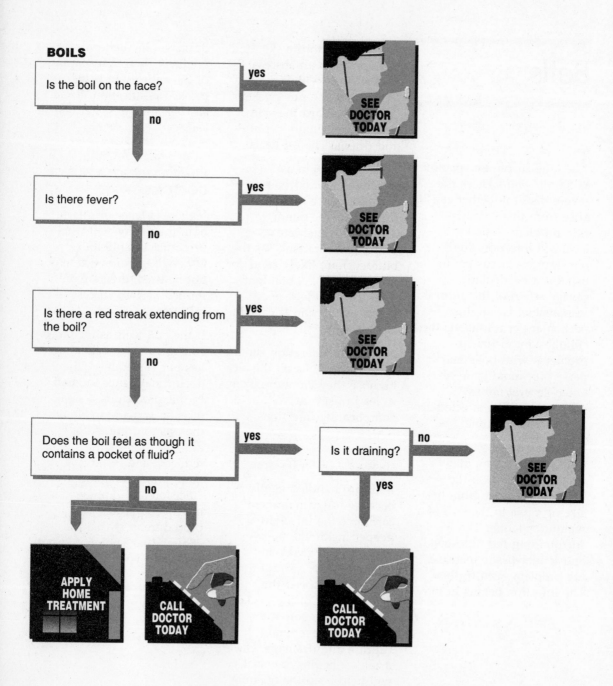

BOILS

Is the boil on the face? — yes → **SEE DOCTOR TODAY**

no ↓

Is there fever? — yes → **SEE DOCTOR TODAY**

no ↓

Is there a red streak extending from the boil? — yes → **SEE DOCTOR TODAY**

no ↓

Does the boil feel as though it contains a pocket of fluid? — yes → Is it draining? — no → **SEE DOCTOR TODAY**

no ↓   yes ↓

APPLY HOME TREATMENT   **CALL DOCTOR TODAY**   **CALL DOCTOR TODAY**

Acne

Acne is a superficial skin eruption caused by a combination of factors. It is triggered by the hormonal changes of puberty and is common in children with oily skin. The increased skin oils (sebum) accumulate below keratin plugs in the openings of the hair follicles and oil glands. In this stagnant area below the plug, secretions accumulate, and bacteria grow. These normal bacteria cause changes in the secretions that make them irritating to the surrounding skin. The result is usually a **pimple**, but sometimes may develop into a larger pocket of secretions, or **cyst**.

Blackheads are formed by the deposition of a pigment known as melanin on keratin plugs; skin irritation is minimal.

HOME TREATMENT

Although excessive dirt will aggravate acne, scrupulous cleaning will not prevent it. With acne, the face should be scrubbed several times daily with a warm washcloth and a mild soap (Dove) to remove skin oils and keratin plugs. The rubbing and heat of the washcloth help dislodge keratin plugs.

Acne Medicines

Benzoyl peroxide and tretinoin are the most effective preparations currently available. Benzoyl peroxide acts by causing a mild peeling of the skin and assists in preventing plugging of the sebaceous glands. It also has some activity against bacteria and is able to interfere somewhat with some of the chemical processes involved in the formation of papules and pustules. It can be irritating, especially if used with abrasive soaps, which is why we do not recommend coarse and gritty soaps. As many as 2% of children may be allergic to this product, so it is not for everyone.

It also takes a number of weeks before any results will be seen with benzoyl peroxide. This is important to bear in mind, for the first few weeks can be discouraging. The goal is to produce a mild dryness of the skin. The best way to achieve this is to begin applications every other day. Begin with the 5% lotion. If there is no skin irritation and the acne persists, it is then possible to increase the concentration to the 10% lotions and go to an everyday basis. The most potent forms are available in gels.

Some cosmetics can aggravate acne; it is best to avoid them as much as possible, especially when treating the skin with the agents we have just been discussing.

There has been publicity about zinc treatment of acne; this has not yet been proved effective, and we do not recommend it. Most diets have sufficient zinc for adequate skin nourishment. Your adolescent's acne will not respond readily to topical treatment if the pimples are large and cystic in nature. Moreover, it is these types of pimples that most often cause scarring. If this situation occurs, you should consult your doctor about some of the effective treatments.

Sunlamps

Acne appears to get better in the summer. Although this is most likely due to increased exposure to sunlight, other factors may be involved. We do *not* recommend the routine use of sunlamps, the primary effect of which is to induce peeling of the skin. This can be done much more safely by peeling agents such as benzoyl peroxide and tretinoin. Sunlamps can produce severe burns.

WHAT TO EXPECT AT THE DOCTOR'S OFFICE

If acne continues as a problem after home treatment, your doctor may prescribe tretinoin. Its actions are similar to benzoyl peroxide, but it is slightly more irritating. In resistant cases, an oral antibiotic (tetracycline or occasionally erythromycin) may be prescribed. Some doctors use these antibiotics applied to the skin.

> Diet is not an important factor in most cases of acne. But if certain foods tend to aggravate the problem, avoid them. There is scant evidence that chocolate aggravates acne.

"Acne surgery" is a term generally applied to the doctor's removal of blackheads with a suction device and an eyedropper. Large cysts are sometimes arrested with injection of steroids. Such procedures should be required only in severe cases and are more usually performed on the back than on the face.

ACNE

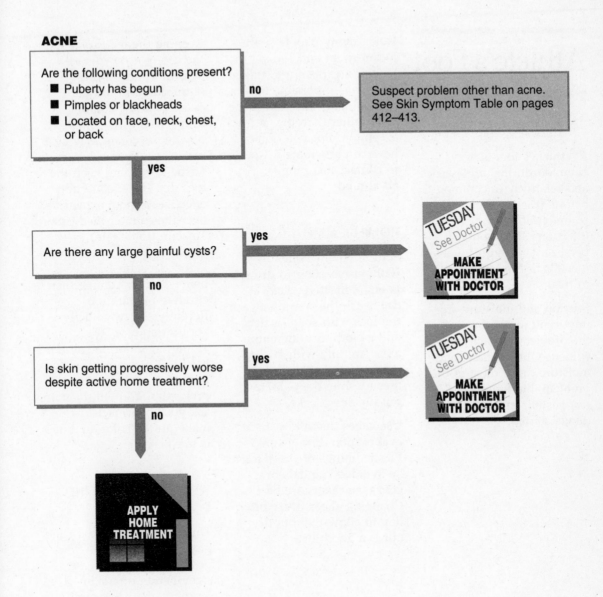

Are the following conditions present?
- Puberty has begun
- Pimples or blackheads
- Located on face, neck, chest, or back

no → Suspect problem other than acne. See Skin Symptom Table on pages 412–413.

yes ↓

Are there any large painful cysts?

yes → TUESDAY See Doctor — MAKE APPOINTMENT WITH DOCTOR

no ↓

Is skin getting progressively worse despite active home treatment?

yes → TUESDAY See Doctor — MAKE APPOINTMENT WITH DOCTOR

no ↓

APPLY HOME TREATMENT

Athlete's Foot

Athlete's foot is very common during adolescence and relatively uncommon before. It is the most common of the fungal infections and is often persistent. When it involves toenails, it can be difficult to treat.

Friction and moisture are important in aggravating the problem. In fact, there is evidence that bacteria and moisture cause most of this problem; the fungus is responsible only for getting things started.

When many people share locker-room and shower facilities, exposure to this fungus is impossible to prevent; and infection is the rule, rather than the exception. But you don't have to participate in sports to contact this fungus; it's all around.

HOME TREATMENT

Scrupulous hygiene, without resorting to drugs, is often effective. *Twice a day,* wash the space between the toes with soap, water, and a cloth; dry the entire area carefully with a towel, particularly between the toes (despite the pain); and put on clean socks.

Use shoes that allow evaporation of moisture. Plastic linings of shoes must be avoided. Sandals or canvas sneakers are best. Changing shoes every other day to allow them to dry out is a good idea.

Keeping the feet dry with the use of a powder is helpful in preventing re-infection. In difficult cases, over-the-counter drugs such as Desenex powder or cream may be used. The powder has the virtue of helping keep the toes dry. If these are not effective, a more expensive over-the-counter medication, Tinactin (tolnaftate), is available in either cream or lotion. Tinactin powder is better as a preventive than curative preparation. Recently, the twice-daily application of a 30% aluminum chloride solution has been recommended for its drying and antibacterial properties. You will have to ask your pharmacist to make up the solution, but it is inexpensive.

WHAT TO EXPECT AT THE DOCTOR'S OFFICE

Through history and physical examination, and possibly laboratory examination of a skin scraping, the doctor will establish the diagnosis. Several other problems, notably a condition called **dyshydrosis**, may mimic athlete's foot. If home treatment has failed,

ATHLETE'S FOOT

Are both of the following conditions present?
- Redness and scaling between toes (may have cracks and small blisters)
- Itching

no → Suspect problem other than athlete's foot. Check Skin Symptom Table on pages 412–413.

yes ↓

Have any of the following occurred?
- Blisters or pus oozing from toes
- Difficulty walking
- Progression despite home treatment

yes → CALL DOCTOR TODAY

no ↓

APPLY HOME TREATMENT

alternative medications may be given, including miconazole, haloprogin, clotrimazole, or ciclopirox. An oral drug, griseofulvin, may be used for fungal infections of the nails but is not recommended for athlete's foot.

Jock Itch

We might wish for a less picturesque name for this condition than "jock itch," but the medical term, *tinea cruris*, is understood by relatively few. Jock itch is a fungus infection of the pubic region. It is aggravated by friction and moisture. It usually does not involve the scrotum or penis, nor does it spread beyond the groin area. For the most part, this is a male disease. Frequently, the fungus grows in an athletic supporter turned old and moldy in a locker room far from a washing machine. The preventive measure for such a problem is obvious.

HOME TREATMENT

The problem should be treated by removing the contributing factors, friction and moisture. This is done by wearing boxer-type shorts rather than briefs, by applying a powder to dry the area after bathing, and by frequently changing soiled or sweaty underclothes. It may take up to two weeks to completely clear this problem, and it may recur. The "powder-and-clean-shorts" treatment will usually be successful without any medication. Tinactin (tolnaftate) will eliminate the fungus if the problem persists.

WHAT TO EXPECT AT THE DOCTOR'S OFFICE

Occasionally, a yeast infection will mimic jock itch. By examination and history, the doctor will attempt to establish the diagnosis and may make a scraping in order to identify a yeast. Medicines used for this problem are virtually always applied to the affected skin; oral drugs or injections are rarely used. Halotex (haloprogin), Lotrimin (clotrimazole), miconazole, and ciclopirox are prescription creams and lotions effective against both fungi and certain yeast infections.

JOCK ITCH

Are all of the following conditions present?
- Involves only groin and thighs
- Redness, oozing, or some peripheral scaling
- Itching

no →

Suspect problem other than jock itch. Check Skin Symptom Table on pages 412–413.

yes ↓

Are there signs of infection?
- Pus
- Crusting

yes →

CALL DOCTOR TODAY

no ↓

APPLY HOME TREATMENT

Sunburn

Sunburn is common, painful, and avoidable. Very rarely, people with sunburn have difficulty with vision; if so, they should be seen by a doctor. Otherwise, a visit to the doctor is unnecessary unless the pain is extraordinarily severe or unless extensive blistering (not peeling) has occurred. Blistering indicates a **second-degree burn** and rarely follows sun exposure.

The pain of sunburn is worst between 6 and 48 hours after sun exposure. Peeling of injured layers of skin occurs later, between three and ten days after the burn.

Prevention

Sunburn is better prevented than treated. Infants in particular should be shielded because they can burn rapidly. Gradual exposure is sensible. Start with only 15 minutes daily and increase exposure time gradually. In the early morning, the sun and its heat are less intense. Remember that water, snow, and high altitude intensify sun exposure. Cloud covering offers only a modest decrease in sun exposure. Protective clothing and sunscreen offer the best protection for young children against the effects of the sun.

Sunscreen products are now evaluated by a number known as the sun protective factor (SPF), with 2 being minimal and over 30 being maximal, which does not permit tanning. For children we recommend sunscreens with an SPF of at least 15. Some sunscreens contain para-aminobenzoid acid (PABA). Effective preparations contain at least 5% PABA. Because of allergic reactions to PABA, new products now dominate the shelves. Benzopherone is another sunscreen that shields a wider spectrum of light rays though it is less effective against damaging ultra-violet light than PABA. There are opaque white sunscreens that contain zinc oxide, quite effective but not very attractive.

HOME TREATMENT

Cool compresses or cool baths with Aveeno (one cup to a tubful of water) may be useful. Ordinary baking soda (one-half cup to a tub) is nearly as effective. Lubricants such as Vaseline feel good to some children but retain heat and should not be used the first day. Avoid products that contain benzocaine. These may give temporary relief but can cause irritation of the skin and may actually prolong healing. Acetaminophen by mouth may ease pain and thus help sleep.

SUNBURN

Are any of these conditions present following prolonged exposure to sun?

- Fever
- Fluid-filled blisters
- Dizziness
- Visual difficulties

yes → CALL DOCTOR TODAY

no → APPLY HOME TREATMENT

WHAT TO EXPECT AT THE DOCTOR'S OFFICE

The doctor will direct the history and physical examination toward determination of the extent of burn and the possibility of other heat-related injuries like sunstroke. If only first-degree burns are found, a prescription steroid lotion may be prescribed. This is not of particular benefit.

The rare second-degree burns may be treated with antibiotics in addition to analgesics or sedation. There is no evidence that steroid lotions or antibiotic creams help at all in the *usual* case of sunburn; most doctors prescribe the same therapy that is available at home.

Frostbite is rare in children, and easy to prevent. Have your children wear warm clothing, including mittens and windproof garments, in cold weather. Don't forget face masks on especially bitter days. When one's torso is warm, blood flows better to the fingers and toes, where frostbite usually occurs. Teach your children to come in from the cold whenever their fingers, nose, or toes start to hurt.

If your child's extremities begin to numb, they are starting to get frostbite. In the extreme case, the affected tissues turn black; see a doctor immediately. Otherwise, warm up the frostbitten area as quickly as possible with gentle rubbing. As the blood flow resumes, the frostbitten part will begin to hurt, sometimes a lot. This is a good sign, since the tissues are obviously still alive. Your child may have leftover numbness for months after minor frostbite, but this does not require medical attention.

Lice

Lice are found in the best of families. Lack of prejudice with respect to social class is as close as lice come to having a virtue. At best, they are a nuisance; at worst, they can cause real disability.

Lice themselves are very small and are seldom seen without the aid of a magnifying glass. There are three types of lice that infect children's hair, bodies, or pubic areas. Usually, it is easier to find the "nits," which are clusters of **louse eggs**. Without magnification, nits will appear as tiny white lumps on hair strands. The louse bite leaves only a pinpoint red spot, but scratching makes things worse. Itching and occasionally small, shallow sores at the base of hairs are clues to the disease.

Pubic lice are not a "venereal disease," although they may be spread from person to person during sexual contact. Unlike syphilis and gonorrhea, lice can be spread by toilet seats, infected bedding, and other sources. Pubic lice bear some resemblance to crabs, hence the use of the name "crabs" to indicate a lice infestation of the pubic hair. A different species of lice can inhabit the scalp or other body hair. Lice like to be close to a warm body all the time and will not stay for long periods of time in clothing not being worn, bedding being used, and so on.

Head lice are quite common in day-care centers and schools. During epidemics, children should be cautioned about sharing hats. Clothing should not be hung in closely crowded areas.

HOME TREATMENT

Over-the-counter preparations containing pyrethrins are effective against lice; these include A200, RID, and R&C shampoo. RID has the advantage of supplying a fine-tooth comb, a rare item these days.

Instructions that come with these drugs must be followed carefully. A synthetic variation of the natural pyrethrin is permethrin (Nix). The advantage of Nix is that only one treatment is required compared to a repeat treatment seven to ten days later with other products. Bedding and clothing must be changed simultaneously and should be washed in water hotter than 125°F or stored in plastic bags for ten days. Sexual partners should be treated at the same time.

LICE

Is either of the following conditions present?
- Lice seen on skin or in clothing
- Nits seen on hair shafts

yes → APPLY HOME TREATMENT

no ↓

Consider a problem other than lice. Check Skin Symptom Table on pages 412–413.

WHAT TO EXPECT AT THE DOCTOR'S OFFICE

If lice are the suspected problem, the doctor will make a careful inspection to see if he or she can find nits or the lice themselves. Doctors almost always use gamma benzene hexachloride (Kwell) for lice. It may be somewhat more effective than the over-the-counter preparations. It is more expensive and has more side effects.

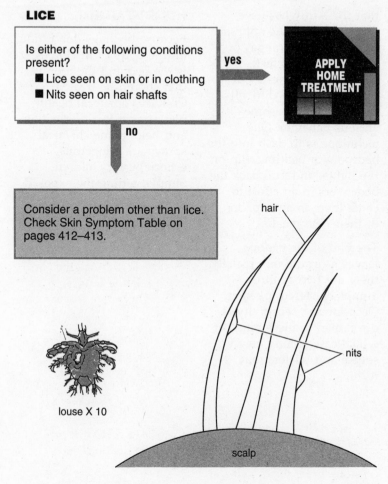

louse X 10

hair

nits

scalp

Bedbugs

Bedbug bites cause itchy red bumps. The adult bedbug is flat, wingless, oval in shape, reddish in color, and about one-quarter inch in length. Like lice, they stay alive by sucking blood. Unlike lice, they feed for only 10 or 15 minutes at a time and spend the rest of the time hiding in crevices and crannies.

Bedbugs feed almost entirely at night because that is when bodies are in bed and because of a real aversion to light. They have such a keen sense of the nearness of a warm body that the Army has used them to detect the approach of an enemy at ranges of several hundred feet! Catching these pests out in the open is very difficult and may require some curious behavior. One technique is to dash into the bedroom at bedtime, flip on the lights, and pull back the bedcovers in an effort to catch them in anticipation of their next meal.

The bite of the bedbug leaves a firm bump; usually, there are two or three bumps clustered together. Occasionally, sensitivity is developed to these bites, in which case itching may be severe and blisters may form.

HOME TREATMENT

Because bedbugs don't hide on the body or in clothes, it is the bed and the room that should be treated. Contact your local health department for information and help. Some chemical sprays are dangerous, especially for children. Simply getting the exposed bedding outdoors and exposed to sun and air for several days works, too.

BEDBUGS

Have bedbugs been seen on or near bed?

yes → APPLY HOME TREATMENT

no ↓

Consider a problem other than bedbugs. Check Skin Symptom Table on pages 412–413.

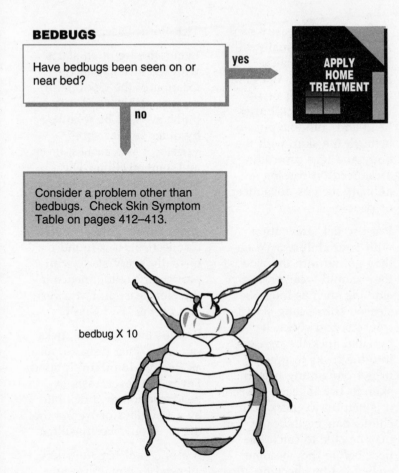

bedbug X 10

WHAT TO EXPECT AT THE DOCTOR'S OFFICE

The doctor will be hard-pressed to make a certain diagnosis of bedbug bites without information from you that bedbugs have been seen in the house. However, the bumps may be suggestive, and it may be decided to assume that the problem is bedbugs initially. If this is the case, treatment with an insecticide will be recommended (see Home Treatment).

Ticks

Outdoor living has its dangers. While bears, mountain lions, and vertical cliffs can usually be avoided, shrubs and tall grasses hide tiny insects eager for a blood meal from a passing animal or person. Ticks are among the most common of the small hazards. The tick achieved celebrity status with its appearance on the cover of *Newsweek*. Notoriety is a more appropriate term for the tick's role in transmitting an organism causing Lyme disease.

Ticks are rather easily seen, and a tick bite usually still has the obvious cause sticking out of it. Ticks are about one-quarter of an inch long. The tick buries its head and crab-like pincers beneath the skin, with the body and legs protruding. Ticks feed on passing animals such as dogs, deer, or people.

Practice tick prevention with your children. When they go out into the woods, they should wear protective clothing such as long-sleeved shirts, long slacks or trousers, and socks. If possible, tuck the trousers into the socks to prevent ticks from getting near the skin. In tick-infested areas, it is helpful to check your child's hair periodically. You may be able to catch the ticks before they become embedded by checking after hikes.

Tick-Borne Diseases

Lyme disease often starts with red swelling at the location of the tick bite. It may resemble a red ring or bull's eye and be followed by other similar spots, a variety of skin rashes, fever, and joint swelling. The nervous system and heart may also become involved. Lyme disease is no longer confined to Lyme, Connecticut; it is found in virtually every state, with particular predominance in the Northeast and Midwest and on the West Coast.

Besides Lyme disease, ticks can carry other diseases, such as **Rocky Mountain Spotted Fever.** If a fever, rash, or headache follow a tick bite by a few days or weeks, the doctor should be consulted.

If a pregnant female tick is allowed to remain feeding for several days, under certain circumstances a peculiar condition called **tick paralysis** may develop. The female tick secretes a toxin that can cause temporary paralysis, clearing shortly after the tick is removed. This complication is quite rare and can happen only if the tick stays in place many days.

TICKS

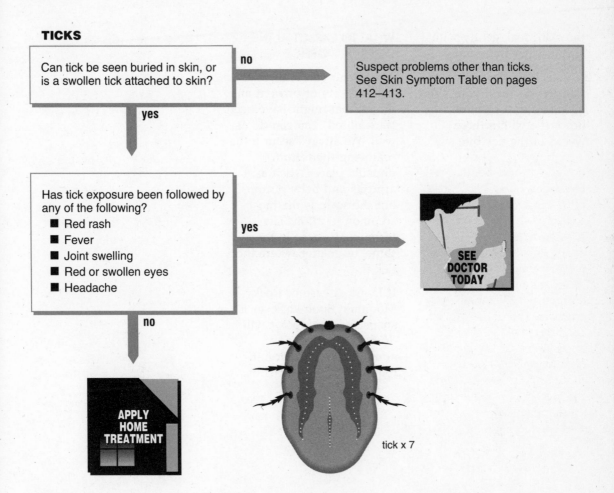

| Can tick be seen buried in skin, or is a swollen tick attached to skin? | **no** → | Suspect problems other than ticks. See Skin Symptom Table on pages 412–413. |

yes ↓

| Has tick exposure been followed by any of the following?
■ Red rash
■ Fever
■ Joint swelling
■ Red or swollen eyes
■ Headache | **yes** → | **SEE DOCTOR TODAY** |

no ↓

APPLY HOME TREATMENT

tick x 7

HOME TREATMENT

Ticks should be removed, although they will eventually "fester" out and complications are unusual. The trick is to get the tick to "let go" with the pincers before removal and not to kill the tick before getting it out. If the mouth parts and pincers remain under the skin, healing may require several weeks.

Make the tick uncomfortable with its new home. Gentle heat from a heated paper clip, alcohol, acetone, or oil will cause the tick to wiggle its legs and begin to withdraw. Grasp the tick (with a tissue if you're squeamish) and remove it quickly with a twisting motion. Tradition says to twist counterclockwise, but we seem to be able to get them out equally well in either direction. If the head is inadvertently left under the skin, soak gently with warm water twice daily until healing is complete.

Call the doctor if your child gets a fever, a rash, or a headache within three weeks of the tick bite.

WHAT TO EXPECT AT THE DOCTOR'S OFFICE

The doctor can remove the tick but cannot prevent any illness that might have been transmitted. You can do as well. We always seem to be removing them from unusual places, such as armpits and belly buttons, but the scalp is the most common location. The technique is exactly the same, no matter where the tick is.

If Lyme disease or Rocky Mountain Spotted Fever is suspected, the doctor will do blood tests and administer antibiotics, if required.

Chiggers

Chiggers are small red mites, sometimes called "redbugs." Their bite contains a chemical that eats away at the skin, causing a tremendous itch. Usually, the small, red sores are around the belt line or other openings in clothes. Careful inspection may reveal the tiny red larvae in the center of the itching sore. They also live on grasses and shrubs.

HOME TREATMENT

Chiggers are better avoided than treated. Using insect repellents, wearing appropriate clothing, and bathing after exposure help cut down on the frequency of bites. Once you get them, they itch, often for several weeks. Keep the sores clean and soak with warm water twice daily. A200 and RID applied the first few days will help kill the larvae, but the itch will persist.

WHAT TO EXPECT AT THE DOCTOR'S OFFICE

Doctors will usually prescribe Kwell for chiggers, which is perhaps slightly more effective than A200 or RID, but does not stop the itching either. Antihistamines make the child drowsy and are not frequently used unless intense itching persists despite home treatment with aspirin, warm baths, oatmeal soaks, and calamine lotion.

CHIGGERS

Are any of the following present?
- Itching red sores around belt line or other opening in clothes
- Itching red sores following contact with grass or shrubs
- Small, red mites seen on skin or red spot in center of sore

no → Suspect problems other than chiggers. See Skin Symptom Table on pages 412–413.

yes ↓

APPLY HOME TREATMENT

Scabies

Scabies is an irritation of the skin caused by a tiny mite related to the chigger. The itching is intense. No one knows why, but scabies seems to be on the rise in this country. As with lice, scabies is no longer related to hygiene. It occurs in the best of families and in the best of neighborhoods. The mite is easily spread from person to person or by contact with items, such as clothing and bedding, that may harbor the mite. Epidemics often spread through schools despite strict precautions against contacts with known cases.

Scabies affects children differently depending on their age:

- Young children—head, neck, shoulders, palms, and soles
- Older children—hands, wrists, and belly

The mite burrows into the skin to lay eggs. These burrows may be evident, especially at the beginning of the problem. However, the mite soon causes the skin to have a reaction to it so that redness, swelling, and blisters follow within a short period. Intense itching causes scratching so that there are plenty of scratch marks, and these may become infected from the bacteria on the skin. Thus, the telltale burrows are often obscured by scratch marks, blisters, and secondary infection.

If you can locate something that looks like a burrow, you might be able to see the mite with the aid of a magnifying lens. This is the only way to be absolutely sure that the problem is scabies, but it is often not possible. The diagnosis in an individual child is most often made on the basis of a problem that is consistent with scabies and the fact that scabies is known to be in the community.

HOME TREATMENT

There is no effective home remedy to eliminate the scabies mite. However, symptoms can be minimized.

For itching, we recommend cool soaks, calamine lotion, or aspirin. Chlorpheniramine (Chlor-Trimeton), an antihistamine, is available without prescription but often causes drowsiness. Follow the directions that come with the package. As in the case of poison ivy, warmth makes the itching worse by releasing histamine, but if all the histamine is released, then relief may be obtained for several hours. (See **Poison Ivy and Poison Oak**, page 428.)

It will take some time before the skin becomes normal, even with effective treatment, but at least some improvement should be noted within 72 hours. If this is not the case, then a visit to the doctor should be made.

SCABIES

Are the following conditions present?

- Intense itching
- Raised red skin in a line (represents a burrow) and possibly blisters or pustules
- Located on hands, especially between the fingers, elbow crease, armpit, groin crease, or behind the knees
- Exposure to scabies

no →

Consider a problem other than scabies. Check Skin Symptom Table on pages 412–413.

yes ↓

TUESDAY
See Doctor

MAKE APPOINTMENT WITH DOCTOR

WHAT TO EXPECT AT THE DOCTOR'S OFFICE

The doctor should examine the entire skin surface for signs of the problem and may examine the area with a magnifying lens in an attempt to identify the mite. He or she may scrape the lesion to examine under the microscope. Most of the time the doctor will be forced to make a decision based on the probability of various kinds of diseases and then treat much as you would at home. The proof of the pudding will be whether or not the treatment is successful.

Gamma benzene hexachloride (Kwell or Scabene) will most often be prescribed for older children. Because of the potency of this medication, it should not be used more than two times, a week apart. Kwell is usually effective in one treatment. The lotion is applied to the entire body below the neck, allowed to remain, and then thoroughly washed off.

Crotamiton creme (Eurax) is most often used for younger children. It is effective and does not have some of the potential side effects associated with gamma benzene hexachloride. Sulfur ointments are also used in infants. Hydroxyzine (Atarax, Vistaril) may be prescribed for intense itching.

Dandruff and Cradle Cap

Although they look somewhat different, cradle cap and dandruff are really part of the same problem; its medical term is **seborrhea**. It occurs when the oil glands in the skin have been stimulated by adult hormones, leading to oiliness and flaking of the scalp. This occurs in the infant because of exposure to the mother's hormones and in older children when they begin to make their own adult hormones. However, it does occur between these two ages; once a child has the problem, it tends to recur. Children will frequently have redness and scaling of the eyebrows and behind the ears as well.

Seborrhea itself is a somewhat ugly but relatively harmless condition. However, it may make the skin more susceptible to infection with yeast or bacteria.

Lookalike Problems

Occasionally, this condition is confused with **ringworm** of the scalp. Careful attention to the conditions listed in the decision chart will usually avoid this confusion. Remember also that ringworm would be unusual in the newborn and very young child.

Psoriasis often stops at the hairline. Furthermore, the scales of psoriasis are on top of raised lesions called "plaques," which is not the case in seborrhea. Psoriasis requires a doctor's help to remedy.

HOME TREATMENT

Shampoos

The heavily advertised anti-dandruff shampoos, such as Head and Shoulders, contain zinc pyrithione and are helpful in mild-to-moderate cases of dandruff. However, because of toxic effects on the nervous system demonstrated in animal studies, we advise against its use in infants and young children.

One of the most effective ingredients is selenium sulfide. Selsun (available by prescription only) and Selsun Blue are brand names of shampoos that contain selenium sulfide; Selsun Blue is available over the counter and, while weaker, is just as good if you apply more of it more frequently. When using these shampoos, it is important that directions be followed carefully, because oiliness and yellowish discoloration of the hair may occur with their use.

Sebutone and Pragmatar contain salicylic acid–sulfur combinations that are not quite as effective as the previously mentioned ingredients. Least effective are products that only contain tar.

DANDRUFF/CRADLE CAP

In an infant, are all of the following conditions present?
- Thick, adherent, oily, yellowish, scaling or crusting patches
- Located on the scalp, behind the ears, or (less frequently) in the skin creases of the groin
- Only mild redness in involved areas

no → Suspect problem other than cradle cap. Check Skin Symptom Table on pages 412–413.

In an older child (especially an adolescent), are all of the following conditions present?
- Fine, white, oily scales
- Confined to scalp and/or eyebrows
- Only *mild* redness in involved areas

no → Suspect problem other than dandruff. Check Skin Symptom Table on pages 412–413.

yes ↓

APPLY HOME TREATMENT

Cradle Cap

Cradle cap is best treated with a scrub brush. If the cradle cap is thick, then rub in warm baby oil, cover with a warm towel, and soak for 15 minutes. Use a fine-tooth comb or scrub brush to help remove the scale; then shampoo with Sebulex or other preparations listed above. Be careful to avoid getting shampoo in the child's eyes.

No matter what you do, the problem will often return, and you may have to repeat the treatment. If the problem gets worse despite home treatment over several weeks, see the doctor.

WHAT TO EXPECT AT THE DOCTOR'S OFFICE

Severe cases of seborrhea may require more than the medications given above; a cortisone cream is most often prescribed. Most often, a trip is made to the doctor in order to clear up some confusion concerning the diagnosis. The doctor usually makes the diagnosis on the basis of the appearance of the rash. Occasionally, scrapings from the involved areas will be looked at under the microscope. Drugs by mouth or by injection are not indicated for seborrhea unless bacterial infection has complicated the problem.

Patchy Loss of Skin Color

Children are constantly getting minor cuts, scrapes, insect bites, and other minor skin infections. During the healing process, it is common for the skin to lose some of its color. With time, the skin coloring generally returns.

Occasionally, **ringworm**, a fungal infection discussed on page 423, will begin as a small round area of scaling with associated loss of skin color.

In the summertime, many children are noticed to have small round spots on the face in which there is little color. The white spots have probably been present for some time, but the tanning of the skin does not occur in these areas. This condition is known as **pityriasis alba**; the cause is unknown, but it is a mild condition of cosmetic concern only. It may take many months to disappear and may recur, but are virtually never any long-term effects.

If there are lightly scaled, tan, pink, or white patches on the neck or back, the problem is most likely due to a fungal infection known as **tinea versicolor**. This is a very minor and superficial fungal infection.

HOME TREATMENT

Waiting is the most effective home treatment for loss of skin color. Tinea versicolor can be treated by applying Selsun Blue shampoo to the affected area, once every day or so until the lesions are gone. Unfortunately, this condition almost always comes back eventually.

WHAT TO EXPECT AT THE DOCTOR'S OFFICE

A history and careful examination of the skin will be performed. Scrapings of the lesions will be taken because tinea versicolor can be identified from these scrapings. Pityriasis alba should be distinguished from more severe fungal infections that may occur on the face. Again, scrapings will help identify the fungus.

LOSS OF SKIN COLOR

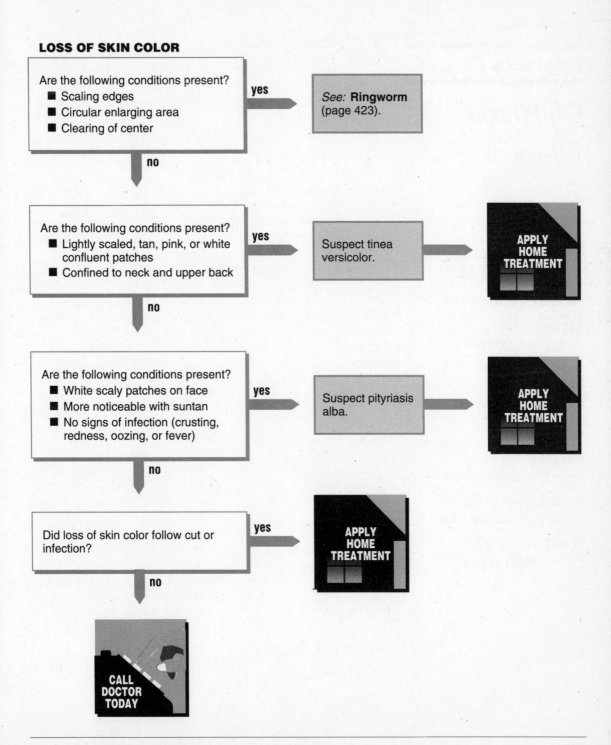

Are the following conditions present?
- Scaling edges
- Circular enlarging area
- Clearing of center

yes → *See:* **Ringworm** (page 423).

no

Are the following conditions present?
- Lightly scaled, tan, pink, or white confluent patches
- Confined to neck and upper back

yes → Suspect tinea versicolor. → APPLY HOME TREATMENT

no

Are the following conditions present?
- White scaly patches on face
- More noticeable with suntan
- No signs of infection (crusting, redness, oozing, or fever)

yes → Suspect pityriasis alba. → APPLY HOME TREATMENT

no

Did loss of skin color follow cut or infection?

yes → APPLY HOME TREATMENT

no

CALL DOCTOR TODAY

CHAPTER K

Childhood Diseases

Mumps

Mumps is a viral infection of the salivary glands. The major salivary glands are located directly below and in front of the ear. Before any swelling is noticeable, the child may have a low fever, complain of a headache or an earache, or experience weakness. Fever is variable; it may be only slightly above normal or as high as 104°F. After several days of these symptoms, one or both salivary glands (parotid glands) may swell.

It is sometimes difficult to distinguish mumps from swollen lymph glands in the neck; in mumps, you will not be able to feel the edge of your child's jaw that is located beneath the ear. With mumps, chewing and swallowing may produce pain behind the ear. Sour substances, such as lemons and pickles, may make the pain worse. When swelling occurs on both sides, children take on the appearance of chipmunks! Other salivary glands besides the parotid may be involved, including those under the jaw and tongue. The openings of these glands into the mouth may become red and puffy.

Approximately one-third of all patients who have mumps do not demonstrate any swelling of glands whatsoever. Therefore, many persons who are concerned about exposure to mumps will already have had the disease without realizing it.

Mumps is quite contagious during the period from 2 days before the first symptoms to the complete disappearance of the parotid gland swelling (usually about a week after the swelling has begun). Mumps will develop in a susceptible exposed person approximately 16 to 18 days after exposure to the virus. In children, it is generally a mild illness.

The decision chart is directed toward detection of the rare complications, which include encephalitis (viral infection of the brain), pancreatitis (viral infection of the pancreas), kidney disease, deafness, and involvement of the testicles or the ovaries. Complications are far more frequent in adults than they are in children.

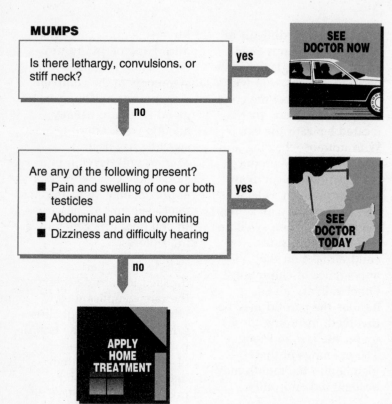

MUMPS

Is there lethargy, convulsions, or stiff neck?

yes → SEE DOCTOR NOW

no ↓

Are any of the following present?
- Pain and swelling of one or both testicles
- Abdominal pain and vomiting
- Dizziness and difficulty hearing

yes → SEE DOCTOR TODAY

no ↓

APPLY HOME TREATMENT

The rare complication of a right ovarian infection may be confused with appendicitis, and blood tests may be required.

Because mumps is a viral disease, there is no medicine that will directly kill the virus. Supportive measures may be necessary for some of the complications; fortunately, these occur rarely, and permanent damage to hearing or other functions is unusual. Mumps very rarely produces sterility in men or women, even with complicating involvement of the testes or ovaries. Mumps vaccination is discussed on page 203.

HOME TREATMENT

The pain may be reduced with acetaminophen. There may be difficulty eating, but adequate fluid intake is important. Sour foods should be avoided, including orange juice. Adults who have not had mumps should avoid exposure to the child until complete disappearance of the swelling. Many adults who do not recall having mumps as a child may have had an extremely mild case and consequently are not at risk of developing mumps.

WHAT TO EXPECT AT THE DOCTOR'S OFFICE

If a complication is suspected, then a visit to the doctor's office may be necessary. The history and physical examination will be directed at confirming the diagnosis or the presence of a complication.

Chicken Pox

It is valuable to know the signs of chicken pox because you can then avoid taking your child to the doctor for these symptoms. As soon as you recognize chicken pox, do *not* give your child aspirin during the course of the disease. Use acetaminophen instead.

How to Recognize Chicken Pox

Before the Rash. Usually there are no symptoms before the rash appears, but occasionally there is fatigue and mild fever 24 hours before the rash is noted.

The Rash. The typical rash goes through the following stages.

1. It may appear as flat red splotches.

2. They become raised and may resemble small pimples.

3. They develop into small blisters, called vesicles, which are very fragile. They may look like drops of water on a red base. The tops are easily scratched off.

4. As the vesicles break, the sores become "pustular" and form a crust. (The crust is made of dried serum, not true pus.) This stage may be reached within several hours of the first appearance of the rash. Itching is often severe in the pustular stage.

5. The crust falls away between the 9th and the 13th day. The vesicles tend to appear in crops, with two to four crops appearing within two to six days. All stages may be present in the same area. They often appear first on the scalp and the face and then spread to the rest of the body, but they may begin anywhere. They are most numerous over shoulders, chest, and back. They are occasionally found on the palms of the hands or the soles of the feet. They may also be found in the mouth or in the vagina. There may be only a few sores, or there may be hundreds.

Fever. After most of the sores have formed crusts, fever usually subsides.

How Chicken Pox Spreads

Chicken pox spreads very easily—over 90% of brothers and sisters catch it! It may be transmitted from 24 hours before the rash appears up to about 6 days after. It is spread by droplets from the mouth or throat or by direct contact with contaminated articles of clothing. It is not spread by dry scabs. The incubation period is from 14 to 17 days. Chicken pox leads to lifelong immunity with rare exceptions.

However, the same virus that causes chicken pox also causes **shingles**, and the individual with a history of chicken pox may develop shingles (herpes zoster) later in life.

Chicken pox can be a serious problem in children with cancer or those receiving drugs that affect the immune system. See page 205 for information about immunizations for chicken pox.

Most of the time, chicken pox should be treated at home. Complications are rare and far less common than with measles. The specific questions on the chart deal with complications that may require more than home treatment: encephalitis (viral infection of the brain), pneumonia, and severe bacterial infection of the lesions. Encephalitis and pneumonia are rare.

HOME TREATMENT

The major problem in dealing with chicken pox is control of the intense itching. Warm baths containing baking soda (one-half cup to a tubful of water) frequently help. Calamine lotion may temporarily be helpful. The use of antihistamines is sometimes necessary and may require contact with your doctor; check by phone before exposing other children in a doctor's office. **Caution:** *Aspirin should not be used, because of the associated risk of Reye's syndrome.* Use acetaminophen for fever.

Cut fingernails or use gloves to prevent scarring from scratching. When lesions occur in the mouth, gargling with salt water (one-half teaspoon salt to an eight-ounce glass) may help. Hands should be washed three times a day, and skin should be kept gently but scrupulously clean in order to prevent a complicating bacterial infection. Minor bacterial infection will respond to soap and time; if it becomes severe and results in return of fever, then see the doctor.

WHAT TO EXPECT AT THE DOCTOR'S OFFICE

Do not be surprised if the doctor is willing and even anxious to treat the case "over the phone." If it is necessary to bring the child to the doctor's office, then attempts should be made to keep him or her separate from the other children. In healthy children, chicken pox has few lasting ill effects; but in children with other serious illnesses, it can be a devastating or even fatal disease. A visit to the doctor's office may not be necessary unless a complication seems possible.

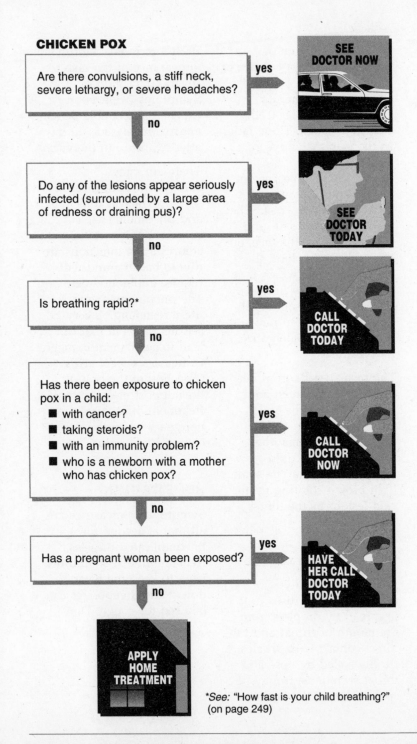

CHICKEN POX

Are there convulsions, a stiff neck, severe lethargy, or severe headaches?

yes → **SEE DOCTOR NOW**

no ↓

Do any of the lesions appear seriously infected (surrounded by a large area of redness or draining pus)?

yes → **SEE DOCTOR TODAY**

no ↓

Is breathing rapid?*

yes → **CALL DOCTOR TODAY**

no ↓

Has there been exposure to chicken pox in a child:
- with cancer?
- taking steroids?
- with an immunity problem?
- who is a newborn with a mother who has chicken pox?

yes → **CALL DOCTOR NOW**

no ↓

Has a pregnant woman been exposed?

yes → **HAVE HER CALL DOCTOR TODAY**

no ↓

APPLY HOME TREATMENT

*See: "How fast is your child breathing?" (on page 249)

Measles

This type of measles is also called "red measles," "seven-day measles," or "ten-day measles," as opposed to German, or three-day, measles (see page 468). It is a preventable disease. Unlike some of the, other childhood illnesses, measles can be quite severe. The results of over two decades of immunizing children are remarkable, with a reduction of 99% of cases. Make sure your child is immunized against measles.

How to Recognize Measles

Early Signs. Measles is a viral illness that begins with fever, weakness, a dry "brassy" cough, and inflamed eyes that are itchy, red, and sensitive to the light. These symptoms begin three to five days before the appearance of the rash.

Another early sign of measles is the appearance of fine white spots on a red base inside the mouth opposite the molar teeth (Koplik's spots). These fade as the skin rash appears.

Rash. The rash begins on about the fifth day as a pink, blotchy, flat rash. The rash first appears around the hairline, on the face, on the neck, and behind the ears. The spots, which fade on pressure early in the illness, become somewhat darker and tend to merge into larger red patches as they mature.

The rash spreads from head to chest to abdomen and finally to the arms and legs. It lasts from four to seven days and may be accompanied by mild itching. There may be some light brown coloring to the skin lesions as the illness progresses.

How Measles Spread

Measles is a highly contagious viral disease. It is spread by droplets from the mouth or throat and by direct contact with articles freshly soiled by nose and throat secretions. It may be spread during the period from 3 to 6 days before the appearance of the rash to several days after. Symptoms begin in an exposed susceptible person approximately 8 to 12 days after exposure to the virus.

There are many complications of measles; sore throats, ear infections, and pneumonia are all common. Many of these complicating infections are due to bacteria and will require antibiotic treatment. The pneumonias can be life-threatening. A very serious problem that can lead to permanent damage is **measles encephalitis** (infection of the brain); life-support measures and treatment of seizures are necessary when this rare complication occurs.

HOME TREATMENT

Symptomatic measures are all that is needed for uncomplicated measles. Acetaminophen should be used to keep the fever down, and a vaporizer can be used for cough. Dim

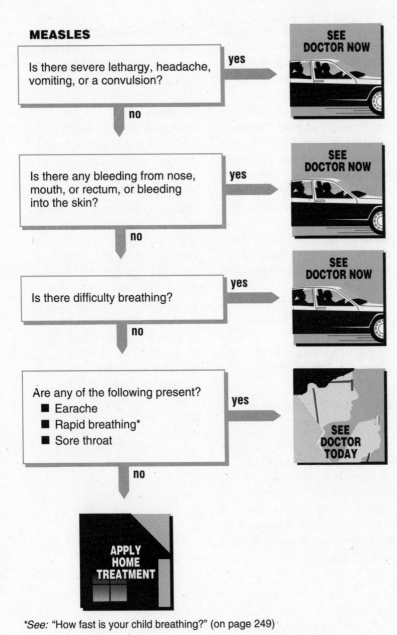

MEASLES

Is there severe lethargy, headache, vomiting, or a convulsion?

yes → **SEE DOCTOR NOW**

no ↓

Is there any bleeding from nose, mouth, or rectum, or bleeding into the skin?

yes → **SEE DOCTOR NOW**

no ↓

Is there difficulty breathing?

yes → **SEE DOCTOR NOW**

no ↓

Are any of the following present?
- Earache
- Rapid breathing*
- Sore throat

yes → **SEE DOCTOR TODAY**

no ↓

APPLY HOME TREATMENT

*See: "How fast is your child breathing?" (on page 249)

lighting in the room often makes the child feel more comfortable because of the eye's sensitivity to light. In general, the child feels "measley." He or she should be isolated until the end of the contagious period. All unimmunized children in contact should be brought to be immunized immediately after symptoms begin in one child.

WHAT TO EXPECT AT THE DOCTOR'S OFFICE

The history and physical examination will be directed at determining the diagnosis of measles and the nature of any complications. Bacterial complications, such as ear infections and pneumonia, can usually be treated with antibiotics. The child with symptoms suggestive of encephalitis (lethargy, stiff neck, convulsions) will be hospitalized, and a spinal tap will be performed. Very rarely, there may be a problem with blood clotting so that bleeding occurs, usually first apparent as dark purple splotches in the skin. It is best to avoid all of these problems with measles immunization, which is discussed on page 199.

Rubella (German Measles)

Rubella is also known as "German measles" and "three-day measles." See page 466 for advice on measles.

How to recognize Rubella

Before the Rash. There may be a few days of mild fatigue. Lymph nodes at the back of the neck may be enlarged and tender.

Rash. The rash first appears on the face as flat or slightly raised red spots. It quickly spreads to the trunk and the extremities, and the discrete spots tend to merge into large patches. The rash of rubella is highly variable and is difficult for even the most experienced parents and doctors to recognize. Often, there is *no* rash.

Fever. The fever rarely goes above 101°F and usually lasts less than two days.

Pain. Joint pains occur in about 10 to 15% of older children and adolescents. The pains usually begin on the third day of illness.

How Rubella Spreads

German measles is a mild virus infection that is not as contagious as measles or chicken pox. It is usually spread by droplets from the mouth or throat. The incubation period is from 12 to 21 days with an average of 16 days.

The specific questions on the chart are addressed to possible complications, which are extremely rare. The main concern with German measles is an infection in an unborn child. If three-day measles occurs during the first month of pregnancy, there is a 50% chance that the fetus will develop an abnormality such as cataracts, heart disease, deafness, or mental deficiency. By the third month of pregnancy, this risk decreases to less than 10%, and it continues to decrease throughout the pregnancy. Because of the problem of congenital defects, a vaccine for German measles has been developed; rubella immunization is discussed on page 200.

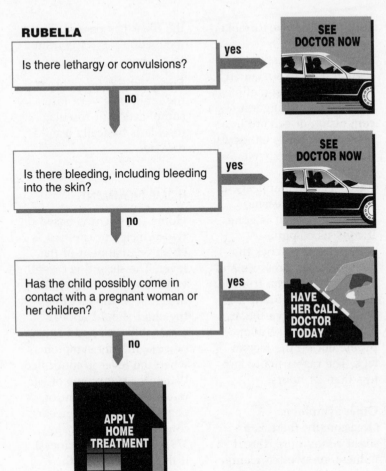

RUBELLA

Is there lethargy or convulsions? — **yes** → **SEE DOCTOR NOW**

no ↓

Is there bleeding, including bleeding into the skin? — **yes** → **SEE DOCTOR NOW**

no ↓

Has the child possibly come in contact with a pregnant woman or her children? — **yes** → **HAVE HER CALL DOCTOR TODAY**

no ↓

APPLY HOME TREATMENT

HOME TREATMENT

Usually no therapy is required. Occasionally fever will require the use of aspirin or acetaminophen. Isolation is usually not imposed.

Avoid any exposure of the child to women who could possibly be pregnant. If a question of such exposure arises, the pregnant woman should discuss the risk with her doctor. Blood tests are available that will indicate whether a pregnant woman has had rubella in the past and is immune, or whether problems with the pregnancy might be encountered.

WHAT TO EXPECT AT THE DOCTOR'S OFFICE

Visits to the doctor's office are seldom required for uncomplicated German measles. Questions about possible infection of pregnant women are more easily and economically discussed over the telephone.

Roseola

Roseola is most common in children under the age of three but may occur at any age. Its main significance lies in the sudden high fever, which may cause a convulsion. Such a convulsion is due to the high temperature and does not indicate that the child has epilepsy. Prompt treatment of the fever is essential (see "Fever," page 299).

How to Recognize Roseola

Fever. There usually are several days of sustained high fever, and sometimes this fever can trigger a convulsion or seizure in a susceptible child. Otherwise, the child appears well.

Rash. The rash appears as the fever is decreasing or shortly after it is gone. It consists of pink, well-defined patches that turn white on pressure and first appear on the trunk. It may be slightly bumpy. It spreads to involve the arms and neck but is seldom prominent on the face or legs. The rash usually lasts less than 24 hours.

Other Symptoms. Occasionally, there is a slight runny nose, throat redness, or swollen glands at the back of the head, behind the ears, or in the neck. Most often, there are no other symptoms.

This disease has been found to be caused by human herpes virus 6. Consequently, this condition, long known as **exanthem subitum**, has become sixth disease. It is contagious. The child should be kept from contact with other children until the fever has passed. The incubation period is from 7 to 17 days.

Encephalitis (infection of the brain) is a very rare complication of roseola; roseola is basically a mild disease.

HOME TREATMENT

Home treatment is based on two principles. The first is effective treatment of the fever. The second is careful watching and waiting. This depends on the fact that the child should appear reasonably well and have no other significant symptoms when the fever is controlled. Watch for symptoms of ear infection (a complaint of pain or tugging at the ear), cough (see **Cough**, page 379), or lethargy (laziness). If these occur, then the appropriate sections of this book should be consulted. If the problem still is not clear, a phone call to the doctor may be necessary.

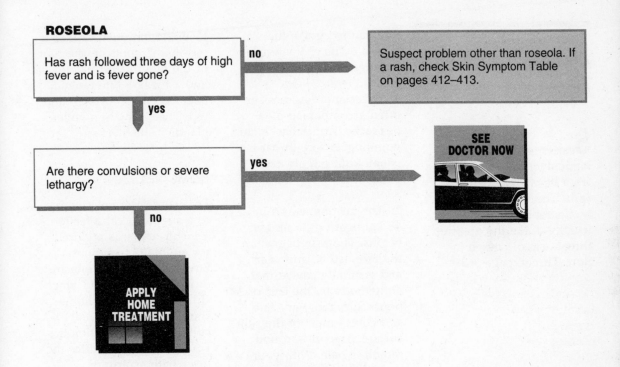

ROSEOLA

Has rash followed three days of high fever and is fever gone? — **no** → Suspect problem other than roseola. If a rash, check Skin Symptom Table on pages 412–413.

yes

Are there convulsions or severe lethargy? — **yes** → SEE DOCTOR NOW

no

APPLY HOME TREATMENT

Remember that roseola should not last more than four or five days; you should consult your doctor about a persistent problem.

WHAT TO EXPECT AT THE DOCTOR'S OFFICE

Children are usually seen soon after the onset of the illness because of the high fever. As noted, at this stage there is little else to be found in roseola. The ears, nose, throat, and chest should be examined. If the fever remains the only finding, then the doctor will recommend home treatment (control of the fever with careful waiting and watching to see if the rash of roseola will appear). There is no medical treatment for roseola other than that available at home.

Scarlet Fever

Scarlet fever derived its name over 300 years ago from its characteristic red rash. The illness is caused by a streptococcal infection, usually of the throat. Strep throats are discussed in **Sore Throat**, page 363.

You can recognize the illness by its characteristic features.

1. Fever and weakness are often accompanied by a headache, stomachache, and vomiting. A sore throat is usually but not always present.

2. The rash appears 12 to 48 hours after the illness begins. The rash begins on the face, trunk, and arms and generally covers the entire body by the end of 24 hours. It is red, very fine, and covers most of the skin surface. The area around the mouth is pale. With your eyes closed, it has the feeling of fine sandpaper. Skin creases, such as in front of the elbow and the armpit, are more deeply red. Pressing on the rash will produce a white spot lasting several seconds.

3. The intense redness of the rash lasts for about five days, although peeling of skin can go on for weeks. It is not unusual for peeling, especially of the palms, to last for more than a month.

Examination often reveals a red throat, spots on the roof of the mouth (soft palate), and a fuzzy white tongue, later becoming swollen and red. There may be swollen glands in the neck.

HOME TREATMENT

Because scarlet fever is due to a streptococcal infection, a medical visit is required for antibiotic treatment. Streptococcal infections are quite contagious, and other children within the home should also have throat cultures. Besides antibiotics, you will wish to reduce the fever with aspirin or acetaminophen, keep up with fluid requirements, and give plenty of cold liquids to help soothe the throat.

SCARLET FEVER

Are both of the following present?
- Fever
- Fine, red rash on trunk and extremities that feels like sandpaper

no →

Suspect a problem other than scarlet fever. If there is a skin rash, check the Skin Symptom Table on pages 412–413.

yes

SEE DOCTOR TODAY

WHAT TO EXPECT AT THE DOCTOR'S OFFICE

There are several rashes that can be confused with scarlet fever, including those of measles and drug reactions. If the rash is sufficiently typical of scarlet fever, the doctor will probably begin antibiotics, usually penicillin (or erythromycin if the child is allergic to penicillin), and take throat cultures from the rest of the family. If the doctor is uncertain of the cause of the rash, a throat culture may be taken before beginning treatment. Treatment that is delayed by a day or two while waiting for culture results will still prevent the complication about which we are most concerned: rheumatic fever.

Fifth Disease

Consider the strange case of the fifth disease, whose only claim to fame is that it might be mistaken for another disease. It is so named because it was always listed last among the five very common contagious rashes of childhood. Its medical name, *erythema infectiosum*, is easily forgotten. The responsible agent for this illness has recently been identified as a virus, parvovirus B19.

Fifth disease comes very close to not being a disease at all for children. It has no symptoms other than rash, has few complications, and needs no treatment. Rare complications include joint pains and nervous system involvement.

The disease causes a characteristic "slapped cheek" appearance in children. The rash often begins on the cheeks and is later found on the backs of the arms and legs. It often has a very fine, lacy, pink appearance. It tends to come and go and may be present one moment and absent the next. It is prone to recur for days or even weeks, especially as a response to heat (warm bath or shower) or irritation. In general, however, the rash around the face will fade within 4 days of its appearance, and the rash on the rest of the body within 3 to 7 days of its appearance.

While the discovery of a virus responsible for roseola has caused it to be named sixth disease, it is for another serious reason we changed the subheading for fifth disease (page 474) to "No longer last or least." Pregnant women exposed to parvovirus B19 can have serious problems. Call your physician immediately if you are exposed to this disease while pregnant.

This rash could worry you or cause you to make an avoidable trip to the doctor's office. It is very contagious; epidemics of fifth disease have resulted in unnecessary school closings. The incubation period is usually from 4 to 14 days.

HOME TREATMENT

There is none. Just watch and wait to make sure you are dealing with fifth disease. Check that there is no fever; fever is very unusual with fifth disease. No restrictions on activities are necessary.

FIFTH DISEASE

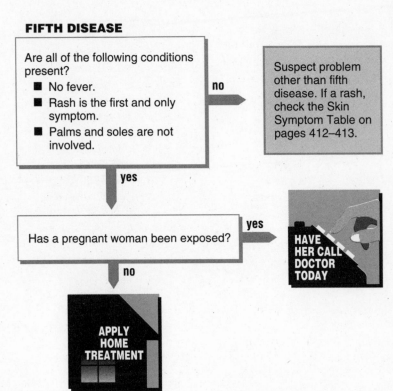

Are all of the following conditions present?
- No fever.
- Rash is the first and only symptom.
- Palms and soles are not involved.

no → Suspect problem other than fifth disease. If a rash, check the Skin Symptom Table on pages 412–413.

yes ↓

Has a pregnant woman been exposed?

yes → HAVE HER CALL DOCTOR TODAY

no ↓

APPLY HOME TREATMENT

WHAT TO EXPECT AT THE DOCTOR'S OFFICE

The doctor may be able to distinguish fifth disease from other rashes. If the rash fits the description given in this section, the doctor is going to make the same diagnosis that you might have. Taking the child's temperature and looking at the rash can be expected. Because there are no tests for the known cause, laboratory tests are unlikely. Waiting and watching is the means of dealing with fifth disease.

CHAPTER L

Bones, Muscles, and Joints

Pain in the Limbs, Muscles, or Joints

Arm and leg pains are quite common in school-age children. Physical exertion can stress bones, joints, and muscles and result in pain. Muscle aches and pains are common after strenuous exercise or during mild viral illness. Severe pain of sudden onset often signifies a muscle cramp.

"Growing pains" are found in the mid-portion of the upper and lower legs, generally at night, in children from 6 to 12 years of age. The cause is unknown, but millions of children have had them, and they are of absolutely no medical significance.

Joint Pains

In children, very seldom does pain in the joint mean arthritis. The word **arthritis** comes from "arth," meaning "joint," and "itis," denoting that the joint is red, warm, swollen, and painful to move. The word **arthralgia** means pain in the joint *without* redness, warmth, or swelling. Because injuries are so common in children, they are the most common cause of joint pain and are discussed under the specific injury.

Most true arthritis in children is caused by infection. Occasionally, arthralgia may precede the arthritis by a day or two. This is often true in **gonococcal arthritis**, which may occur in adolescents.

Arthritis accompanied by a fever occurs in **rheumatic fever**. Arthritis accompanied by abdominal pain and a blotchy, dark rash most prominent on the legs can be part of a rare illness in children that temporarily affects small blood vessels. Arthritis may also occur in children with **sickle-cell disease**. In addition, arthritis somewhat similar to the rheumatoid arthritis of adults occurs occasionally in children.

Arthralgia (pain without swelling) is a common temporary complaint in many viral illnesses. It can occur with **German measles** (rubella) and may be accompanied by swollen lymph glands behind the ears as well as the rash. If rubella is suspected, special attention must be taken to prevent exposure of pregnant women.

A cause of knee pain in active adolescents is **Osgood–Schlatter's disease**. This problem is caused by a strong set of thigh muscles pulling at their insertion on the tibial bone in the knee. Resting the leg is the treatment of choice. Some diseases of children and adolescents that cause pain in the knees actually may have their origins in the hips. Persistent arthralgia should therefore be attended to by a doctor.

HOME TREATMENT

Heat will often relax sore muscles caused by strenuous exercise. Stretching exercises may prevent cramps, especially in the calves. Acetaminophen is often effective in relieving the muscle pain of viral illness. Ibuprofen also has a role (see page 232). Aspirin should not be given to relieve the muscle ache of influenza because of the association with Reye's syndrome. Aspirin is effective in muscle and joint injuries.

A combination of the above remedies may be of comfort in severe "growing pains." Home treatments for various injuries are discussed in those particular sections. All arthritis in children must be seen by a doctor immediately because of the potential serious consequences of the underlying infections. In the arthralgia that accompanies viral illness such as rubella, or that follows rubella immunization, aspirin may afford relief of pain.

WHAT TO EXPECT AT THE DOCTOR'S OFFICE

A history and physical examination of the joints will be made. A more complete examination will depend on the specific history. Depending on the nature of the condition, blood tests and X-rays will often be obtained. If a joint contains fluid, the fluid may be removed and tested. Treatment of serious infections will very often require hospital administration of antibiotics.

PAIN IN LIMBS

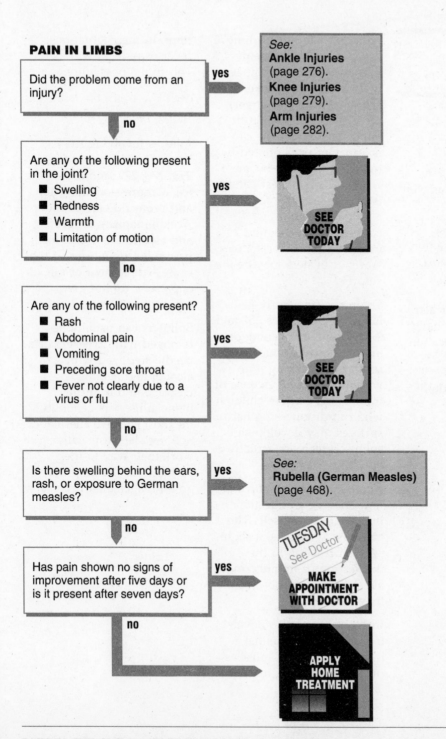

Did the problem come from an injury? — **yes** →

See:
Ankle Injuries (page 276).
Knee Injuries (page 279).
Arm Injuries (page 282).

↓ **no**

Are any of the following present in the joint?
- Swelling
- Redness
- Warmth
- Limitation of motion

yes → **SEE DOCTOR TODAY**

↓ **no**

Are any of the following present?
- Rash
- Abdominal pain
- Vomiting
- Preceding sore throat
- Fever not clearly due to a virus or flu

yes → **SEE DOCTOR TODAY**

↓ **no**

Is there swelling behind the ears, rash, or exposure to German measles? — **yes** →

See:
Rubella (German Measles) (page 468).

↓ **no**

Has pain shown no signs of improvement after five days or is it present after seven days? — **yes** →

TUESDAY See Doctor
MAKE APPOINTMENT WITH DOCTOR

↓ **no**

APPLY HOME TREATMENT

Limp

Many children will experience periods when they are not walking as well as usual.

When children are first learning to walk, there may be several setbacks. Most children will improve steadily; but on rare occasions, a child particularly favoring one leg may have a congenital hip dislocation that was not detected during routine child health-supervision visits.

Although the initial glimpse of an offspring suddenly limping conjures up images of Tiny Tim (the Dickens character—not the singer), there is seldom reason to panic. Sometimes the cause is apparent or is offered by the child ("Megan bit me"). Usually, some investigating is warranted. Bare feet have a knack for acquiring splinters and becoming irritated from other sources. Warts or blisters may be the culprits.

Trauma is frequently involved (see pages 252 and 273). Often, an evening of jumping off the couch may result in a morning limp without obvious evidence of trauma. Very active children who run distances on hard surfaces may develop shin splints, which are actually very tiny fractures.

Another common source of limp may be an injection into the thigh muscle. The muscle may develop a slight swollen reaction to the medication or immunizing agent. Rarely, an abscess may develop.

Infections may also develop in bones (osteomyelitis), joints (arthritis) or muscles (myositis), leading to a limp. In these situations, fever or redness will alert you to the possibility.

HOME TREATMENT

Injuries involving the leg have been discussed (pages 273, 276, 279, and 477). Rest, ice, compression, elevation, and protection are in order. Acetaminophen, ibuprofen, and aspirin are useful for pain relief in injuries. If the cause is unknown or a fever is present, do not give aspirin to children.

Splinters can usually be removed with tweezers or a needle sterilized with alcohol and heating. Distraction during this home surgery is generally the most effective anesthetic. Splinters left in usually work their way to the surface, but the limp may go on longer than necessary.

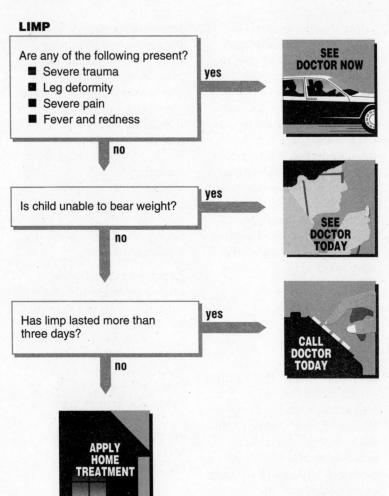

LIMP

Are any of the following present?
- ■ Severe trauma
- ■ Leg deformity
- ■ Severe pain
- ■ Fever and redness

yes → SEE DOCTOR NOW

no ↓

Is child unable to bear weight?

yes → SEE DOCTOR TODAY

no ↓

Has limp lasted more than three days?

yes → CALL DOCTOR TODAY

no ↓

APPLY HOME TREATMENT

WHAT TO EXPECT AT THE DOCTOR'S OFFICE

A careful history and examination will provide probable diagnoses for the majority of limps. A good part of the visit will be spent on inspecting, prodding, and rotating your child's feet, ankles, legs, knees, thighs, and hips. X-rays are often ordered to detect possible fractures or dislocations.

Blood tests are possible if fever or redness accompanies the limp. In certain instances, a joint (or bone) may have a needle placed inside to remove fluid in order to evaluate a potentially serious infection.

Low-Back Pain

Low-back pain is uncommon before adolescence. If a young child complains of back pain and there is accompanying fever or pain on urination, a **urinary tract infection** is often to blame. **Falls** on the tailbone are another fairly frequent cause of this symptom. Although children's bones are far more resilient than those of adults, spinal injuries do occur; persistent pain or weakness in the lower extremities should alert parents to this possibility. Weakness without pain is unusual. However, if it is clearly present, see the doctor.

Some developmental **disorders of the hip** may present themselves first as back complaints; often these children will have a limp or other problems as well. Another cause of back complaints is a developmental (congenital) **abnormality of the spine**. A spine that appears crooked either in a side-to-side or back-to-front direction should be discussed with a doctor during a regularly scheduled visit.

In Adolescents

Low-back pain in adolescents is similar to that in adults and usually results from muscular strain. The strain causes spasm of the supportive muscles alongside the spine. Any injury to the back may produce such spasms, resulting in severe pain and stiffness. Pain may be immediate or may occur some hours after the exertion or injury; often the cause is not clear.

Muscular problems of the back due to exertion or lifting must heal naturally; give them time. The pain due to muscular back strain is usually in the low back; if it extends beyond the low-back area, a more serious problem may exist. Pain that travels down one leg (**sciatica**) suggests pressure on the nerves as they leave the spinal cord. Such complications require medical attention.

Backache may also be caused by the daily strain of supporting an obese body, by the rupture of small fat sacs, or by the loose ligaments following rapid weight reduction. Obesity is not good for the back. Low-back ache is common during the menstrual period. This is discussed further in **Menstrual Problems**, page 532.

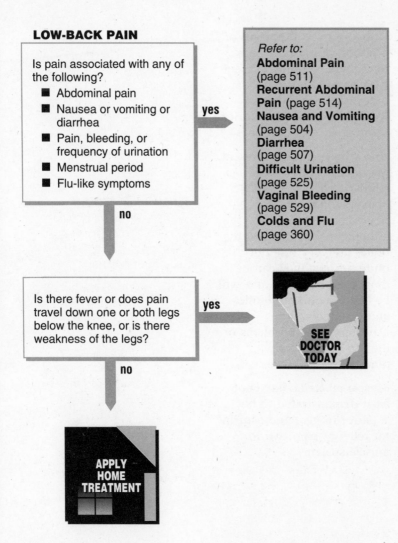

LOW-BACK PAIN

Is pain associated with any of the following?
- Abdominal pain
- Nausea or vomiting or diarrhea
- Pain, bleeding, or frequency of urination
- Menstrual period
- Flu-like symptoms

yes →

Refer to:
Abdominal Pain (page 511)
Recurrent Abdominal Pain (page 514)
Nausea and Vomiting (page 504)
Diarrhea (page 507)
Difficult Urination (page 525)
Vaginal Bleeding (page 529)
Colds and Flu (page 360)

no ↓

Is there fever or does pain travel down one or both legs below the knee, or is there weakness of the legs?

yes →

SEE DOCTOR TODAY

no ↓

APPLY HOME TREATMENT

Severe muscle spasm pain usually lasts for 48 to 72 hours and is followed by days or weeks of less severe pain. Complete recovery will generally take six weeks; strenuous activity during that six weeks can bring the problem back and delay complete recovery.

Slow improvement is the rule with back pain. But if there is no improvement within 48 hours, see the doctor.

After healing, an exercise program can help prevent re-injury. No drugs will hasten healing; they only reduce symptoms. The child should sleep pillowless on a very firm mattress, with a bedboard under the mattress. Some individuals report being most comfortable on the floor. A folded towel beneath the low back may increase comfort.

Heat applied to the affected area will provide some relief. Acetaminophen, ibuprofen (for children over 12), or two aspirin every four hours (in children over 10) may be continued for as long as there is significant pain.

HOME TREATMENT

When a muscle strain is present, resting the injured part will help healing; the spasm of the muscle itself helps rest the part. When the pain first appears, having the child rest flat on his or her back for at least 24 hours may help. After the complete bed rest, a gradual increase in activity, carefully avoiding re-injury, is begun.

When standing or sitting, at least one foot can be elevated and placed on an adjacent chair with the knee flexed. This position helps straighten the lower back and increase comfort.

WHAT TO EXPECT AT THE DOCTOR'S OFFICE

The physical examination will be directed toward determining whether the problem has its origins in the urinary tract, the skeleton, or the muscular system. The examination will center on the back, the abdomen, the arms, and legs, with special attention to testing the nerve function of the legs.

If the injury is the result of a fall or blow to the back, or if there is a limp present, X-rays are indicated; otherwise, usually not. X-rays do not show injuries to muscles, only to bones.

If low-back strain is present, the doctor's advice will be similar to that described above. Urinary tract infections will require antibiotics. Infections of the bones are very rare but quite serious; they can be diagnosed by X-ray and will be treated with antibiotics and hospitalization. Developmental problems of the hip or spine may require special braces.

Muscle relaxants have not been demonstrated to be superior to heat and aspirin for relief of pain due to muscle spasm.

Neck Pain

A variety of problems can be responsible for children having pain or difficulty moving their necks.

Newborn infants often have considerable pressure applied to their necks during delivery. A small amount of bleeding in one of the neck muscles can produce both swelling and difficulty moving the neck; this may not become apparent in a baby for several weeks.

Other types of muscle problems may affect older children, especially spasm that may follow either an injury or occur spontaneously. These neck problems are known as **torticollis**, or wry neck.

Infection

Often, children will complain of neck pain or refuse to move their neck because of an infection. Pneumonia or pharyngitis (sore throat) are two of the more common infections that produce this symptom. A swelling that feels like a lima bean represents a swollen lymph node and usually signals an infection. On occasion, inability to move the neck represents a very serious infection, meningitis.

If your child experiences difficulty touching chin to chest or is irritable, drowsy, or vomiting, an immediate doctor visit is necessary. Unwillingness to move the neck may be a protective reaction to a rare infection causing swelling of the opening portion of the trachea (**epiglottitis**). Difficulty breathing in and drooling are other symptoms of this problem that merit immediate medical attention.

HOME TREATMENT

For minor aches and muscle spasms, heat can be soothing and pain relievers such as acetaminophen, ibuprofen, or aspirin quite helpful. If your infant has developed torticollis, you will probably receive instructions from your doctor on gentle stretching exercises.

WHAT TO EXPECT AT THE DOCTOR'S OFFICE

After careful examination of the neck muscles and movement of the head, your doctor's course of action will depend on whether muscle spasm, trauma, or infection seems most likely. If trauma is suspected, X-rays of the neck, collarbones, and shoulder may be taken. Depending on the location of the suspected infection, a throat culture, chest X-ray, or evaluation of swollen lymph glands (see **Swollen Glands**, page 386) may be done.

NECK PAIN

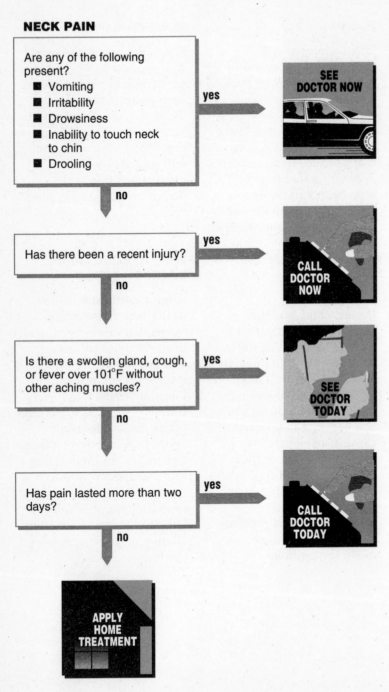

Are any of the following present?
- Vomiting
- Irritability
- Drowsiness
- Inability to touch neck to chin
- Drooling

yes → SEE DOCTOR NOW

no

Has there been a recent injury?

yes → CALL DOCTOR NOW

no

Is there a swollen gland, cough, or fever over 101°F without other aching muscles?

yes → SEE DOCTOR TODAY

no

Has pain lasted more than two days?

yes → CALL DOCTOR TODAY

no

APPLY HOME TREATMENT

If meningitis is suspected, your doctor will need to immediately do a lumbar puncture (spinal tap). While this is a terrifying procedure to contemplate, it is extraordinary how low the risks are; but, more important, it is a crucial procedure that must be done without delay if meningitis is suspected.

Bowlegs and Knock-Knees

These problems are quite common and almost never need treatment. Practically all normal infants and toddlers have some bowing of the legs. The bowing is worst at about one year of age and usually disappears by age two. The strengthening of the leg muscles that occurs in the first year of walking appears to be responsible for correcting the leg bowing. Bowed legs are often more apparent than real and may disappear merely by placing the ankles together.

Knock-knee usually develops a year or so later. It appears the worst at about age three and is often accompanied by pigeon toeing, which is actually a compensating balancing response. Knock-knee tends to correct itself without any treatment by age six.

Many years ago, vitamin D deficiency (**rickets**) sometimes caused bowleg, but this is virtually never the case in an adequately nourished child in this country today. In some parts of the eastern United States, a hereditary form of rickets still occurs.

Rarely, there are cases of bowleg that require medical treatment, but in these the bowing is severe and tends to get even worse as time goes by. Knock-knee that requires medical treatment is also rare. A proper decision to treat is made after careful measurements indicate that the problem is getting worse and that the deformity is more than mild. This problem can almost always be discussed incidental to another examination and seldom requires a separate appointment with the doctor.

HOME TREATMENT

Watching is usually all that is required. Remember that bowleg tends to be most apparent during the first year of life and tends to resolve thereafter. Adequate walking and other exercise that will strengthen the leg muscles are important for the overall health of your child, as well as for proper leg development.

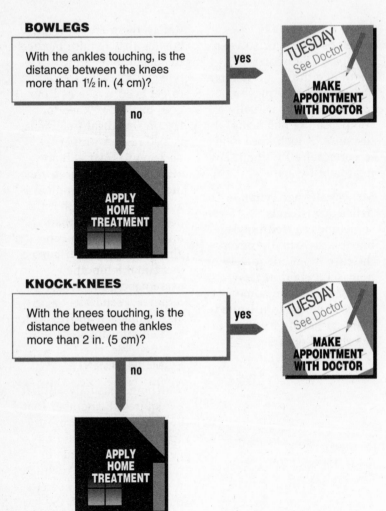

BOWLEGS

With the ankles touching, is the distance between the knees more than 1½ in. (4 cm)?

yes → TUESDAY See Doctor — MAKE APPOINTMENT WITH DOCTOR

no ↓ APPLY HOME TREATMENT

KNOCK-KNEES

With the knees touching, is the distance between the ankles more than 2 in. (5 cm)?

yes → TUESDAY See Doctor — MAKE APPOINTMENT WITH DOCTOR

no ↓ APPLY HOME TREATMENT

WHAT TO EXPECT AT THE DOCTOR'S OFFICE

Careful examination will be made of the back, hips, and legs with appropriate measurements. Observation of the child's gait will be made. If there is excess bowing, an X-ray may be taken, and a referral to an orthopedic surgeon may be made. Treatment usually consists of using a nighttime brace (Dennis–Browne splint). Surgery is rarely needed. Surgical intervention in a severe case of knock-knee is seldom recommended before age eight or nine.

Pigeon Toes and Flat Feet

As a nation, we have had a fascination with feet. Many readers may remember shoe stores with fluoroscopes, where you could stare with wonder at the bones in your feet. The radiation was dangerous and, of course, not necessary for shoe fitting; yet it was tolerated for some time.

Infants

Most concerns about feet can be allayed by understanding normal development. When children are born, their feet often are turned in because of the cramped uterine environment. You should be able to straighten the foot easily by manipulating it gently with your own fingers. If it cannot be easily straightened, discuss this with your doctor. In this situation, the bones of the foot may not be aligned perfectly, causing a problem known as **metatarsus adductus**. This problem is simple to correct if detected early in the newborn.

All infants have fat feet, not to be confused with flat feet. Unless there is an obvious bony deformity of the foot, with ankle bones protruding on the inside of the foot, you need not be concerned.

Toddlers

When children begin to walk, they keep their legs far apart and have their feet pointing out. This "duck-walking" provides the most stable base for the new, unsteady walker.

Many toddlers appear to have bowed legs and later walk with their toes pointing in—"pigeon toes." Rarely, **toeing in** can be caused by the leg's being set into the hip at the wrong angle; with the child lying on his or her back, the feet should be able to be turned outward. (If this is not possible, check it with your doctor on the next visit.) Toeing in can also be caused when one of the bones in the lower leg is twisted in excessively; as the child grows, the bone naturally twists outward. Excessive twisting can be detected by checking the child's ankle bones when the child is sitting on a table. If the outer ankle bone is in front of the inner, this condition is present. Pigeon toeing can also be caused by metatarsus adductus.

Most children outgrow their pigeon toeing by the age of four. A severe problem causing frequent tripping or any of the findings discussed in the previous paragraph suggest the need for discussion with a doctor.

Older Children

In the older child, a flat foot can be checked for by examining the shoe. A worn inner edge of the heel indicates a possible flat foot. Children with severe flat feet may complain of foot pain and have no visible arch even if they stand on their toes. Most flat feet are more imagined than real.

HOME TREATMENT

Most children will grow up to have straight feet and legs, and most of the variations are normal developmental occurrences. The best home treatment is to encourage physical activities because muscles assist in the growing and straightening process. You need *not* spend large sums on shoes; the purpose of a shoe is to protect the bottom of the child's foot from scrapes and cuts. Except for the first few months of walking when high-top shoes may assist the walking process, sneakers are a good choice. They are inexpensive (which is especially important for a rapidly growing foot), have a straight last, and won't cramp the toes.

WHAT TO EXPECT AT THE DOCTOR'S OFFICE

A thorough examination of the foot, ankle, knee, and hips will be done, and the doctor will observe the child's gait. If a foot deformity such as metatarsus adductus is suspected, X-rays may be ordered. Treatment will depend on the child's age and the extent of the problem. Finally, you should be reassured that perfectly straight feet and legs, although cosmetically more satisfying, do not offer any functional advantages. Many sports are played with greater ease by people with feet that are slightly turned in (tennis) or out (fencing).

PIGEON TOES/FLAT FEET

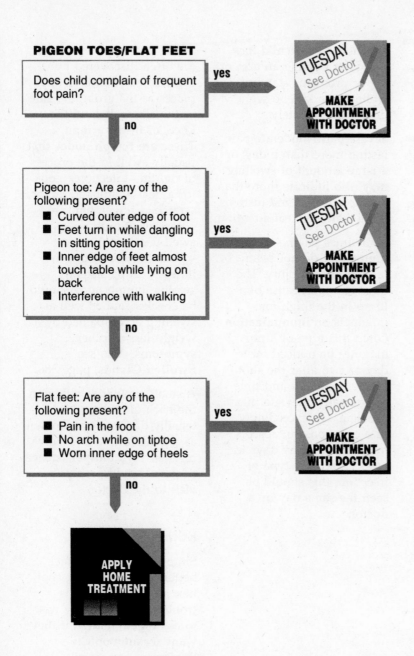

Does child complain of frequent foot pain?

yes → TUESDAY See Doctor — **MAKE APPOINTMENT WITH DOCTOR**

no ↓

Pigeon toe: Are any of the following present?
- Curved outer edge of foot
- Feet turn in while dangling in sitting position
- Inner edge of feet almost touch table while lying on back
- Interference with walking

yes → TUESDAY See Doctor — **MAKE APPOINTMENT WITH DOCTOR**

no ↓

Flat feet: Are any of the following present?
- Pain in the foot
- No arch while on tiptoe
- Worn inner edge of heels

yes → TUESDAY See Doctor — **MAKE APPOINTMENT WITH DOCTOR**

no ↓

APPLY HOME TREATMENT

Arm and Leg Lumps

Active children have countless ways of entangling, compressing, bruising, and sometimes shattering arms and legs. While we are all accustomed to seeing a slight amount of swelling that accompanies a minor injury, the presence of other symptoms can help indicate whether or not the problem is serious. When making decisions after an accident, always consider the severity of the impact; also, be sure to read Chapter D, "Common Injuries."

Whenever your child has difficulty moving an arm or leg or problems standing or walking, be sure to see a doctor immediately.

Severe pain, moderate pain lasting more than a day, or a large amount of swelling may also indicate more than a simple bruise and justify immediate medical attention.

Sometimes parents notice lumps that do not seem related to any injury. The most common cause of a lump in the thigh for infants is an **immunization**. Often this type of lump may not be noticed for a day or two after the shot. Home treatment and time will usually alleviate this problem. On exceedingly rare occasions, an abscess may follow many days to weeks after any type of injection; this should be seen the same day by a doctor.

Small, bean-shaped lumps are often discovered by children or parents in such places as the groin, the side of the elbow, behind the knee, or in the armpit. These are **lymph nodes** that usually swell in the process of fighting infections. Look for a nearby stubbed toe or scratched finger. On rare occasions, lymph nodes swell because of more serious infections or other illnesses. A persistently swollen lymph node should be checked out as well as swelling accompanied by weight loss or other symptoms. Also see **Swollen Glands**, page 386.

A small circular swelling on the back of the hand is usually due to a cyst known as a **ganglion**. Most resolve on their own but persistence or pain should prompt a visit to the doctor.

HOME TREATMENT

Simple bruises often feel better with something cool next to them. If pain is troublesome, you may wish to ask older children if they want acetaminophen, ibuprofen, or aspirin.

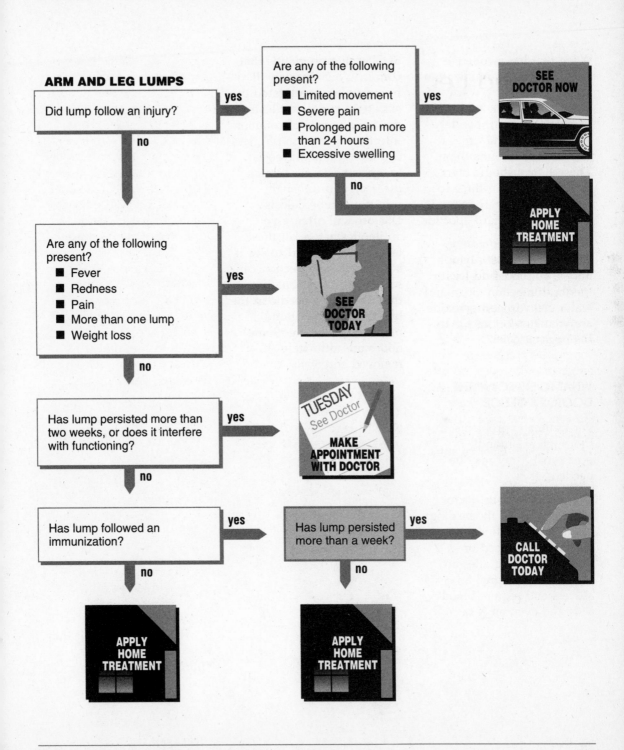

ARM AND LEG LUMPS

Did lump follow an injury? — **yes** → Are any of the following present?
- Limited movement
- Severe pain
- Prolonged pain more than 24 hours
- Excessive swelling

yes → SEE DOCTOR NOW

no → APPLY HOME TREATMENT

no →

Are any of the following present?
- Fever
- Redness
- Pain
- More than one lump
- Weight loss

yes → SEE DOCTOR TODAY

no →

Has lump persisted more than two weeks, or does it interfere with functioning? — **yes** → TUESDAY See Doctor — MAKE APPOINTMENT WITH DOCTOR

no →

Has lump followed an immunization? — **yes** → Has lump persisted more than a week? — **yes** → CALL DOCTOR TODAY

no → APPLY HOME TREATMENT

no → APPLY HOME TREATMENT

A leg that has swollen because of an immunization may be helped by warm compresses. Soak a washcloth in warm water and apply several times a day to your infant's thigh. These lumps should start getting smaller in three to five days, but there may still be a small hard spot for weeks.

When dealing with lymph nodes, you must deal with the local infection. Soap and water or hydrogen peroxide are your first options. Also see pages 386–387.

WHAT TO EXPECT AT THE DOCTOR'S OFFICE

For traumatic problems, a thorough exam and occasional X-ray may be needed.

There is little your doctor can do to speed the healing of an immunization reaction. Prolonged or massive swelling after an injection may signal an abscess and require blood tests and even an X-ray.

Swollen lymph glands that concern your doctor will probably be investigated by skin tests and blood tests seeking a variety of common infections (strep, staph), less common (mononucleosis, toxoplasmosis, cat scratch disease), and unusual infections (tuberculosis). Doctors will often give antibiotics before all these tests are completed to see if a presumptive infection can be treated. Occasionally, they will put a needle in the lymph node to obtain material for testing. Some nodes are ultimately removed surgically.

CHAPTER M

Chest Pain, Shortness of Breath, and Palpitations

Chest Pain

Chest pain can come from the chest wall (including muscles, ligaments, ribs, and rib cartilage), the lungs, the outside covering of the heart (pericardium), the gullet (esophagus), the diaphragm, the spine, the skin, and the organs in the upper part of the abdomen. Heart pain almost never occurs under 30 years of age.

Chest pain in children frequently results from severe **coughing**, which leads to a pain or burning sensation underneath the breastbone (sternum).

Chest pain is a very common problem in adolescents. Often it is difficult even for a doctor to determine the precise origin of the pain. In general, pains that are made worse by breathing, coughing, or movement of the chest are due to problems of the chest wall or the lungs. You can check for **chest-wall pain** by pressing a finger on the chest at the spot of discomfort and reproducing or aggravating the pain; it usually becomes worse with movement of the chest.

A shooting pain lasting a few seconds is common in healthy young people and means nothing. It is probably due to a trapped **gas** bubble of the stomach. A sensation of a "catch" at the end of a deep breath is also trivial and does not need attention. The hyperventilation syndrome (see **Stress, Anxiety, and Depression**, page 326) is a frequent cause of chest pain, particularly in adolescents. If there is dizziness or a tingling in the fingers, suspect this problem.

Adolescent males will often complain of chest pain when they experience tenderness in a swollen breast (see gynecomastia, pages 74–75). Adolescent females may also complain of chest pain when they are experiencing breast problems.

One of the serious problems that can present itself as chest pain is a **pneumothorax**. A pneumothorax is a problem in which the outside lining of the lung has ruptured and air begins to accumulate around the outside of the lung, compressing it. This problem often occurs in patients with a history of asthma but can occur spontaneously in patients without any prior lung problems. Breathing becomes progressively difficult; if chest pain is accompanied by shortness of breath, the doctor should be consulted immediately.

CHEST PAIN

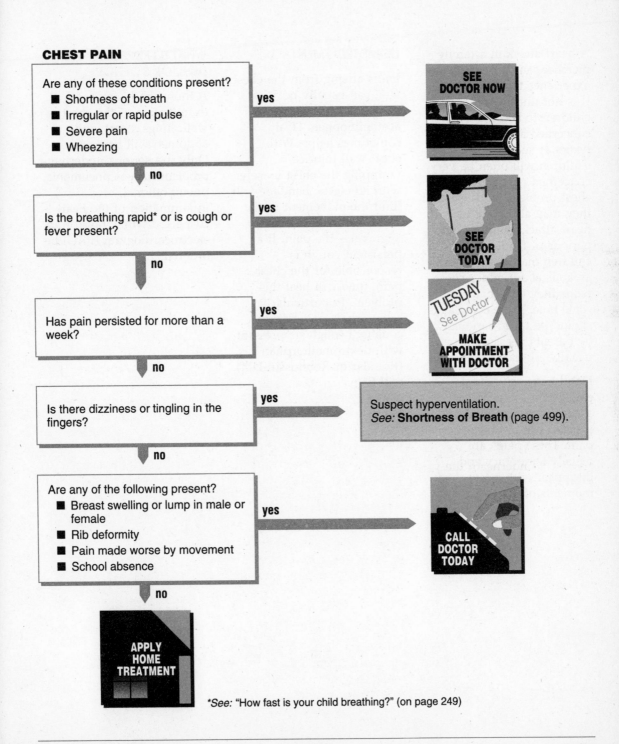

Are any of these conditions present?
- Shortness of breath
- Irregular or rapid pulse
- Severe pain
- Wheezing

yes → **SEE DOCTOR NOW**

no ↓

Is the breathing rapid* or is cough or fever present?

yes → **SEE DOCTOR TODAY**

no ↓

Has pain persisted for more than a week?

yes → TUESDAY See Doctor — **MAKE APPOINTMENT WITH DOCTOR**

no ↓

Is there dizziness or tingling in the fingers?

yes → Suspect hyperventilation. *See:* **Shortness of Breath** (page 499).

no ↓

Are any of the following present?
- Breast swelling or lump in male or female
- Rib deformity
- Pain made worse by movement
- School absence

yes → **CALL DOCTOR TODAY**

no ↓

APPLY HOME TREATMENT

**See:* "How fast is your child breathing?" (on page 249)

A heart attack in a family member can be a frightening experience for everyone. It is not uncommon for children to become concerned about their own hearts at such a time. Children will often be very concerned about minor chest problems and fear that they may also be having a heart attack. During other periods of family crisis, children may also become anxious about minor chest pains they have had for some time. These complaints should be taken seriously by parents; reassurance is usually all that is required.

Finally, excessive hard exercise in someone not accustomed to prolonged exercise can produce chest pain. These pains are usually temporary and disappear with a few moments of rest.

HOME TREATMENT

Pains arising from the chest wall can usually be managed at home with acetaminophen. Heat sometimes helps. With chest-wall injuries, wrapping the chest loosely with an elastic bandage will limit the movement of the chest wall, which often aggravates the pain. If a persistent cough is responsible for the chest pain, time will heal this problem. In particularly severe chest pain due to coughs, a cough suppressant with dextromethorphan (Romilar or Robitussin-DM) will help.

WHAT TO EXPECT AT THE DOCTOR'S OFFICE

A thorough history and examination of the chest wall, lungs, heart, and abdomen will be made. Only if a serious underlying problem such as pneumonia, pneumothorax, or inflammation of the heart is suspected will X-rays or an electrocardiogram (EKG) be ordered.

Shortness of Breath

When children run hard and long, they become short of breath; this symptom is certainly normal under such circumstances. Medical use of the term "shortness of breath" does not include shortness of breath after such heavy exertion.

Shortness of breath can be due to several different problems. Your child may experience difficulty breathing in. This symptom is common in **croup** (page 379), where it is usually accompanied by a barking cough.

If a child is having difficulty breathing while gasping for breath, drooling, or breathing with the head tilted forward, a serious obstruction of the airway passages is likely; the child should immediately be brought to the doctor.

If a child is having difficulty breathing out, the expiration will take longer than normal. Often difficulties in breathing out are accompanied by **wheezing**. More often, difficulty breathing out is accompanied by wheezing that cannot be heard unless you place your ear to your child's chest. Wheezing is discussed further on page 382.

Rapid Breathing

The term "shortness of breath" is frequently used to mean rapid breathing. Respiratory rates in infants are usually high; breathing rates of 50 to 60 times per minute are common. As infants grow older, the normal respiratory rate declines. By about one year of age, the respiratory rate of a child is between 25 and 35 while resting, although in an active but not exercising child it may be as high as 45. It is therefore important to assess the breathing rate when the child is resting. Here are resting breathing rates that should cause concern:

- In children over one year of age: 40 or more
- In children over six: 30 or more

The most common reasons for elevated respiratory rates are **fevers** and **pneumonia**. One of the body's mechanisms for lowering temperature is to increase the respiratory rate. If you reduce the fever with acetaminophen or cool baths and the child's respiratory rate is still elevated, you should be concerned about possible pneumonia.

Another cause for an elevated respiratory rate is an overdose of aspirin; if you suspect this, call the doctor.

More unusual causes of an elevated respiratory rate include diabetes and metabolic disturbances that may accompany severe diarrhea. These need the doctor's attention.

SHORTNESS OF BREATH

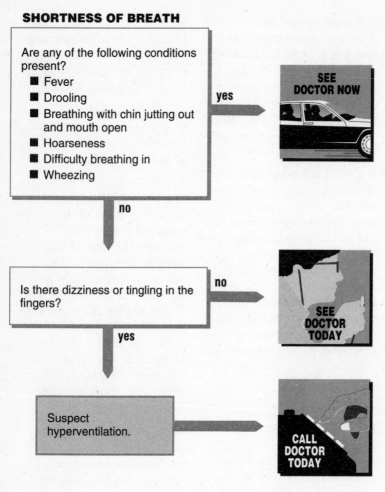

Are any of the following conditions present?
- Fever
- Drooling
- Breathing with chin jutting out and mouth open
- Hoarseness
- Difficulty breathing in
- Wheezing

yes → **SEE DOCTOR NOW**

no ↓

Is there dizziness or tingling in the fingers?

no → **SEE DOCTOR TODAY**

yes ↓

Suspect hyperventilation.

→ **CALL DOCTOR TODAY**

HOME TREATMENT

Most causes of true shortness of breath require medical attention. Mist will usually relieve croup (see page 379). Children with asthma will already be under medical supervision, and the usual regimen should be begun. All other cases should be brought to the doctor.

WHAT TO EXPECT AT THE DOCTOR'S OFFICE

A thorough history and examination of the lungs, heart, and upper airway passages will be done. Depending on the nature of the problem, chest and neck X-rays may be ordered. Severe obstruction of the airway will almost always require hospitalization. Pneumonia, asthma, and croup can usually be managed without hospitalization.

Palpitations

Pounding of the heart is brought on by strenuous exercise or intense emotion and is seldom associated with serious disease. Most of us have experienced what is known as "bent-bumper syndrome"; after a near collision with another car, the heart seems almost to stop and then pounds with such force that you feel as though you're being punched in the chest. Simultaneously, the knees become wobbly and the palms sweaty. These events are due to a large discharge of adrenalin from the adrenal glands. Almost no one is concerned by such pounding of the heart. But if there is no obvious exertion or frightening event, many people become worried.

Most people who complain of palpitations do not have heart disease but are overly concerned about the possibility of such disease and thus overly sensitive to normal heart actions. Often this is because of heart disease in parents, other relatives, or friends.

Taking a Pulse

An irregular or very fast pulse may be more serious. There is a normal variation in the pulse with respiration (faster when breathing in, slower when breathing out). Even though the pulse may speed or slow, the normal pulse has a regular rhythm. Occasional extra heartbeats occur in nearly everyone. Consistently irregular pulses, however, are usually abnormal. The pulse can be felt on the inside of the wrist, in the neck, or over the heart itself. Ask the nurse to check you out on taking pulses on your next visit. Take your own pulse and those of your children, noting the variation with respiration.

The most common time for palpitations to occur is just before going to sleep. If the pulse rate is under 120, relax.

Causes

Hyperventilation may also cause pounding and chest pain, but the heart rate also remains less than 120 beats per minute. Refer to **Shortness of Breath**, page 499.

In older children and adolescents, a resting heart rate greater than 120 beats per minute (without exercise) is a cause to check with the doctor. Young children may have normal heart rates in that range, but they rarely complain of the heart pounding. If one should, check the situation with your doctor. Keep in mind that the most frequent causes of rapid heartbeat (other than exercise) are anxiety and fever. The presence of **shortness of breath** (page 499) or **chest pain** (page 496) increases the chances of a significant problem.

While most soft drinks have only a fraction of the amount of caffeine contained in coffee, excess intake can lead to palpitations.

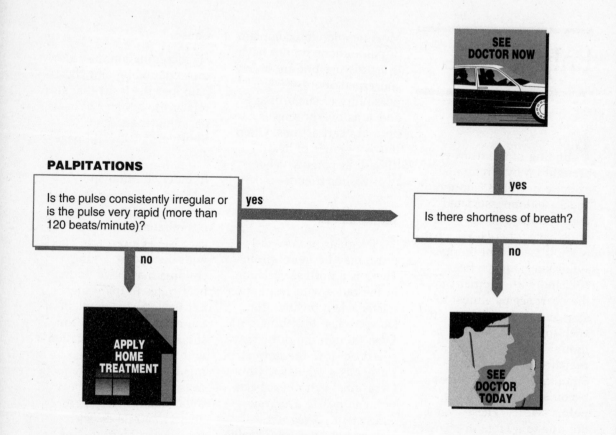

PALPITATIONS

Is the pulse consistently irregular or is the pulse very rapid (more than 120 beats/minute)?

yes

no

APPLY HOME TREATMENT

SEE DOCTOR NOW

Is there shortness of breath?

yes

no

SEE DOCTOR TODAY

HOME TREATMENT

If a child seems stressed or anxious, focus on this feeling rather than on the possibilities of heart disease. If anxiety does not seem likely and the child has none of the other symptoms on the chart, discuss the problem with the doctor at your next visit.

WHAT TO EXPECT AT THE DOCTOR'S OFFICE

Tell the doctor the exact rate of the pulse and whether the pulse rhythm was regular. Usually, the symptoms will have disappeared by the time the doctor is consulted, so that your accuracy is important. The doctor will examine the heart and lungs. An electrocardiogram (EKG) is unlikely to help if the problem is not present when it's being done. A chest X-ray is seldom needed.

CHAPTER N

Digestive Tract Problems

Nausea and Vomiting

Vomiting is another way that the body expels germs or other irritants. Unfortunately, it's a "drastic measure" that can be frightening and exhausting for children. Prolonged or excessive vomiting can cause more serious problems.

Dangers of Vomiting

A child who is vomiting and losing fluid in other ways, such as diarrhea or sweating, may suffer **dehydration**. Children become dehydrated more quickly than adults, and infants under six months of age can become dehydrated very easily, especially in warm summer weather or while running a fever. Signs of dehydration are:

- Marked thirst, dry mouth
- Sunken eyes
- Dry skin that wrinkles easily (Gently pinch the skin on the stomach using all five fingers. If the skin

does not spring back normally after you release it but remains tented up, dehydration is indicated.)
- Urine that is scant or deep yellow in color

If you see any of these signs in a small child, see the doctor immediately.

If vomiting is bloody, if it is accompanied by flecks of blood in the stool (feces), and/or if intermittent abdominal pain is present, intestinal blockage is possible, and professional medical care is necessary.

Newborns

Most newborns and infants will spit up a small amount of food after feeding. Even with proper burping, small amounts of vomitus can be expected at this age. Often, placing an infant face down for a nap after meals may help. Persistent or violent vomiting or the failure to gain weight are signs that this may be more than simple spitting up. Fever is unusual in the newborn. If it is present along with vomiting, the child should be brought to the doctor immediately.

Older Children

Infants and toddlers will often vomit with viral infections of the gastro-intestinal tract. With many gastro-intestinal infections, diarrhea accompanies the vomiting. The younger the child, the more serious this combination can be. When abdominal pain is present, parents are often concerned about the possibility of appendicitis or other serious abdominal problems. These are discussed further in **Acute Abdominal Pain**, page 511.

In younger children, the possibility of accidental poisoning or medication intake should not be overlooked. In older children, excess alcohol ingestion can cause vomiting.

Hepatitis may begin with nausea and vomiting; often children with hepatitis will have dark urine. Yellow jaundice is not always present in hepatitis, but abdominal tenderness over the liver (the upper right quarter of the abdomen underneath the rib cage) usually is. If you suspect hepatitis, a visit to the doctor is important. Other members of the family can obtain gamma globulin shots, which will reduce the severity of the symptoms in the event that they come down with the disease.

Nausea and vomiting can also accompany urinary tract infections. Fever, problems with urination, and occasionally abdominal or back pain will also be present. See **Difficult Urination**, page 525.

A serious infection that produces vomiting, and one that is fortunately not very common, is **meningitis**. This infection of the covering of the brain and spinal cord will also produce either irritability or lethargy, and fever is almost always present. In young children, the soft spot (fontanel) will be bulging; older children will have a stiff neck that prevents them from touching their chin to their chest. Meningitis is a medical emergency that must be attended immediately. Head injuries can cause vomiting and are discussed in further detail on page 285.

Excessive excitement or emotional stress can also cause vomiting in young children. Vomiting, when it accompanies headaches, may be a sign of migraine in children, discussed further in **Headache**, page 309.

Nausea usually precedes or accompanies vomiting in children with mild gastrointestinal infections. Often, it occurs by itself, especially in children who have problems with **motion sickness**. Car sickness is aggravated by looking out the side windows of a moving vehicle, which forces the children to continuously focus on objects rapidly moving past them. The eye movements required bring about the car sickness.

Many medications can cause nausea. Remember also that if your child is taking a medication for another purpose, the vomiting will interfere with the absorption of the medicine. Call the doctor.

HOME TREATMENT

Avoid solid foods. Frequent, small feedings of clear liquids should be given instead. A tablespoon of clear fluid every few minutes will usually stay down. Often, Popsicles or iced fruit bars will work if nothing else will stay down.

As the condition improves, larger amounts of fluids and then Jell-O and applesauce may be given. Sometimes, sucking on hard candy or chewing ice chips helps. In younger children, you may wish to give commercially available electrolyte solutions (Pedialyte, Ricelyte, Lytren). These are effective in keeping children from becoming dehydrated but are of very little caloric value.

After a period of 8 to 12 hours without vomiting, your child can begin to eat bland foods: bananas, rice, and applesauce for infants; bread and soups can be added for older children.

WHAT TO EXPECT AT THE DOCTOR'S OFFICE

A history and physical examination will direct attention to whether the child is dehydrated and to the abdomen. With girls, a urinalysis will often be obtained. If a serious underlying condition is suspected, blood tests and abdominal X-rays will be ordered.

NAUSEA AND VOMITING

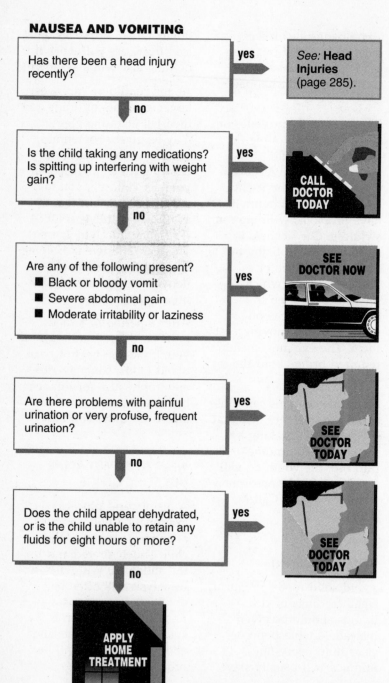

Has there been a head injury recently? — **yes** → *See:* **Head Injuries** (page 285).

no

Is the child taking any medications? Is spitting up interfering with weight gain? — **yes** → **CALL DOCTOR TODAY**

no

Are any of the following present?
- Black or bloody vomit
- Severe abdominal pain
- Moderate irritability or laziness

— **yes** → **SEE DOCTOR NOW**

no

Are there problems with painful urination or very profuse, frequent urination? — **yes** → **SEE DOCTOR TODAY**

no

Does the child appear dehydrated, or is the child unable to retain any fluids for eight hours or more? — **yes** → **SEE DOCTOR TODAY**

no

APPLY HOME TREATMENT

For particularly severe vomiting some doctors may give trimethobenzamide (Tigan), promethazine (Phenergan), or prochlorpromazine (Compazine) suppositories. These suppositories are not recommended for treating uncomplicated vomiting. Although useful in preventing vomiting in children who are taking anti-cancer medicines (known to cause vomiting), their effectiveness in controlling other types of vomiting is unknown. The side effects of these drugs include nervous system disturbances that may confuse a diagnostic evaluation.

If dehydration is a problem, intravenous fluids and hospitalization may be required; this is more usual for children under the age of two. Some doctors will give intravenous fluid therapy in their offices.

Diarrhea

Diarrhea is another common problem in children of all ages. Like constipation, it may be just a normal variation in bowel habits rather than a disease. Newborn infants who are exclusively breast-fed have more frequent and softer stools than bottle-fed infants do. Having more than a dozen of these soft stools a day is not uncommon; but despite the frequency, this is not diarrhea. Diarrhea in infants consists of a liquid, runny stool. In older children, two or three soft to runny stools a day may be considered a sign of diarrhea.

Diarrhea can be potentially dangerous, especially in young infants. Both water and body salts can be lost, leading to dehydration. It is always important to start giving your child the proper fluids when diarrhea begins.

Causes

The most common cause is **viral gastroenteritis**. Often there are other symptoms, such as fever, runny nose, and fatigue. Like most viral infections, the problem should end in three to four days. The diarrhea is often accompanied by vomiting, which may aggravate the child's fluid losses. Because of this outpouring of fluids from both ends, it is crucial to maintain adequate hydration. Assessing dehydration is discussed on pages 504 and 508.

Diarrhea can also be caused at times by **bacterial infections**. The bacteria usually come from contaminated food or water and can be particularly severe in infants. Profuse, watery diarrhea, especially accompanied with flecks of blood, should arouse a parent's suspicion of a bacterial infection. The intestines respond to the presence of viruses and bacteria by increasing their movement. The intestines are trying to get rid of the infection, and it is a natural defense mechanism of the body that produces the rapid, frequent bowel movements. There can be irritation of the lining of the intestines, and some of the cells that produce enzymes necessary to digest food are temporarily damaged. Much of the ordinary digestive processes are hampered because the food does not remain long enough in the intestines to be digested.

Antibiotics, by altering the normal bacterial pattern in the intestines, often cause diarrhea. If this seems to be the case, the prescribing doctor should be contacted.

Milk allergy is often blamed for diarrhea in children but is seldom the cause. However, some children may have a limited amount of an enzyme (lactase) necessary for digesting milk. This may cause several symptoms, including diarrhea, bloating, and abdominal pain.

Chronic diarrhea is unusual in children; but when it occurs, it can be a serious problem. It requires professional medical care, especially when accompanied by weight loss.

HOME TREATMENT

Management of diarrhea is primarily by fluid and dietary manipulation. All children have sufficient caloric reserve to withstand several days of no food intake. However, no child has sufficient *fluid* reserves to withstand several days of diarrhea without fluid intake.

Treatment of diarrhea begins with giving the child special liquids known as **oral electrolyte solutions**. Solutions available in supermarkets and pharmacies, such as Lytren, Ricelyte, Pedialyte, and Rehydralyte, are essential for infants because they replace fluid *and* the right amount of salt and sugar. Juices or flat sodas may actually increase diarrhea in infants and young children and should be avoided.

If the child seems to be tolerating the fluids, give the constipating foods that spell BRAT:

- Bananas
- Rice
- Applesauce
- Toast

This is commonly referred to as the "brat diet." Milk should be avoided, because very often the enzyme lactase is lost during a bout of diarrhea; milk contains a great deal of lactose and cannot be digested properly. Fats should be avoided for several days because they will not remain in the intestines long enough to be digested. Although there is no real evidence that the presence of undigested fat in the digestive tract is harmful, it does make the bowel movement smell bad.

Infants can become dehydrated very quickly. Consult the decision chart for indications when to call your doctor. Don't hesitate to do so with young infants. If an older child is still having watery diarrhea after three days of clear liquid therapy, consult your doctor by phone. Often, the recommendation will be to begin solid foods because continuous liquid diets can also eventually result in diarrhea.

There are no medications that we consider safe and effective for use in children. The narcotic or narcotic-like preparations (paregoric, Parelixir, Lomotil) used by adults to control diarrhea should be avoided for children. Over-the-counter preparations that include kaolin and pectin will change the consistency of the stool from a liquid to a semi-solid state, but they will not reduce the amount or frequency of the bowel movements. Pepto-bismol has recently been shown to help moderate to severe dehydration when used with oral rehydration. However, the frequency of doses required is cumbersome.

As soon as diarrhea begins, you may wish to protect the diaper area with petroleum jelly (Vaseline). If skin breakdown has occurred and sores are present, ointments should be avoided, and efforts should be made to keep the diaper area as dry as possible.

DIARRHEA

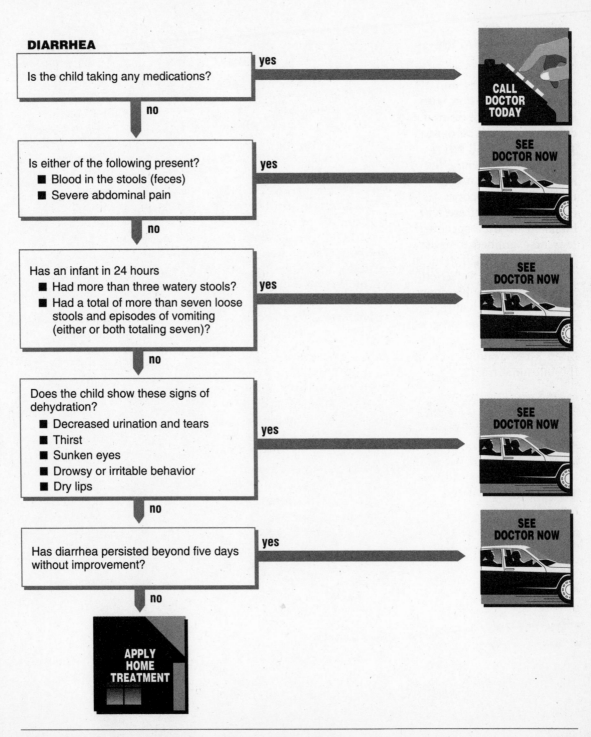

Is the child taking any medications? — **yes** → CALL DOCTOR TODAY

no ↓

Is either of the following present?
- Blood in the stools (feces)
- Severe abdominal pain

— **yes** → SEE DOCTOR NOW

no ↓

Has an infant in 24 hours
- Had more than three watery stools?
- Had a total of more than seven loose stools and episodes of vomiting (either or both totaling seven)?

— **yes** → SEE DOCTOR NOW

no ↓

Does the child show these signs of dehydration?
- Decreased urination and tears
- Thirst
- Sunken eyes
- Drowsy or irritable behavior
- Dry lips

— **yes** → SEE DOCTOR NOW

no ↓

Has diarrhea persisted beyond five days without improvement? — **yes** → SEE DOCTOR NOW

no ↓

APPLY HOME TREATMENT

WHAT TO EXPECT AT THE DOCTOR'S OFFICE

A thorough history and physical examination with special attention to assessing dehydration will be done. The abdomen will be examined. Frequently, the stools will be examined under the microscope; occasionally, a culture will be taken. A urine specimen may be examined to assist in assessing dehydration. In cases of bacterial infection, an antibiotic may be given. Chronic diarrhea will require more extensive evaluation of the stools, blood tests, and often X-ray examinations of the intestinal tract.

Abdominal Pain

Abdominal pain is one of the most common concerns of parents, and the causes for abdominal pain in children change with age of the child. During the first few months of life, **colic** is the most common cause and is discussed on page 516.

As children become older, they often need to be reminded to have a bowel movement, and this will relieve their abdominal pain.

Abdominal pain associated with vomiting and diarrhea is common with gastro-enteritis or **stomach flu**. These problems are of greater concern in infants (see pages 504 and 507) than in older children. Vomiting without diarrhea and especially without bowel movements is of concern because of the possibility of an obstructed or blocked intestine. Obstruction can also produce severe intermittent pain and especially violent vomiting. All these signs signal the need for a trip to the doctor.

In older children, infections commonly produce abdominal pain, even though the infection itself may not be in the abdomen. Sore throats, ear infections, and excessive coughing from a cold can all cause bellyaches. Pneumonia can also cause abdominal pain and is usually accompanied by a rapid breathing rate and fever.

Hepatitis may cause abdominal pain (usually in the upper right corner of the abdomen) and is usually accompanied by nausea, vomiting, and sometimes yellow jaundice. Household members can be given potential protection from the symptoms of hepatitis by gamma globulin shots.

Urinary tract infections can also be present with abdominal pain, along with discomfort or frequency in passing urine. See page 525.

Appendicitis occurs less frequently than any of the problems discussed so far. While appendicitis most commonly occurs in young adults, it can occur at any age. The diagnosis of appendicitis is especially hard to make in toddlers, who cannot describe the problem but may look very sick. Older children may begin by complaining of pain in the center of the abdomen. In some, this pain will then move to the lower right part of the belly. Some children will have pain only in the lower right belly; about 20% will have pain elsewhere. There is usually fever and often nausea and vomiting. Because of abdominal tenderness, children may rest with their right leg bent. Sometimes they will not bear weight on the right leg. The abdomen may be sensitive to touch. If these symptoms are present, see the doctor. If you are uncertain, waiting a few hours does not substantially increase the risk of the appendix bursting. However, waiting a long period of time (more than 12 hours) does increase the risk.

Abdominal pain with unusual accompanying symptoms like joint pain and rash suggest rare illnesses like rheumatic fever and Henoch–Shonlein purpura, which require medical attention.

Ulcers are more common in adults but do occur in children. Vomiting dark material or blood should signal this as a possibility. Pain in the upper abdomen recurrent for many days needs a check.

Finally, do not overlook the possibility that your child may have swallowed some object or medicine. Always be careful with medicines and dangerous products.

HOME TREATMENT

Watchful waiting in the first few hours is the best approach. For diarrhea, see page 507. For nausea and vomiting, see page 504. Small amounts of clear liquid can be given. Aspirin should not be given, but fever can be treated with acetaminophen or ibuprofen if the child is uncomfortable from the fever. If pain and fever are the only problems, be sure to evaluate your child's appearance every few hours and do not hesitate to call the doctor for further advice.

WHAT TO EXPECT AT THE DOCTOR'S OFFICE

A thorough history and examination of ears, throat, chest, and abdomen will be made. A rectal examination will be done if appendicitis is suspected. Often a blood test and a urinalysis will also be performed. If intestinal obstruction is suspected, X-rays will be taken; they are not helpful in most cases of abdominal pain.

ABDOMINAL PAIN

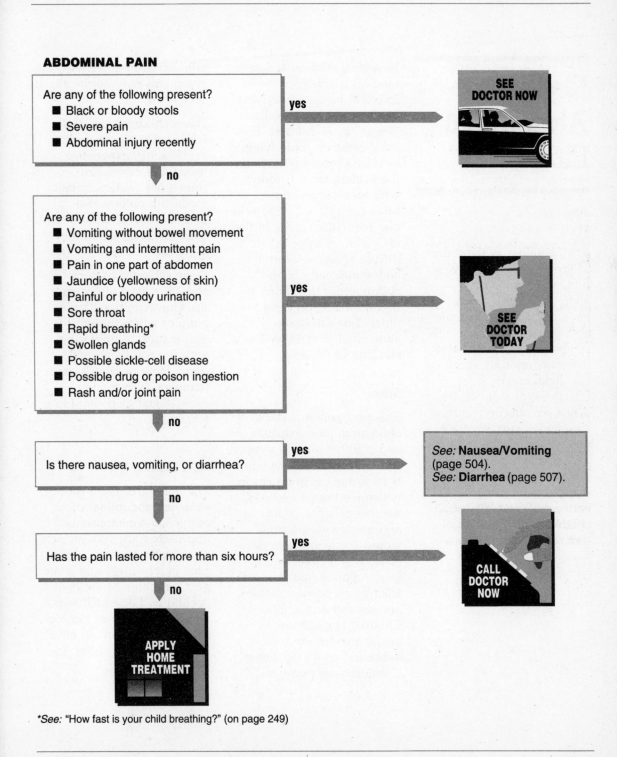

Are any of the following present?
- Black or bloody stools
- Severe pain
- Abdominal injury recently

yes → SEE DOCTOR NOW

no

Are any of the following present?
- Vomiting without bowel movement
- Vomiting and intermittent pain
- Pain in one part of abdomen
- Jaundice (yellowness of skin)
- Painful or bloody urination
- Sore throat
- Rapid breathing*
- Swollen glands
- Possible sickle-cell disease
- Possible drug or poison ingestion
- Rash and/or joint pain

yes → SEE DOCTOR TODAY

no

Is there nausea, vomiting, or diarrhea?

yes → *See:* **Nausea/Vomiting** (page 504). *See:* **Diarrhea** (page 507).

no

Has the pain lasted for more than six hours?

yes → CALL DOCTOR NOW

no → APPLY HOME TREATMENT

See: "How fast is your child breathing?" (on page 249)

Recurrent Abdominal Pain

All children in the course of growing up will have a number of bouts of abdominal pain. As we have discussed with acute abdominal pain, these symptoms are often related to respiratory tract infections, coughing, urinary tract infections, or stomach flu. When we refer to recurrent abdominal pain, we are talking about repeated bouts of abdominal pain for which no explanation is obvious. Somewhere between 10 and 20% of all children will complain of these recurrent problems.

In several studies, serious medical problems were found to be the underlying cause of pain in less than 10% of these children. The most common underlying medical problems were in the urinary tract. Children with medical problems often complained of pain in one particular corner of the abdomen, as opposed to diffuse or vague central abdominal pain. Some children lose the ability to digest milk as they get older. This can cause abdominal pain as well as a swelling of the belly.

Stress

The most common cause of abdominal pain in adults and children alike is stress. It is only natural for us to react to the environment in which we live. A bad day, the loss of a pet, or an argument with a friend would create stress in all of us. It is a common misconception that children's feelings are not as complex and sensitive as adults. Children become just as angry, anxious, and depressed as we do. Often, if children are permitted to talk about these feelings, the symptoms may not be as severe.

Other children and adults when subjected to similar stresses may experience headaches. Stress can sometimes cause serious medical problems like ulcers; while ulcers are less common in children, they do occur.

Being subjected to numerous stresses is part of growing up. However, if abdominal pain, or any symptom for that matter, is interfering with the child's normal functioning in school, at home, or with friends, professional help should be sought.

HOME TREATMENT

The home treatment for recurrent abdominal pain requires a commonsense approach. Care may include clear-liquid feedings, some time in a hot tub, and some gentle rubbing of the belly. A child will often tell you what makes him or her feel better. In addition, an effort should be made to link the

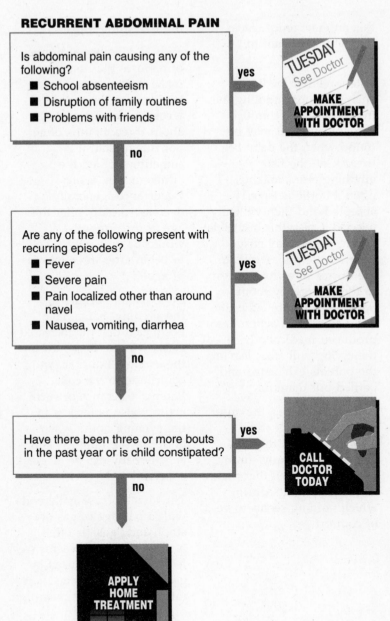

RECURRENT ABDOMINAL PAIN

Is abdominal pain causing any of the following?
- School absenteeism
- Disruption of family routines
- Problems with friends

yes → **MAKE APPOINTMENT WITH DOCTOR**

no ↓

Are any of the following present with recurring episodes?
- Fever
- Severe pain
- Pain localized other than around navel
- Nausea, vomiting, diarrhea

yes → **MAKE APPOINTMENT WITH DOCTOR**

no ↓

Have there been three or more bouts in the past year or is child constipated?

yes → **CALL DOCTOR TODAY**

no ↓

APPLY HOME TREATMENT

cause of the abdominal pain with the symptom. Talking with your child about what is going on in school and what may be upsetting him or her can often be profitable. Children sometimes exploit abdominal pain in order to get attention or get their own way. Your own judgment will guide you in these cases.

WHAT TO EXPECT AT THE DOCTOR'S OFFICE

The greatest amount of time and attention should be directed toward the medical history. A thorough physical examination should be performed, including examination of the head, chest, abdomen, and genitalia. A urinalysis and most likely an examination of the stool will be performed. Often an additional trip to the office can be avoided by finding out beforehand whether or not a stool specimen will be required. X-rays are necessary in only a few circumstances; do not expect them as part of the routine evaluation.

Colic

The word *colic* means "of the colon," but the term is commonly used to mean a prolonged period of unexplained crying in infants. Abdominal pain is not necessarily responsible for these bouts of crying.

The first episode of colic can be a disturbing experience. New babies seldom demonstrate colic symptoms while in the hospital, for simple reasons:

- Colic is very unusual within the first few days of life.
- Infants are often removed to the nursery if they are crying in the room.

Once the child is home, the baby who had always seemed perfectly well behaved suddenly begins to scream. This usually occurs in the evening with both parents at home, and neither is able to supply an explanation. Is the baby having terrible pain? What can you do if the baby can't tell you where it hurts? Is this an emergency? Does the baby just want to be fed?

Often, neither feeding, changing, nor cuddling the infant provides comfort. Temporary relief may be found when the baby begins to suck on the fists or anything else available; often a bottle is eagerly accepted and then violently rejected. Given this insatiable crying, panic may ensue with frantic phone calls to the doctor or a frenzied trip to the emergency room. Sometimes before either of these courses of action has produced medical advice, the crying will stop, leaving the parents exhausted and baffled but thankful.

In most instances, the onset and resolution of this problem are not this dramatic. The baby simply has a crying spell that may last several hours during which nothing seems to be of comfort.

Signs of Colic

Typical colic has a number of features that usually make it easy to recognize. It begins after the second week of life and peaks at about three months of age. For this reason, it is sometimes called the "three-month" colic. Occurrences generally decrease rapidly after the age of three months. It would be unusual for colic to begin after three months of age.

Colic usually occurs in the evening. Some have postulated that the increased activity in the home during the evening hours may contribute to the colic. Parents' tolerance of such activity also is far less in the evening hours when both may be fatigued and a colicky baby can be of great concern.

The attack occasionally ends with a passage of gas or stool, and "gas" is often blamed for the problem. We feel that colic is caused by gas problems in some infants, but not the majority of them.

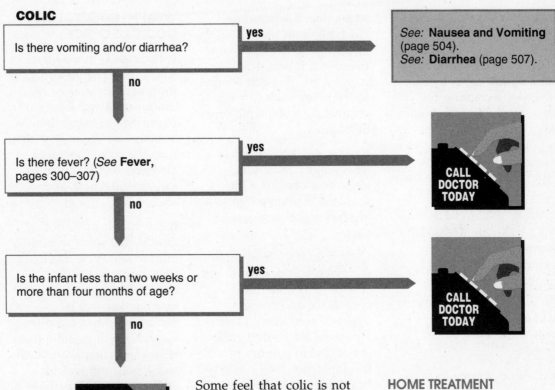

COLIC

Is there vomiting and/or diarrhea?	**yes** → *See:* **Nausea and Vomiting** (page 504). *See:* **Diarrhea** (page 507).

no ↓

Is there fever? (*See* **Fever,** pages 300–307)	**yes** → CALL DOCTOR TODAY

no ↓

Is the infant less than two weeks or more than four months of age?	**yes** → CALL DOCTOR TODAY

no ↓

APPLY HOME TREATMENT

The baby will seem perfectly well before and after these attacks and there should be no fever, vomiting, or diarrhea. Repeated episodes are the rule and may occur with great regularity at the same time each day.

Some feel that colic is not related to abdominal pain at all but is merely a characteristic that is part of the normal development of some children. Although all children cry, some seem to cry longer and are more difficult to console than others. Colic can be better appreciated by a reading of the section on crying (see pages 88–89).

HOME TREATMENT

Colic has been treated in many ways, all of which work some of the time and none of which work all of the time. This leads us to believe that most of the time the problem cures itself and probably requires no treatment.

With true colic, there is no reason for any medication to be given to the child. Parents have found that cuddling, rocking, soothing music, walking with the baby, taking the baby for a car ride, wrapping a child snuggly, soothing talk, pacifiers, backrubs, and placing the baby on his or her stomach are sometimes effective. Increasing the amount of time you hold your baby during quiet periods has been effective in reducing crying in some babies, although many truly "colicky" babies are characteristically inconsolable.

Many home remedies include giving a small amount of an alcoholic beverage. However, we discourage relying on alcohol, or any other drug for that matter, for the relief of any but the most disturbing symptoms. Another popular home remedy is an ounce or two of warm, weak tea. Tea seems more likely to stimulate the bowel than to put it to rest, but occasionally it may provide relief.

In children who have insufficient lactase, switching to a soy formula may help. Breast-fed babies are not immune to colic and a trial of a milk-free diet (though tough on a lactating mother) may help. Switching from one formula

to another is commonly tried although the effectiveness is unknown. Because it is safe, it is worth a try. But don't forget to try going back to your original formula, which may be cheaper.

Finally, don't be afraid to leave your baby crying alone in a crib for a few moments while you go to another room to rest your ears.

Colic is terribly frustrating for parents to deal with and is known to undermine confidence and cause depression in many parents. However, be assured colic is not due to parent incompetence. You can take heart in knowing that your child will outgrow this stage in a few months.

If any one attack persists beyond four hours, give the doctor a call. If the attacks do not seem to be diminishing by four months of age, discuss them during a regular doctor visit.

We do not recommend any medication for colic, and suggest avoiding them, for their effectiveness has not been demonstrated.

WHAT TO EXPECT AT THE DOCTOR'S OFFICE

If the visit is made for a single long-lasting attack, a thorough physical examination will be performed to see if there is a reason for the crying other than the colic. Particular attention will be given to ears, throat, chest, and abdomen.

If the visit is made for attacks that seem to be colic but are still occurring after four months of age, a careful history and physical examination will be performed. Most often the diagnosis will be colic that is lasting longer than usual, and a policy of watching and waiting with perhaps a dietary change may be recommended. It is very rare for serious medical problems to be present.

Eating Difficulty

Most children will lose their appetite during an illness, occasionally even during the most minor ones. When children feel bad, they often have neither the energy for nor an interest in eating. This is a feeling we all have experienced.

If an ill child appears interested in eating but experiences difficulties once the food is in the mouth, a sore throat or mouth sores may be the culprit. On rare occasions, the entrance to the child's windpipe (trachea) becomes infected and so swollen that a child will refuse to eat and will often drool and hold his or her head forward in order to alleviate difficulty breathing. This is a clear-cut emergency.

Toddlers often go through stubborn periods when other activities are more compelling than eating. At this stage, it is often helpful to offer "finger" foods at frequent intervals and fluids to keep them hydrated.

Older children may refuse to eat for longer periods of time. This may represent depression or herald anorexia. In these circumstances, a doctor should be consulted.

HOME TREATMENT

Most children can withstand a dramatic change in their diet during a short illness. It is not necessary to consume all the basic food groups every meal each day.

What is critical is that children receive a sufficient amount of liquids every day so that they will not become dehydrated. Cold juices seem to be time-honored favorites. With mouth sores, keep away from acidic juices such as orange or pineapple or lemonade. Although sodas are not nutritionally commendable, during illnesses we become expedient and approve any form of fluid or calories that may be acceptable to a child. Sodas without caffeine are preferable. Commercially available electrolyte solutions, such as Pedialyte or Lytren, are helpful in decreasing the likelihood of dehydration, although they are of little caloric value.

Most parents quickly learn that toddlers and younger children have variable appetites. Major struggles over food generally intensify meal wars. However, reasonable rules should be established—"If you're not hungry tonight, you still need to be at the dinner table for our family talk, but you don't have to eat"—and infractions dealt with immediately and appropriately—"You ignored our warning not to take your sister's dessert, so you will not be allowed to watch television tonight."

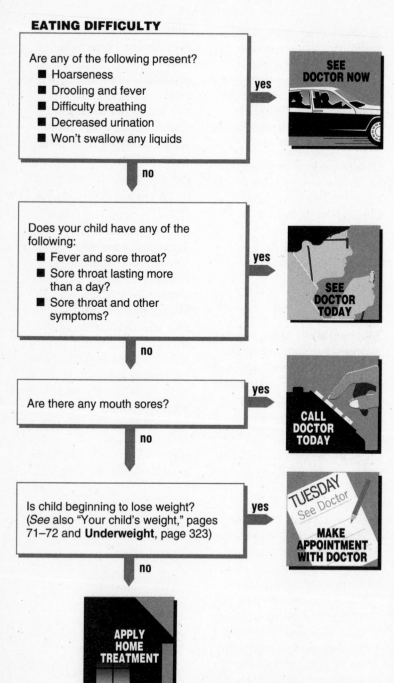

EATING DIFFICULTY

Are any of the following present?
- Hoarseness
- Drooling and fever
- Difficulty breathing
- Decreased urination
- Won't swallow any liquids

yes → **SEE DOCTOR NOW**

no ↓

Does your child have any of the following:
- Fever and sore throat?
- Sore throat lasting more than a day?
- Sore throat and other symptoms?

yes → **SEE DOCTOR TODAY**

no ↓

Are there any mouth sores?

yes → **CALL DOCTOR TODAY**

no ↓

Is child beginning to lose weight? (*See* also "Your child's weight," pages 71–72 and **Underweight**, page 323)

yes → **TUESDAY See Doctor — MAKE APPOINTMENT WITH DOCTOR**

no ↓

APPLY HOME TREATMENT

Once children are old enough to eat, they should not be "force fed" unless as part of a prescribed regimen for a feeding disorder. Children who are picky eaters should have firm routines, few distractions during mealtime, defined time periods for eating (about 20 minutes), and minimal snacking and fluids between meals. Confrontations and negative comments should be avoided.

WHAT TO EXPECT AT THE DOCTOR'S OFFICE

After a history and careful exam, your doctor may investigate certain infections with appropriate lab tests, such as a throat culture. A neck X-ray may be ordered if there is a breathing difficulty. If dehydration is suspected, a urinalysis and/or blood test may be done. If the problem has been chronic, you should be prepared for a number of visits and involvement of other professionals if there is depression or anorexia. Treatment of depression in childhood always involves counseling; drug therapy is seldom used.

Rectal Problems

Rectal bleeding is not a very common problem in children. It is most often seen during a diaper change when a few streaks of blood are noticed, usually on the surface of the stool. In newborn infants, this is most often due to a tiny tear in the rectum. This tear will usually heal by itself so long as the stools are not hard. In older infants and children, rectal bleeding is usually due to **constipation** which is discussed on page 318.

If abdominal pain accompanies rectal bleeding, this may be a sign of a blocked intestine or a bacterial gastro-enteritis; both of these require medical attention quickly.

Pinworms

Often a child will suddenly awaken crying in the early evening with rectal pain; there may be intense itching as well. This almost always means pinworms. Though these small worms are seldom seen, they are quite common. They live in the rectum, and the female emerges at night and secretes a sticky and irritating substance around the anus into which she lays her eggs. Occasionally, the worms move into the vagina, causing pain and itching in that area. The scratching can lead to vaginal infections in girls.

You can confirm the diagnosis of pinworms by checking for them with a flashlight several hours after the child's bedtime. They are about one-quarter inch long and look like white threads. While infestations with pinworms often resolve without medication, several prescription drugs that kill the pinworms will speed the process.

Bacterial infections can also cause rectal redness and itching. This is a consideration when pinworms are not seen.

HOME TREATMENT

If rectal bleeding is due to constipation, the stool should be softened. This can be accomplished by including more fruit (especially prunes or prune juice), fiber (bran, celery, whole-wheat bread), and fluids in the diet. Seldom are over-the-counter laxatives (Colace, Metamucil, Maltsupex) necessary.

Temporary relief of itching may be accomplished by giving aspirin or acetaminophen. If itching is present in girls, good hygiene and tub baths will help avoid vaginal infections.

WHAT TO EXPECT AT THE DOCTOR'S OFFICE

Through an examination of the anus and rectum, small tears and fissures can easily be detected. Seldom are pinworms noticeable during the day. If you have seen them at night, the doctor may rely on your observation.

Often, you may be asked to remove some of the pinworms with a piece of Scotch tape and bring them in for evaluation. After you have collected the pinworms, fold the Scotch tape over so that only the non-sticking surface is exposed. Even if you don't see the pinworms, the Scotch tape applied to the area around the anus that is itching will collect the eggs, which can be identified under the microscope.

Pinworms will usually be treated with one of several effective oral drugs. Treatment of the entire family is often necessary.

A moderate degree of redness around the rectum may signify a streptococcal infection, requiring antibiotics.

RECTAL PROBLEMS

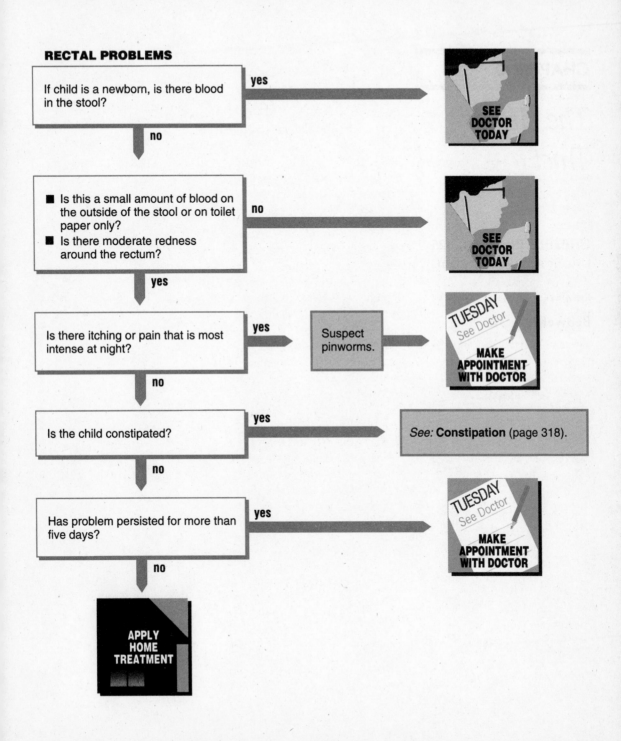

If child is a newborn, is there blood in the stool? — yes → **SEE DOCTOR TODAY**

↓ no

- Is this a small amount of blood on the outside of the stool or on toilet paper only?
- Is there moderate redness around the rectum?

— no → **SEE DOCTOR TODAY**

↓ yes

Is there itching or pain that is most intense at night? — yes → Suspect pinworms. → **TUESDAY See Doctor — MAKE APPOINTMENT WITH DOCTOR**

↓ no

Is the child constipated? — yes → *See:* **Constipation** (page 318).

↓ no

Has problem persisted for more than five days? — yes → **TUESDAY See Doctor — MAKE APPOINTMENT WITH DOCTOR**

↓ no

APPLY HOME TREATMENT

CHAPTER O

The Urinary Tract

Urination Problems

Painful, frequent, or bloody urination, usually indicating a bladder infection, is much more common in girls than in boys. In addition, many girls have episodes of frequency or burning on urination in which there is no infection but only irritation of the end portion of the urethra.

This is characterized by frequent passage of small amounts of urine.

Chemical irritants such as bubble baths have been incriminated in these outside irritations of the urethra (urethritis). Both boys and girls are susceptible to urethral irritation from trauma, chronic itching from pinworms, or masturbation.

Boys seldom have episodes of painful or frequent urination, but those that occur are more likely to be related to an infection. Sometimes infections involve not only the lower urinary tract (urethra and the bladder) but the kidneys as well. With a **kidney infection**, the child is likely to appear much more ill, fever is significant, and there may be nausea and vomiting, abdominal pain, back pain, or true shaking chills.

The resumption of bedwetting in a previously dry child may provide a clue to a urinary tract infection (see page 525). If there is a fever and the bedwetting is not accompanied by any other stressful psychological event, a urinary tract infection should be suspected.

Not uncommonly, a complaint of burning on urination is accompanied by a vaginal discharge in girls. In these instances, it is likely that the vaginal irritation has involved the urethra or urinary opening. If the doctor has investigated this problem before, then following the treatment recommended by the doctor is advisable. This may include using vaginal suppositories or soaking in a bath with vinegar added.

Blood in Urine

Blood in the urine can mean a problem with the bladder or the kidneys, and the child should always be brought to the doctor. This can be a sign of infection or kidney stone, or it can be due to injury. Red urine does not always signify blood; beets can produce a pink urine color in some children.

HOME TREATMENT

Home treatment is useful in relieving symptoms but not in eliminating infection. All new episodes of urinary burning, frequency, pain, or blood should be investigated by a doctor.

Drinking lots of liquids helps. Cranberry juice is better than some drinks because it contains a chemical known as quinic acid, which is transformed in the body to another chemical having antibacterial properties. However, cranberry juice does *not* provide enough of these chemicals to make it a reliable therapy. It is *not* adequate treatment for a bacterial infection.

Acetaminophen may help relieve pain. If a vaginal discharge is present, see page 529.

WHAT TO EXPECT AT THE DOCTOR'S OFFICE

The doctor will perform a urinalysis and examine the back, the abdomen, the vaginal opening, and the urinary opening or urethra. The urinalysis may indicate the need for a urine culture and/or antibiotics.

Pyridium, a medication that is helpful in relieving urethral pain, is sometimes prescribed. If there is an infection present, then X-rays of the kidney may be ordered. Most doctors will order an X-ray (I.V.P. or sonogram) for boys at any age and girls less than three years old after an initial infection. Do not be surprised or discouraged if your child has a second urinary tract infection. Eighty percent of children with one infection will develop a second, but only a small minority will have long-term problems.

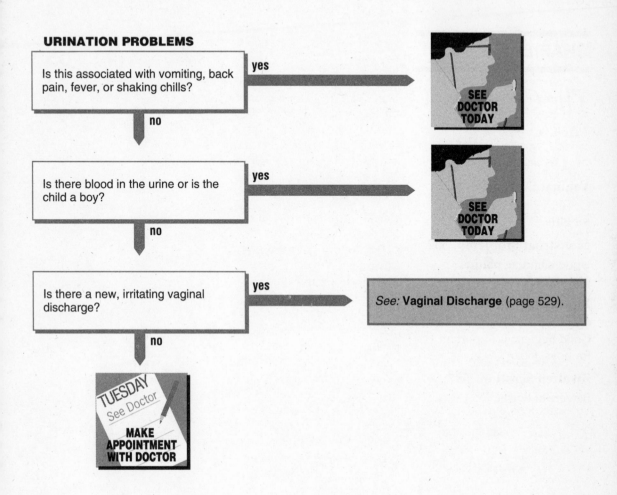

URINATION PROBLEMS

Is this associated with vomiting, back pain, fever, or shaking chills?

yes → SEE DOCTOR TODAY

no

Is there blood in the urine or is the child a boy?

yes → SEE DOCTOR TODAY

no

Is there a new, irritating vaginal discharge?

yes → *See:* **Vaginal Discharge** (page 529).

no

TUESDAY See Doctor — MAKE APPOINTMENT WITH DOCTOR

CHAPTER P

The Genitals

Vaginal Discharge

Doctors usually use the term "discharge" to mean something abnormal, and this can be confusing. Girls have vaginal secretions (discharges) that are normal, although the amount is much less than an adult's. Adult hormones increase the amount of secreted material; children have naturally increased amounts of secretions in two situations.

The first is in the newborn baby during the first week of life. Often there is some vaginal discharge or bleeding during this time due to stimulation by the mother's adult hormones.

After these first two weeks, the child will have only a small amount of clear secretions until about one year before the onset of menstrual periods. At this time, the girl begins to make her own adult hormones; the amount of secretion increases, and the secretions become thicker. These normal secretions are sometimes called a "normal discharge." Some secretion is normal in mid-cycle after menstruation has begun.

Abnormal Discharges

Poor hygiene may contribute to vaginal discharges in children. Scratching in the genital region often accompanies the rectal itching common

with **pinworms** and can lead to a vaginal infection; pinworms occasionally reach the vagina and cause itching directly as well.

Another cause of abnormal discharges in girls is a **yeast infection** (*Candida*), just as in adults. Yeast infections are likely when a child is taking an antibiotic.

Other causes of discharges in children are unusual in women. Just as children stick things in their ears and noses, they may also put objects in the vagina. This may lead to a bacterial infection and a discharge with a particularly bad smell. A forgotten tampon can cause the same reaction. Sand and toilet paper particles can also lead to infection and discharge. For this reason we recommend that girls be taught to wipe with toilet tissue from front to back, rather than back to front. Sometimes the chemical irritation of bubble baths will begin an itch–scratch–infection cycle.

Occasionally, a discharge can result from gonorrhea or other **sexually transmitted diseases**. These are always a result of sexual contact. Children are often too frightened to discuss these experiences with parents. In judging whether venereal disease is possible, it is important not to jump to conclusions in either direction. This is an area in which your doctor can be of great help.

HOME TREATMENT

Frequently, attention to the hygiene of the outside genital structures will be sufficient to clear the discharge. Gentle washing with soap and water, as well as warm baths, are helpful. Eliminate bubble baths if they had been used.

If you suspect a foreign object in the vagina, it is possible to look for yourself. This can be accomplished by having your child lie with her chest on the ground or a table while she is on her knees. In a moment or two, the vaginal opening will relax enough for you to see inside; a flashlight will help.

Yeast does not grow well in an acid environment, and minor infections often respond to vinegar (3% acetic acid) sitz baths. Fill the tub with enough water to cover the child's bottom and then add one cup of white vinegar. Soak for at least 15 minutes and do this twice a day if possible. This will often clear up the problem, and vaginal suppositories can be avoided. If the problem does not improve within five days, then make an appointment with the doctor.

WHAT TO EXPECT AT THE DOCTOR'S OFFICE

The abdomen and outside of the vagina (vulva) will be examined. The discharge will be looked at under a microscope. In adolescents, a pelvic examination may be needed. In younger children, this will be avoided if possible. If the outside of the vagina is very irritated, then an antifungal cream or ointment (Mycolog, miconozole, Candeptin) may be used. If there is an infection inside the vagina, the doctor will have a presumptive diagnosis after performing a few immediate laboratory tests. Appropriate therapy, usually with an oral antibiotic, will be implemented until a definitive laboratory diagnosis is made.

VAGINAL DISCHARGE

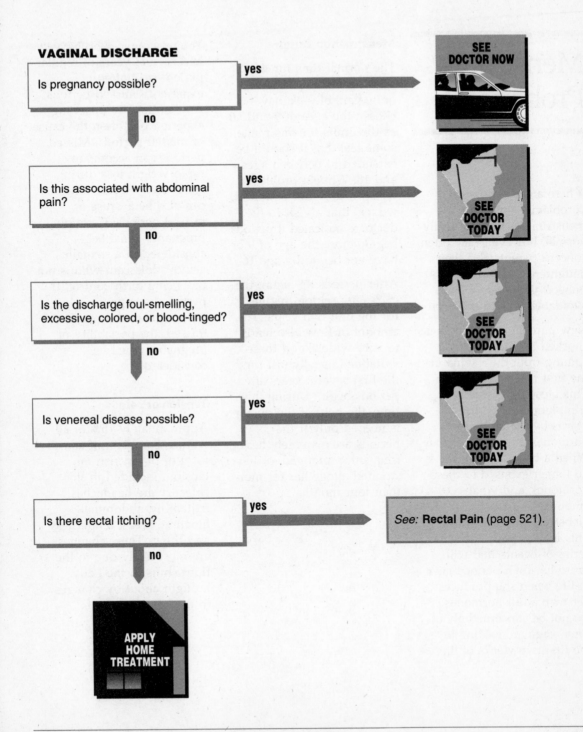

Is pregnancy possible?

— yes → **SEE DOCTOR NOW**

no ↓

Is this associated with abdominal pain?

— yes → **SEE DOCTOR TODAY**

no ↓

Is the discharge foul-smelling, excessive, colored, or blood-tinged?

— yes → **SEE DOCTOR TODAY**

no ↓

Is venereal disease possible?

— yes → **SEE DOCTOR TODAY**

no ↓

Is there rectal itching?

— yes → *See:* **Rectal Pain** (page 521).

no ↓

APPLY HOME TREATMENT

Menstrual Problems

There are a number of "problems" involving menstruation that are really normal; however, they often concern parents and their daughters and can lead to unnecessary anxiety and avoidable trips to the doctor.

New parents are sometimes shocked to find blood coming from the vagina in the first two weeks of life. This bleeding is due to stimulation of the baby's uterus by the mother's hormones during pregnancy. When a baby is born, she is no longer exposed to these hormones, and what amounts to a small menstrual period follows. This is essentially the same series of events that will cause her own periods later in life when she produces her own adult hormones. Do not be concerned about some vaginal bleeding in the first two weeks of life.

Menstruation Patterns

The normal time for the first menstrual period is quite variable. We have chosen the ages of 9 and 16 as the limits for this range; some feel that it should be extended to between ages 8 and 18. While a problem will probably not be found, we feel that a visit to the doctor is indicated if periods begin before the age of 9 or have not begun by age 16.

After periods do begin, the cycles are seldom regular for the first two years. The amount of flow also tends to vary widely, and these variations usually last for the first several years after periods begin. During this time, the help of the doctor is needed only if the periods are extremely heavy, frequent, prolonged, or have stopped altogether for more than four months.

Young women who have had several years of regular periods will often experience missed periods. An emotionally upsetting experience is often the cause of missed periods. Missed periods can accompany rapid weight loss during crash diets. Severe illnesses can also be a cause of missed periods. Certainly, pregnancy must be considered in a sexually active adolescent who is not practicing birth control. If periods have previously been regular and one is missed, the possibility of pregnancy must be considered.

Tension or pain

Approximately 15% of all women complain of some form of pre-menstrual tension. Included in this category are headaches, irritability, abdominal bloating, breast tenderness, and thirst. These changes are most likely due to the fluid shifts in the body brought about by changes in the hormone cycle.

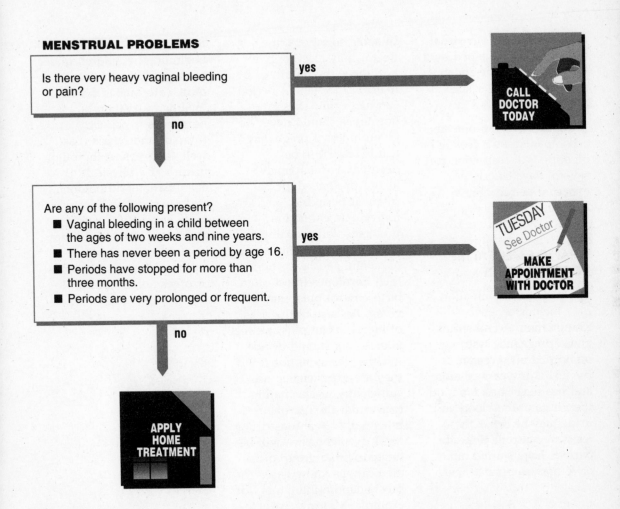

MENSTRUAL PROBLEMS

Is there very heavy vaginal bleeding or pain?

yes → CALL DOCTOR TODAY

no ↓

Are any of the following present?
- Vaginal bleeding in a child between the ages of two weeks and nine years.
- There has never been a period by age 16.
- Periods have stopped for more than three months.
- Periods are very prolonged or frequent.

yes → MAKE APPOINTMENT WITH DOCTOR

no ↓

APPLY HOME TREATMENT

Approximately 5% experience **dysmenorrhea,** or severe pain during menstruation; these symptoms most often begin during the first few years of the menstrual cycles. These crampy lower abdominal and back pains usually begin shortly before the onset of the period and last for about 24 hours. Occasionally, the pain may begin two days before the period and may last up to a total of about four days. Less than 5% of the women with this problem have any abnormality of their reproductive system.

HOME TREATMENT

If the problem is irregularity alone, then no treatment is necessary other than reassuring your daughter that this is normal for the first few years and even after. An occasional heavy period may be helped by bed rest to decrease the amount of flow. Menstrual cramps may be helped by a heating pad on the abdomen and a pain reliever such as aspirin.

If menstrual symptoms are accompanied by a feeling of bloating, salt restriction for several days before the expected period may be helpful.

WHAT TO EXPECT AT THE DOCTOR'S OFFICE

The physical examination may include a pelvic examination. A Pap smear and a pregnancy test may be needed on occasion. If the girl is 16 years or more and has never had a period, then tests of the blood and urine may be done; these tests may also be done if periods have started but have now stopped.

In some situations, the doctor may elect to give hormones (by mouth or by injection) to see if a period can be begun. These hormones should never be given unless a pregnancy test is known to be negative. For extremely heavy, prolonged, or frequent periods, blood tests to evaluate endocrine problems will probably be done. If the problem is one of painful and heavy periods, then hormones (most often birth control pills) may be given. Because of the risks of birth control pills, we feel adolescents should decide whether the symptoms they are experiencing are sufficiently incapacitating to warrant the use of potentially hazardous drugs. Most dysmenorrhea will be treated with a group of medications known as prostaglandin inhibitors (for example, Motrin).

Evaluation of periods beginning in children under the age of nine is usually quite extensive and will include evaluation of hormones (pituitary, thyroid, and ovarian) as well as X-rays. A thorough history and physical examination will be done, and questions will be asked to determine if it was possible that the child may have taken some of her mother's birth control pills or other hormones.

Problems with the Penis

Skin oils and secretions tend to accumulate underneath the foreskin of the uncircumcised penis. This accumulation may cause irritation and may lead to infection. With a severe infection the foreskin may swell and prevent the passage of urine.

To avoid these problems, parents can begin a program of foreskin hygiene. The foreskin in very young infants cannot be pulled back, but by the end of the first year the underlying glans of the penis should be visible. Parents can begin to gently pull the foreskin back and carefully wash the area as part of every infant's bath. Retraction should never be forceful and can begin in late infancy. Once-a-week retraction is sufficient. Do not worry if the foreskin is not fully retractable by age four or five years because this is common in many boys. Return the foreskin to its normal position after washing. As the child becomes older, this should become a part of his bath routine as well.

Sometimes the foreskin is so tight that it cannot be pulled back; this problem requires the help of the doctor. If the foreskin is retracted and cannot be returned to its former position, the blood supply to the end of the penis may be impaired; this problem requires the prompt attention of the doctor.

Discharges from the urinary opening of the penis are rare before adolescence, but at any age they require the help of the doctor. Sometimes an infection under the foreskin will produce enough pus so that there appears to be a discharge. If this is the case, the infection is bad enough to need the help of your doctor. For minor irritation underneath the foreskin, use home treatment.

Another common problem is getting the skin of the penis caught in a zipper. This most often occurs when parents are in a hurry to zip up a younger child. Little boys seldom zip fast enough to cause this problem themselves.

HOME TREATMENT

Careful cleaning of the area under the foreskin is essential. This is most easily accomplished with a gentle washcloth dipped in warm water. Remember to put the foreskin back in its normal position after washing. Soaking in a warm tub is also useful with foreskin problems.

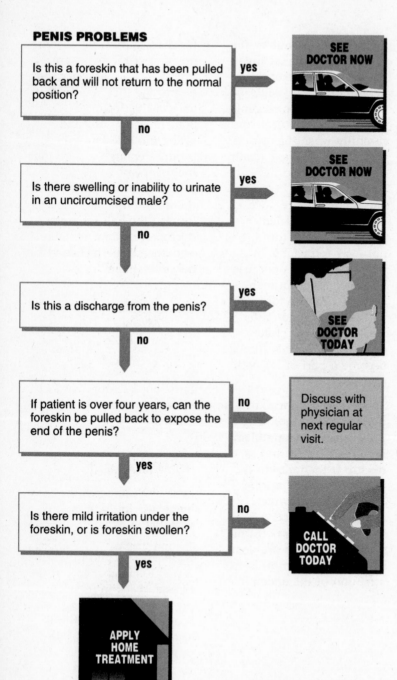

PENIS PROBLEMS

Is this a foreskin that has been pulled back and will not return to the normal position? → **yes** → SEE DOCTOR NOW

↓ **no**

Is there swelling or inability to urinate in an uncircumcised male? → **yes** → SEE DOCTOR NOW

↓ **no**

Is this a discharge from the penis? → **yes** → SEE DOCTOR TODAY

↓ **no**

If patient is over four years, can the foreskin be pulled back to expose the end of the penis? → **no** → Discuss with physician at next regular visit.

↓ **yes**

Is there mild irritation under the foreskin, or is foreskin swollen? → **no** → CALL DOCTOR TODAY

↓ **yes**

APPLY HOME TREATMENT

WHAT TO EXPECT AT THE DOCTOR'S OFFICE

If the foreskin is pulled back and restricting blood flow to the end of the penis, treatment usually will consist of cold compresses and certain medications. Rarely, a minor surgical procedure will be necessary to relieve the constriction. If the foreskin is so tight that it cannot be pulled back in an older boy, then it may be stretched; you will be instructed in a method of stretching the foreskin. This is a gradual process and requires some time. Circumcision is almost never necessary in dealing with foreskin problems.

If a discharge is present, it will be examined under the microscope. A culture will most likely be taken. Boys have been known to put foreign objects inside their penis; this can be a cause of infection. Most discharges do not indicate gonorrhea, but gonorrhea and other sexually transmitted disorders can occur at any age. These infections will be treated with antibiotics.

Swollen Scrotum

Many parents are surprised at the large size of their newborn baby boy's genitals, especially the scrotal sac that contains the testes. The size is actually accounted for by an accumulation of fluid in a sac, called a **hydrocele.** This common condition usually resolves on its own in the first six to nine months of life.

Another common problem that can cause the scrotum to swell at any time in a child's life is a **hernia**. A small opening in the muscles of the abdomen allows a portion of the child's intestine to slide into the scrotum. This usually occurs on one side, although some children will also develop a hernia on the opposite side. Once the muscle opening is present, the intestine may slide in and out so that the swelling is often present one moment and gone the next. Because

it is possible for the intestines to become either twisted or permanently entrapped in the scrotum, it is necessary to repair the hernia surgically.

Adolescents may find painless masses in their scotums. All painless masses should be evaluated by a physician. A firm mass can actually be cancer. Non-firm conditions are common; they are not emergencies but do require medical management.

Pain

Pain in the scrotum is of serious concern. Injury to the scrotum can be painful; but if pain persists more than several hours or is accompanied by swelling, you should see your doctor. If there has been no antecedent trauma, your child may be experiencing either a twisting of the testicle or an infection of the stalk of the testicle (*epididymitis*). Both of these need immediate attention.

HOME TREATMENT

There are no specific actions for you at home. Hydroceles in newborns will resolve on their own; all other problems need professional care.

WHAT TO EXPECT AT THE DOCTOR'S OFFICE

Your doctor will probably use a flashlight when examining the scrotum. It is sometimes possible to distinguish the way fluid lights up compared to more solid structures such as the intestine. In older children, a urinalysis will be done, possibly accompanied by a blood count if infection is suspected. If a serious twisting problem is suspected, blood flow to the testicle may be measured by a Doppler technique or by special X-ray (radionuclide scan). If an infection is diagnosed, antibiotics will be prescribed.

Surgery will be required immediately for twisted testes and eventually to repair a hernia.

SWOLLEN SCROTUM

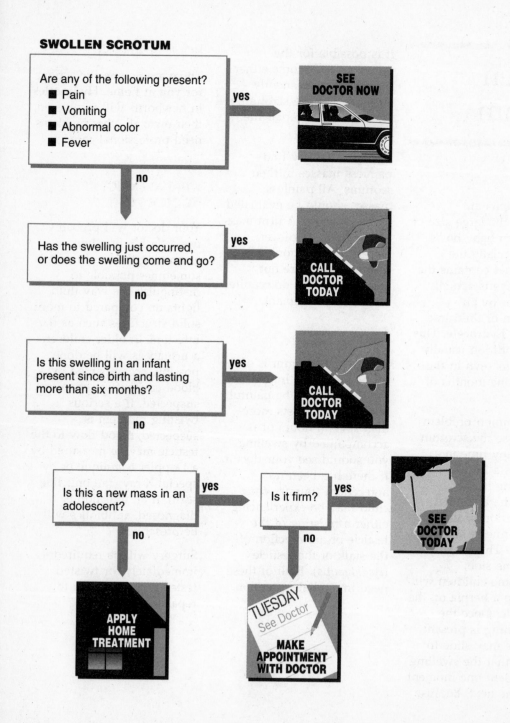

Are any of the following present?
- Pain
- Vomiting
- Abnormal color
- Fever

yes → SEE DOCTOR NOW

no

Has the swelling just occurred, or does the swelling come and go?

yes → CALL DOCTOR TODAY

no

Is this swelling in an infant present since birth and lasting more than six months?

yes → CALL DOCTOR TODAY

no

Is this a new mass in an adolescent?

yes → **Is it firm?**

no ↓

no → TUESDAY See Doctor — MAKE APPOINTMENT WITH DOCTOR

yes → SEE DOCTOR TODAY

APPLY HOME TREATMENT

CHAPTER Q

Adolescent Sexuality

**Preventing Unwanted
Consequences** *540*

Preventing Unwanted Consequences

SPEAKING FRANKLY

Discussions about sexuality between children and their parents are unusual in our society. We anticipate that few readers will have more than cursory knowledge about their mother's sexual experiences, for instance. Some may feel uncomfortable reading this chapter, much less using its information to discuss sexuality with an adolescent. Nevertheless, it is important for parents to discuss sex with children. Frank talk can mean the difference between a child's continuing to grow to adulthood or having a life prematurely encumbered by pregnancy or prematurely ended by AIDS.

Virtually all adolescents have sexual interests; that is what dating is all about. At what age a person's first sexual intercourse occurs depends on many things. Biology is important; some adolescents mature more quickly than others. Personal goals, risk taking, role models, activities of friends, and social pressures are all part of the complex process of sexual decision making. In addition, adolescence is characterized by explorations into new feelings and trying on new roles. Adolescents need to talk with someone about these issues. You can assume they are talking with their friends, and a few are talking with health-care providers. But it is best if they can talk about sexuality with their parents as well. Just telling teenagers "no" is not a conversation. It does not give them the understanding to make a responsible choice at the right time.

To have open conversations with your teenagers about sex, you should start several years in advance. Chapter 5's section on sexuality (see pages 111–115) covers questions that children under the age of ten are likely to ask. By setting up a relationship of trust early, you and your child will feel less awkward when more difficult questions arise.

Parents should encourage adolescents to refrain from sex until they are psychologically mature enough to deal with the responsibilities that accompany it. But encouragement alone is not enough, as our national health records show. The United States is a leader in such biomedical preventive measures as immunizations, but is woefully behind most Western societies in preventing adolescent pregnancy, sexually transmitted diseases (STDs), and the emotional trauma that accompanies both. Our reluctance to discuss sexual behavior in living rooms and schoolrooms has resulted in American teenagers being far less informed than Canadian and many European teenagers and far less reliable in their use of contraception and their practice of "safer sex."

Fortunately, there are signs that our culture is changing. For example, a recent episode of the television series *Northern Exposure* depicted an 80-year-old grandmother openly discussing orgasms with her 30-year-old granddaughter. This realistic depiction of communication about sexual concerns is an improvement over both silence and the exaggerations of such other television programs as *Studs*. If real-life families shared information as responsibly, we would be a much healthier society.

Government Timidity

Until recently the U.S. government was profoundly reluctant to endorse frank discussions of sexual matters. We ran up against this regressive attitude in 1992 when the federal government chose to distribute copies of the previous edition of *Taking Care of Your Child* to 300,000 employees, but insisted that the chapter on adolescent sexuality be removed. An

anonymous official objected, even though the chapter urged parents to give teenagers guidance before they drive off on their first date and to foster age-appropriate decision making throughout their children's lives. Ironically, the book the government wanted could discuss how parents of four- to ten-year-olds might answer questions about sexual behavior but could not advise parents of teenagers. The real heart of the controversy was giving families any information about contraception in an election year.

This attempted censorship was widely reported in the press and came to the attention of prominent members of Congress, including the chairperson of the House Select Committee on Children and Youth, Rep. Pat Schroeder. Under pressure, the executive branch eventually allowed a version of this chapter to be mailed separately to employees—with a "warning label" on the envelope.

The same attitude toward sexual information was evident in other federal actions that canceled research into adolescent sexuality, prevented condoms from being mentioned in AIDS-prevention programs, and limited abortion counseling. Even after basketball star Earvin "Magic" Johnson announced that he had contracted the AIDS virus as a result of his poor judgment, even after teenage birth rates had gone up, the government shied away from giving citizens all the information needed to make responsible choices. The federal government seems now to better understand that open discussions of health issues must not be censored. It is already difficult in this society to have realistic conversations about sex.

FACTS ABOUT ADOLESCENT SEXUALITY

The most recent information released by the National Center for Health Statistics reveals that teenage birth rates have grown to their highest level in 15 years. Between 1986 and 1989 the pregnancy rate for women 15 to 17 years old increased 20%. This statistic demonstrates failure on the part of teenagers to make responsible decisions about sexuality, and failure on the parts of parents, educators, and public-policy makers to encourage informed choices. Unfortunately, no single step will suffice to improve the decision making of teenagers and diminish the consequences of unsafe sex.

The many surveys on adolescent sexuality give slightly different results because of their different methods, but the basic message is the same: The majority of youth today will have had a sexual experience in their teens, with 10 to 15% having an initial experience before age 15, and 50% before age 17, including nearly 40% of young women. Teenage sexual experience extends across all major segments of

our population: among women ages 18 and 19, 78% of African-Americans, 74% of whites, and 70% of Hispanics report having had sexual intercourse.

Sexually Transmitted Diseases (STDs)

When an adolescent becomes sexually active, he or she is at risk for a variety of sexually transmitted diseases, including:

- AIDS
- Syphilis
- Gonorrhea
- Herpes
- *Chlamydia*
- Human papilloma virus (genital warts)
- Trichomonas
- Lymphogranuloma venereum
- Pelvic inflammatory disease
- Lice
- Scabies
- Monilia
- Hepatitis
- Chancroid

Some adolescents face higher risks for these problems, due to early maturation, other risk-taking behaviors besides sex, and the use of alcohol or other drugs. Developmentally, many adolescents are not good planners, and others have difficulties understanding real risks. Adolescent women are more vulnerable to certain diseases for biological factors relating to their cervix. However, it is behavior more than biology that leads to serious health consequences.

Compared to the United States, many Western countries have lower rates of STDs among their teenagers. Clearly, the cost of treating these diseases, added to the cost of teen pregnancies, is one factor in our rising health-care bills. No country in the world spends more than the United States on medical care, and reforming our health-care system requires reversing such trends.

The growing list of serious sexual diseases makes it imperative that your adolescent discuss the full range of contraceptive/ disease prevention options with his or her doctor. While a variety of techniques prevent pregnancy, fewer protect against STDs. The parent's role is to foster open communication and responsible decision making, and to help the adolescent make good use of the health system. The following information may be helpful to you and your child, but the ultimate decision should be shared between your child and his or her physician.

CONTRACEPTION

The most effective method of contraception and STD prevention is, of course, not to have sexual intercourse. There are psychological, social, and developmental arguments against having early intercourse as well. Other than abstinence, only condoms used in

combination with a spermicide can prevent infection. Your adolescent should discuss with his or her doctor the full range of contraceptive/disease-prevention options. The following is a brief discussion of currently available methods. Table 4 lists pregnancy rates for the various contraceptive methods and for no protection at all.

Factors in the Choice of Contraception

Contraception methods vary in risk to the user, in effectiveness, and in convenience. Methods that are most effective unfortunately tend to have the highest risks. Methods that pose little risk, such as diaphragms and condoms, are somewhat less effective for contraception. Some lack of effectiveness of these methods is largely due to inconvenience or aesthetic considerations so that they are used improperly or not at all. The amount of preparation required is especially important to teenagers. Sexual encounters at these ages often occur at erratic intervals and are unplanned. Some

adolescents (and adults) do not always have strong motivation toward planning and control, especially when lovemaking has begun.

Several methods of birth control are particularly likely to fail and are not recommended.

- **Coitus interruptus**, the withdrawal of the penis just before ejaculation, is not totally effective even when practiced faithfully, and the practice can be emotionally difficult.
- The same thing may be said of **douching** immediately after intercourse.
- The **rhythm method** requires fairly regular menstrual cycles, which are often lacking in adolescence, and is often ineffective.

In reality, the only advantage of these three methods is their lack of side effects and, for some, their religious acceptability.

Methods of Contraception

Spermicides, or chemicals that kill sperm, are available for vaginal application in the form of foams, gels, creams, suppositories, and sponges. Used alone they are a poor choice for a contraceptive. There are essentially no side effects, but these methods are effective for only about 60 minutes after insertion, and the instructions for use must be followed exactly if they are to be as effective as possible. Many people find that these preparations are inconvenient or just plain messy. Their cost may represent a significant expense for an adolescent.

Latex condoms have a good deal to recommend them. If used correctly, they are 97% effective. There are no side effects, they are inexpensive and widely available, and they give protection against STDs, especially if used with a spermicide containing nonoxynol 9. But they don't work if they are in the wallet or on the drugstore shelf during intercourse, and that's where they often are.

TABLE 4   *Number of Pregnancies per Year Expected for 100 Women Using Contraceptive Methods*

	Used Correctly and Consistently	*Average Experience*[*]
Combined birth control pills	0.1	3
Progestogen-only pill	0.5	3
IUD	2–8	3
Condom	2–4	12
Diaphragm	2–6	18
Foam (spermicide)	3	21
Implant (Norplant)	.04	.04
Coitus interruptus (withdrawal)	4–7	18
Cervical cap	6	18
Rhythm, mucus method, calendar	1–9	20
No protection	85	85

Source: Robert A. Hatcher, Felicia Stewart, James Trussell, et al., Contraceptive Technology, 1990–1992. *New York: Irvington, 1990. Reprinted by permission.*
[*] *The "Average Experience" group includes women who were using the method inconsistently or incorrectly.*

The combination of a **condom and spermicides** such as nonoxynol 9 is nearly as effective as the birth control pill for contraception and has the added benefit of preventing sexual diseases. This is a highly recommended form of prevention for teenagers, especially for those with infrequent intercourse where daily pills may be less desirable and expensive.

A **diaphragm and spermicidal jelly** is a compromise that is acceptable to some. It is effective in preventing pregnancy, although not as effective as birth control pills or intrauterine devices are. Protection lasts for 12 hours or so after insertion. It requires that intercourse be anticipated; this can be a problem. The diaphragm must be worn for several hours following intercourse as well. Rare occurrences of toxic shock syndrome have been associated with

diaphragm use in situations where the diaphragm has not been removed after a 24-hour period. Diaphragms cannot be used by women who are allergic to latex products. There are no other side effects or complications of diaphragms, and we highly recommend them. Diaphragms must be individually fitted.

Contraceptive sponges have recently been approved for sale as an over-the-counter contraceptive method. The sponges contain a substance that can inhibit or kill sperm. The sponge is inserted into the vagina and up against the cervix by the woman before intercourse and remains there for 6 to 8 hours after intercourse. The effectiveness of the sponge is approximately similar to that of the diaphragm or the condom. Occasionally, some women have difficulty removing the sponge. It is advisable to remove the sponge within 24 hours after the insertion to decrease the chance of getting toxic shock syndrome.

Cervical caps are contraceptive devices that are similar to diaphragms in the way they are used. The basic difference is that the cervical cap has a smaller diameter than the diaphragm and it fits snugly over the cervix. It must be inserted before intercourse and remain over the cervix for six to eight hours after intercourse. The cervical cap is comparable to the diaphragm in its effectiveness, and some women may find it more comfortable. Some women may have difficulty removing the cervical cap, but this is very infrequent. Another rare problem with the cervical cap is the development of cervical discomfort and/or abrasions in women who are using one that does not fit properly. The cervical cap must be fitted by a professional who has been thoroughly trained in the proper techniques. The cautions about toxic shock syndrome included in the discussion of diaphragms apply here also.

An **intrauterine device** (IUD) must be inserted by a doctor but requires no preparation at the time of intercourse. The IUD may cause bleeding and cramps. Sometimes it can be difficult to insert into the uterus. In the past, there were reports that a particular kind of IUD (Dalkon Shield) had been the apparent cause of severe and even fatal infections. This particular type of IUD is no longer used, but these complications might be seen rarely with other types. The IUD may be expelled, but the expulsion not noticed. Pregnancy can occur with the IUD in place, and tubal pregnancies are more common in patients with an IUD.

Birth control pills are a frequent choice for contraception because they are the most effective means of preventing pregnancy, if taken properly, and because they do not require any thought at the time of intercourse. Obviously their use requires that a doctor write a prescription (usually an examination is performed as well), and the patient must remember to take the pill daily. If taken regularly, protection against pregnancy is almost 100%.

TABLE 5 *Risks of Pregnancy, Contraceptives, and Other Dangers*

	Chances of Death in a Year
Motorcycling	1 in 1,000
Playing football	1 in 25,000
Pregnancy	1 in 14,300
Automobile driving	1 in 6,000
Oral contraceptive	
NONSMOKER	1 in 63,000
SMOKER	1 in 16,000
IUD	1 in 100,000
Barrier methods (diaphragm, condom, or spermicide)	
DEATH RESULTING FROM USE	NONE
DEATH RESULTING FROM 10 PERCENT BECOMING PREGNANT	1 in 143,000
Illegal abortion	1 in 3,000
Legal abortion	
LESS THAN 9 WEEKS	1 in 500,000
9–12 WEEKS	1 in 67,000
13–15 WEEKS	1 in 23,000
MORE THAN 15 WEEKS	1 in 10,000

Source: Robert A. Hatcher, Felicia Stewart, James Trussell, et al., Contraceptive Technology, 1990–1992. New York: Irvington, 1990. Reprinted by permission.

The risks of birth control pills can be significant depending on a woman's medical history and/or family history of specific diseases. They may cause blood clots, and these clots have been fatal on occasion. They may cause or contribute to high blood pressure. There are many less dangerous side effects: weight gain, nausea, fluid retention, migraine headaches, vaginal bleeding, and yeast infections of the vagina. Some of the side effects can be eliminated by changing the ratio of the hormones in the pills.

Most birth control pills contain combinations of two hormones: progestin and estrogen. The "mini-pill" is a progestin-only pill. It is nearly as effective as regular birth control pills but usually menstrual periods stop. This usually creates the need for periodic pregnancy testing to see if the user is pregnant. The expense in obtaining the pills may be significant.

Hormonal implants (Norplant) provide contraception that lasts up to five years. Capsules are inserted under the skin of the upper arm and release a small amount of progestin every day. They can be removed at any time. Side effects commonly encountered are bleeding between periods and headaches. An excellent contraceptive, Norplant does nothing to prevent sexually transmitted diseases.

Hormonal injections such as Depo-provera, a progesterone injection given every three months, may be appropriate in special situations.

Safety and Reliability

How safe are birth control pills and IUDs? It is impossible to predict a specific risk for a particular patient. The best information available simply boils down to this: The risk of either of these two methods is less than the risk of not using them if sexual activity is anything more than very occasional. Table 5 demonstrates the relative risks of various contraceptive methods and other activities.

Where to get reliable advice about sex and contraception is sometimes a problem. Teenagers should be able to get advice in an atmosphere that is non-judgmental and supportive. All family doctors and pediatricians should maintain confidentiality and most provide sound and supportive advice. However many young people will feel embarassed discussing sexuality with their childhood physician. Planned Parenthood clinics or public health clinics are alternatives. Planned Parenthood clinics provide competent sex advice in general and contraceptive advice in particular, as well as the medical services necessary for diaphragms, IUDs, and birth control pills. The teenager may be more comfortable in such an atmosphere than at the office of the family doctor. Emergency rooms and walk-in clinics are not good places for this type of help.

PART V

Family Records

Name_____

Birth Information

Date _____ Weight_____ Mother's age _____ Length of pregnancy _____

Complications _____

Medical History	Date	Illness
Hospitalizations	_____	_____
Other medical problems	_____	_____
(Include serious illness	_____	_____
or injury, hearing or vision	_____	_____
problem, positive TB test, etc.)	_____	_____

Allergies

Medicines _____

Other _____

Family Medical History

Allergy/Asthma _____	High blood pressure _____
Heart Disease _____	Tuberculosis_____
Diabetes _____	Other _____

Immunizations

	Date		Date
2 months		**15 months**	
DPT #1	_____	Measles, Mumps,	
OPV #1	_____	Rubella	_____
HIB	_____		
4 months		**18 months**	
DPT #2	_____	DPT booster	_____
OPV #2	_____	OPV booster	_____
HIB	_____	HIB	_____
6 months		**4-6 years**	
DPT #3	_____	DT booster	_____
OPV #3 (optional)	_____	OPV booster	_____
HIB	_____		
		5-18 years	
		Measles, Mumps,	
		Rubella	_____

DPT = Diphtheria, pertussis (whooping cough), and tetanus (lockjaw)
DT = Diphtheria and tetanus (lockjaw)

HIB = Hemophilus influenza B vaccine
OPV = Oral polio vaccine
Rubella = German measles (three-day measles)

Note: Diphtheria and tetanus immunization is recommended every ten years for life, with an additional tetanus booster for contaminated wounds more than five years after the last booster.

*Name*_____

Birth Information

Date _____ Weight ____ Mother's age _____ Length of pregnancy _____

Complications _____ _____

Medical History   *Date*   *Illness*

Hospitalizations

Other medical problems

   (Include serious illness

   or injury, hearing or vision

   problem, positive TB test, etc.)

Allergies

Medicines _____ _____

Other _____ _____

Family Medical History

Allergy/Asthma _____   High blood pressure_____

Heart Disease_____   Tuberculosis _____

Diabetes _____   Other_____

Immunizations   *Date*   *Date*

2 months

DPT #1   _____

OPV #1   _____

HIB   _____

4 months

DPT #2   _____

OPV #2   _____

HIB   _____

6 months

DPT #3   _____

OPV #3 (optional)   _____

HIB   _____

15 months

Measles, Mumps,

Rubella   _____

18 months

DPT booster   _____

OPV booster   _____

HIB   _____

4-6 years

DT booster   _____

OPV booster   _____

5-18 years

Measles, Mumps,

Rubella   _____

DPT = Diphtheria, pertussis (whooping cough), and tetanus (lockjaw)

DT = Diphtheria and tetanus (lockjaw)

HIB = Hemophilus influenza B vaccine

OPV = Oral polio vaccine

Rubella = German measles (three-day measles)

Note: Diphtheria and tetanus immunization is recommended every ten years for life, with an additional tetanus booster for contaminated wounds more than five years after the last booster.

*Name*_____

Birth Information

Date _____ Weight_____ Mother's age _____ Length of pregnancy _____

Complications _____

Medical History

	Date	Illness
Hospitalizations	_____	_____
Other medical problems	_____	_____
(Include serious illness	_____	_____
or injury, hearing or vision	_____	_____
problem, positive TB test, etc.)	_____	_____

Allergies

Medicines _____
Other _____

Family Medical History

Allergy/Asthma _____	High blood pressure _____
Heart Disease _____	Tuberculosis_____
Diabetes _____	Other _____

Immunizations

	Date			Date
2 months			**15 months**	
DPT #1	_____		Measles, Mumps,	
OPV #1	_____		Rubella	_____
HIB	_____			
4 months			**18 months**	
DPT #2	_____		DPT booster	_____
OPV #2	_____		OPV booster	_____
HIB	_____		HIB	_____
6 months			**4-6 years**	
DPT #3	_____		DT booster	_____
OPV #3 (optional)	_____		OPV booster	_____
HIB	_____			
			5-18 years	
			Measles, Mumps,	
			Rubella	_____

DPT = Diphtheria, pertussis (whooping cough), and tetanus (lockjaw)
DT = Diphtheria and tetanus (lockjaw)

HIB = Hemophilus influenza B vaccine
OPV = Oral polio vaccine
Rubella = German measles (three-day measles)

Note: Diphtheria and tetanus immunization is recommended every ten years for life, with an additional tetanus booster for contaminated wounds more than five years after the last booster.

Growth Charts

H ere are four charts that will help you keep a permanent record of your child's growth. This information will be of interest to your child's doctor on routine visits. It is also referred to several times in this book to help you decide whether a doctor's visit is necessary.

Two of the charts are for girls and two for boys. On each chart you can plot your own child's height and weight at various ages. The line for height and the one for weight will be plotted on two different scales on the same chart. As you can see, age is shown at the bottom of each chart with height on the left margin and weight on the right.

To plot your child's height:

1. Find the vertical line for his or her age at the bottom of the chart.

2. Find the horizontal line for the child's height on the left side of the page.

3. Mark the point where the two lines cross.

4. The next time you measure your child, mark the new age and height position on the chart.

5. By connecting the points, you will develop a line that shows your child's growth.

To plot weight:

1. Find the line for the child's age at the bottom of the chart.

2. Locate the horizontal line for the child's weight on the right side of the chart.

3. Mark the point where the two lines cross.

4. Proceed as for the height part of the chart.

If you have more than one girl or boy, assign a different color or symbol to each one so their records can be easily distinguished.

The curved lines with numbers represent "percentile" of normal children. For example, if your boy weighs 31 pounds (about 14 kilograms) at 27 months, he is in the 75th percentile. This means that about 75% of normal boys weigh less than he does and about 25% weigh more. Please remember that these are simply statistical expressions that may be helpful in certain circumstances. Being in a particular percentile does not in itself make a child "normal" or "abnormal."

Girls: Birth to 36 Months

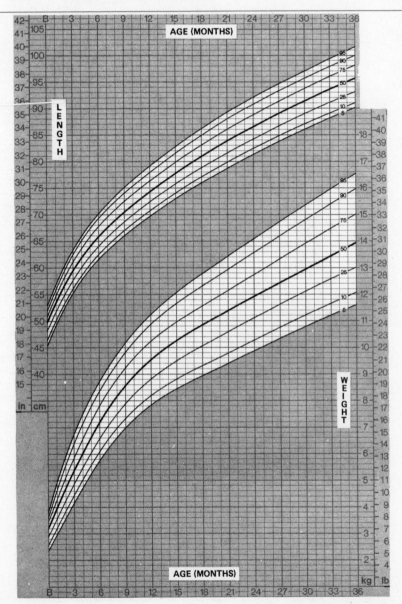

Copyright © 1976 Ross Laboratories. Adapted from National Center for Health Statistics: NCHS Growth Charts, 1976. Monthly Vital Statistics Report, Vol. 25, No. 3, Supp. (HRA) 76-1120. Health Resources Administration, Rockville, Maryland, June 1976. Data from the Fels Research Institute, Yellow Springs, Ohio.

Girls: 2 to 18 Years

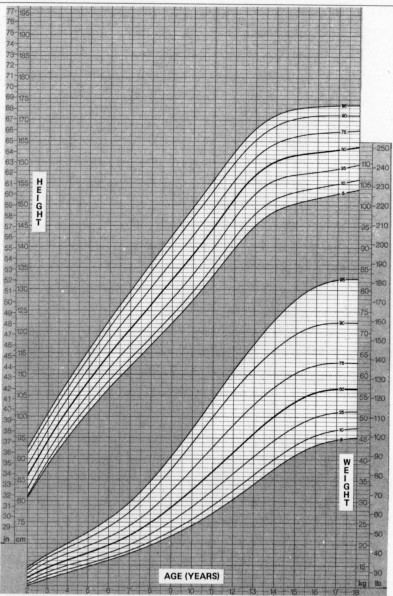

Boys: Birth to 36 Months

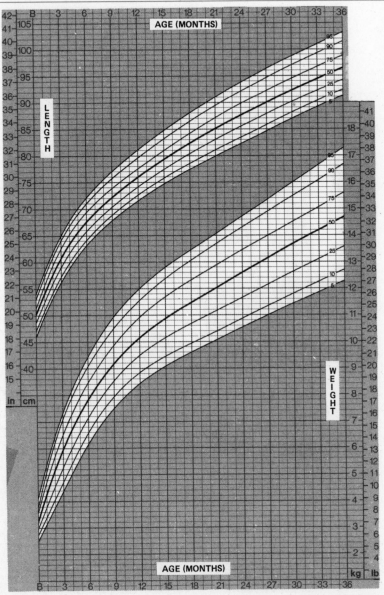

Copyright © 1976 Ross Laboratories. Adapted from National Center for Health Statistics: NCHS Growth Charts, 1976. Monthly Vital Statistics Report, Vol. 25, No. 3, Supp. (HRA) 76-1120. Health Resources Administration, Rockville, Maryland, June 1976. Data from the Fels Research Institute, Yellow Springs, Ohio.

Boys: 2 to 18 Years

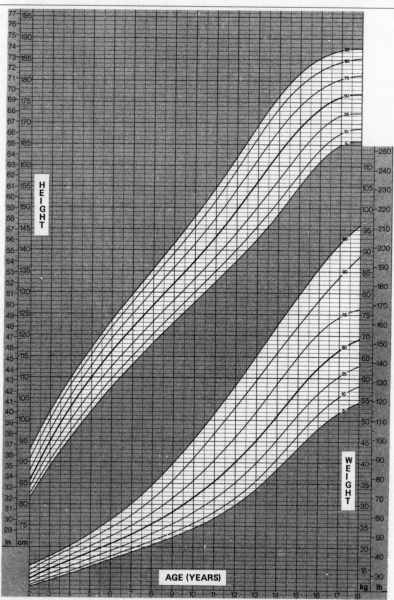

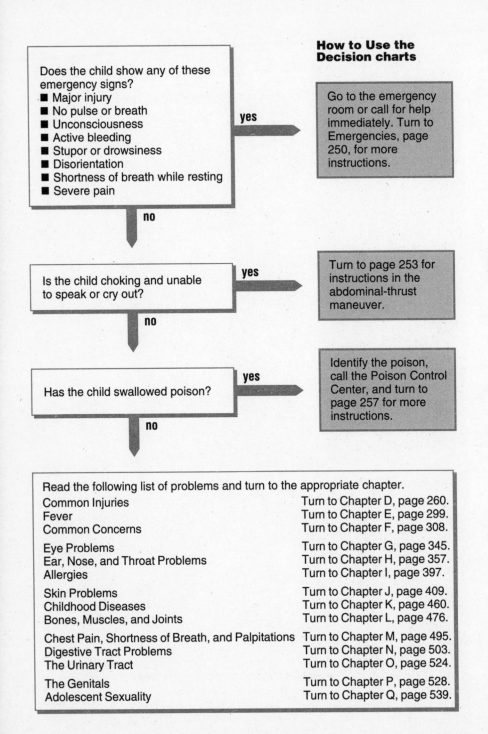

Does the child show any of these emergency signs?
■ Major injury
■ No pulse or breath
■ Unconsciousness
■ Active bleeding
■ Stupor or drowsiness
■ Disorientation
■ Shortness of breath while resting
■ Severe pain

yes

How to Use the Decision charts

Go to the emergency room or call for help immediately. Turn to Emergencies, page 250, for more instructions.

no

Is the child choking and unable to speak or cry out?

yes

Turn to page 253 for instructions in the abdominal-thrust maneuver.

no

Has the child swallowed poison?

yes

Identify the poison, call the Poison Control Center, and turn to page 257 for more instructions.

no

Read the following list of problems and turn to the appropriate chapter.

Index

For advice on a common medical problem, look up the primary symptom in this index. Numbers in **boldface** indicate the pages where you can find the most information on each subject. These are usually the pages with decision charts and advice on home treatment and when to see a doctor.

INDEX

INDEX

INDEX

INDEX

Medications. *See also specific medication or type of medication*
abdominal pain due to, 512
abuse of. *See* Substance abuse
for acne, 437-438
administration of, **236-239**
for asthma, 402-403
breast feeding and, 19
childproof bottles for, 224, 304-305
child's resistance to taking, 237-239
eye, for newborns, 39
headache caused by, 309, 310
household supply of. *See* Home pharmacy
hyperactivity caused by, 312, 314
in labor, **30-31**
nausea caused by, 505
overdose of, 258
in pregnancy, **12-13, 14-15**
for nausea, 7
questions to ask before giving, 237, 238
for runny nose, 226-227, 374
sleep disturbances due to, in toddlers, 94
time vs., 239-240
weight gain related to, 320
in wheezing treatment, 383
Medihaler-Iso. *See* Isoproterenol
Melanoma, 431. *See also* Skin cancer
Membranes, rupture of, 30
Meningitis
headache in, 309
neck in, 485, 486
vomiting in, 505
Meningococcal disease, immunization against, **207**
Menstrual cycle
missed periods and, 532
onset of, 75, 532
patterns of, 532
problems with, **532-534**
teaching school-age girls about, 114
Mental health, help for families and, 136-140
Meperidine (Demerol), in labor, 31
Mercurochrome, 262, 269
Merthiolate, 262
in swimmer's ear, 369
Metabolic disorders, infant blood testing for, 40
Metamucil, 229, 318, 522
Metaprel. *See* Metaproterenol
Metaproterenol, 403
Metatarsus adductus, 161, 489
Methadone, in pregnancy, 15
Metronidazole (Flagyl), in pregnancy, 15
MicaTin. *See* Miconazole
Miconazole

for diaper rash, 418
for ringworm, 423
Migraine, 309, 310
vomiting and, 505
Milia, 414
Miliaria (heat rash), 414, 416
Milk, 63
allergy to, 399-400
diarrhea and, 507
anemia and, 329
bottles of. *See also* Bottle feeding
teething on, 209
toddlers put to sleep with, 94, 366
breast, 60-61. *See also* Breast feeding
diarrhea and, 507, 508
eczema and, 406
evaporated, 63
intolerance of, lactase deficiency and, 399, 400
in poisoning, 258
soybean substitutes for, 400
Milk of Magnesia, 229
Mineral oil, 229, 318
Minerals, in pregnancy, 6
Missing children, 189-192
Mist. *See* Vaporizers
Mistakes, acceptance of, importance of, 116
Mites
chiggers, 453
scabies due to, 454-455
Moles, 55
melanoma, 431
Mongolism. *See* Down syndrome
Monilial infection. *See* Yeast infection
Mononucleosis, 329, **363**
swollen glands in, 387
Moral judgment
in adolescence, 107
in early school years, 104
Morning sickness, 7
Mothers. *See also* Parents
breast feeding advantages for, 18-19
working outside home, 118
Motion. *See also* Activity; Muscle skills
abnormal, of knee, 279
Motion sickness, 505
Mouth. *See also* Dental problems; Lips; Throat; Tongue
bad breath and, **391-392**
dehydration and, 249
of infant, 57
sores in, **393-395**
swallowing difficulty and, 519
Mucus
cough and, 377
nasal
humidity and, 374
post-nasal drip and, 373

Multiple problems, decision charts and, 243
Mumps, 461-462
complications of, 461, 462
immunization against, **203**
Mupirocin
impetigo and, 421
scrapes and, 270
Murder, 218-219
Murine, 230
Muscle pain, 477-479
Muscle skills. *See also* Motion
bladder control and, 82
bowel control and, 80
crossed eyes and, 355
development of, 76
feeding ability and, 65-66
school failure and, 145
sports readiness and, 152
Muscle spasms
headache due to, 309, 310
low-back pain due to, 482, 483
Muscle weakness, headache associated with, 310
Myocardial infarction, 215

Nails. *See* Fingernails; Toenails
Naps, 90
Narcotic preparations. *See also specific narcotics*
for diarrhea, 229
in labor, 31
in pregnancy, 15
Nasal discharge. *See* Runny nose
Nasal sprays, for allergic rhinitis, 404
National Center for Missing and Exploited Children, 189, 192
Nausea, 504-506. *See also* Vomiting
in pregnancy, 7
vertigo and, 332
Navel, hernia of, 59
repair of, **175-176**
Neck. *See also* Throat
in meningitis, 485, 505
pain in, **485-486**
stiff, 505
swollen glands in, 386
wry, 485
Needles, puncture wounds from, 264, 265
Negativism
as developmental milestone, 82
of toddlers, 91-92
Neosporin, scrapes and, 269
Neosynephrine, 226, 374
Nerve injury
fractures and, 273
puncture wounds and, 264

INDEX

of adolescents, 106, 107-108
societal, testing of, 104
Vaporizers, 228, 235
croup and, 380
runny nose and, 374
wheezing and, 382
Variability, in growth and development, 84-85
Venereal diseases. *See* Sexually transmitted diseases (STDs)
Ventilation tubes, **175**, 367
Ventolin. *See* Albuterol
Verbal ability. *See also* Communication; Language
development of
in pre-schoolers, 100
variability in, 85
expression of negative feelings and, temper tantrum prevention through, 92, 95-96
speech development and, hearing loss and, 371
Vergo, for warts, 431
Vernix caseosa, 54
Vertigo, 332, 333
Vesicles, in chicken pox, 463
Viral infections. *See also* Infections; *specific type*
croup caused by, 379
epidemic conjunctivitis due to, 346
frequent, screening for cause of, 170-171
mouth sores due to, 393, 395
syndromes, 358
upper respiratory, 358, 359, **360-362**. *See also* Colds
Visine, 230
Vision. *See also* Eye
infant testing for, 56-57
problems with, **355-356**
school failure and, 145
Visiting, in maternity wards, hospital routine and, 42
Vistaril. *See* Hydroxyzine
Visual motor ability, school failure and, 145
Vitamin D
bowlegs and, 487
breast feeding and, 17
Vitamin K, for newborns, 39
Vitamins
for infants, 44
in pregnancy, 6, 15
supplemental, **236**
Vocal cords, hoarseness and, 384
Vocational counseling, for adolescents, 107
Voiding. *See* Urination
Vomiting, 504-506
bloody, 504

in bulimia, 324
dangers of, 504
diarrhea and, 507
head injury and, 285, 286
induction of, in poisoning, 224, **233-234**, **258-259**
nature of, 248
by newborns, 504
by older children, 504-505
in pregnancy, 7
in vertigo, 332

Walkers, 180
Walking
ankle injuries and, 277
limp and, 480-481
pigeon toes and, 489
Walking ability, development of
individuality in, 84
variability in, 84-85
Warm compresses, in stye treatment, 353
Warts, 431-432
Water. *See also* Fluid intake
scalding with, 288. *See also* Burns
sponging with, in fever reduction, 303-304, 335
Water jets
in dental care, 211
in ear wax removal, 368, 371
Water safety, 186-187
Water supply, fluoridation of, 211
Waterproofing, classes in, 186
Weakness, 329-331
headache associated with, 310
prolonged, 329-330
Weight. *See also* Overweight; Underweight
blood pressure and, 217
growth of child and, **71-72**
measurements of
growth charts for, 553-558
during well-baby visits, 323
sports and, 154
Weight gain
of infants, 71
in pregnancy, 5
Weight loss
recommendations for, 321-322
sports safety and, 154
Well-baby checks, 159-163
height and weight measurements in, 323
Wheezing, 382-383
in asthma, 401
shortness of breath and, 499
Whooping cough, immunization against, 196-197
Wisdom teeth, 212

Wood, puncture wounds from, 266
Wood's lamp, in ringworm diagnosis, 424
Workroom, safety in, 182
Worms, pinworms, 521, 522
Wound care. *See also* Wounds; *specific type of wound*
bandages in, 235
sterilizing agents and antiseptics in, **234-235**, 262
stitches in. *See* Stitches
Wounds. *See also specific type*; Wound care
infected, **290-291**
cuts, 261
puncture wounds, 264
scrapes, 269
normal healing of, 290, 291
tetanus shots and, 271
Wrist, injuries to, **282-283**
Wry neck, 485

X-ray examination, 158
of fractures, 274
in head injury, 286
swallowed objects and, 338
Xylocaine. *See* Lidocaine

Yard, safety in, 182
Yeast infection. *See also* Fungal infection
diaper rash due to, 417
jock itch vs., 442
oral, 393, 395
vaginal, 529, 530
Young children. *See* Infants; Pre-schoolers; Toddlers

Zen macrobiotic diet, anemia and, 329
Zinc, in acne treatment, 438
Zinc oxide ointment, for diaper rash, 417
Zyradryl, 428